THE YEARBOOK OF AGRICULTURE / 1977

Gardening
For Food and Fun

U.S. DEPARTMENT OF AGRICULTURE

Gardening suggestions are keyed to numbers on photos. 3, try to have a sunny site for your garden. Be sure trees or buildings will not shade the garden during most of the day in the growing season. 4, an electrically heated hotbed will let your plants benefit from sunlight even in winter.

5, buy top quality transplants, with healthy green foliage, not wilted or with dry soil, and free of insects or diseases. 7, a transplant should be placed in the garden at the same depth that it was in the original container. 6, seedlings get off to a good start in fertile and loose-textured soil.

8, planting bean seeds in furrow with string as guide for a straight row. 11, hot caps protect young plants on chilly spring days. 10, plastic and other mulches control weeds and conserve moisture. 12, seedlings can be grown in peat moss containers—reducing problems from insects, diseases, or poor soil.

Some insects help control garden pests. 9, larva of lady bug shown attacking aphids. 13, assassin bug feeds on immature insects. 14, bee, here pollinating almond blossoms, is the top garden pollinator. Avoid spraying when helpful insects are around.

Garden structures: A plastic greenhouse, 16, can be used to start plants in the spring. 15, corrugated plastic row cover is held in place with stakes. 17, twine trellises for peas and wire cages for tomatoes.

Some basic garden activities and aids. 18, hoeing when weeds are small makes weed control easier. 19, cans with bottoms cut off are pushed into ground to protect plants from cutworms. 20, young gardeners planting vegetable seeds in various types of containers.

Containers ranging from flower pots to a washtub can be used for growing fruits or vegetables in a sunny window or on a patio.

Trickle irrigation, 23, using tubes that drip water where it will moisten the root zone, is a suggested water-conserving technique. Harvesting, 24, 25, is an important occasion. Children at a botanical garden weigh their produce. This helps any gardener learn which varieties are the most productive, a guide on what to plant next year.

Kohlrabi, 27, and okra, 28, are unfamiliar to many gardeners, but try them for something new and different. Everyone knows the potato: take care not to damage the spuds if you harvest them with your pitchfork.

29

30, thinning beets allows more room for the remaining plants to grow. The removed tops and immature beets can be cooked and eaten. Small Fry is a popular cherry tomato.

Snap beans, turnips, and carrots are popular with home gardeners. The season for these vegetables can be lengthened by successive plantings.

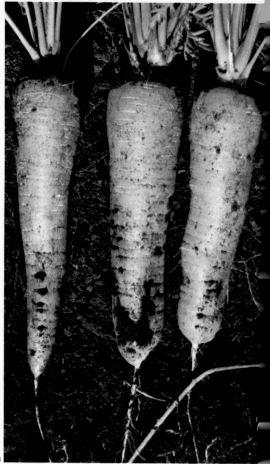

Tying cauliflower leaves helps assure a white head of cauliflower. *Supersweet* variety of watermelon resists such diseases as anthracnose and fusarium wilt. Buy disease-resistant varieties of plants for your garden when they are available.

All the garden work seems worthwhile when the harvest is good, whether it's a variety of produce, pumpkins, sweet shelled peas, or yellow squash and green beans grown in a community garden, 39.

A variety of pests usually are ready to take over the garden and injure your crops. 40, white grub works underground and damages roots of many garden plants. Mexican bean beetle, 41, is shown in adult, larva, and egg stages. The adult and larva are feeding on a bean leaf.

42, aphids—also called plant lice—are tiny, but can stunt plant growth by sucking plant juices and transmitting virus diseases. Lady bugs are helpful in controlling aphids. 43, corn earworm feeds on corn, tomatoes, and many other plants, including vegetable soybeans (shown here). 44, Colorado potato beetle is a common national pest of potatoes and tomatoes.

45, cabbage looper infests not only cabbage but cabbage relatives such as broccoli, cauliflower and collards. 46, codling moth is especially bothersome to apple growers. Harlequin bug, 47, attacks primarily cabbage and related plants such as turnips, horseradish and kale.

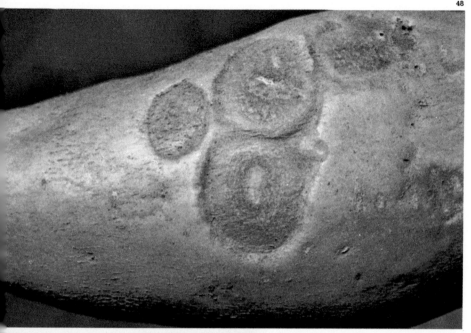

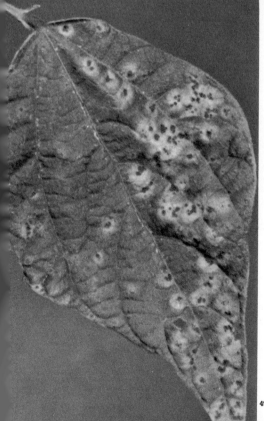

Insects are not the only plant pests—so are disease organisms such as soil rot on sweet potato, 48, and rust on a snap bean leaf, 49.

Among other vegetable plant diseases are scab on squash, 50, and mildew on cucumber, 51.

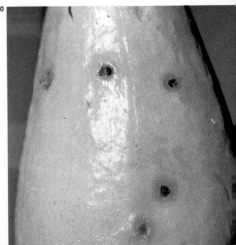

If your tomatoes look like 52, they may have fusarium wilt. Closeup shows anthracnose on a tomato. Brown spot on sweet corn leaf, 54, is at an advanced stage and probably would reduce the yield.

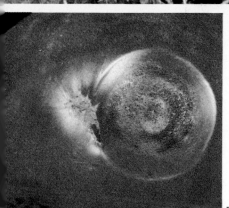

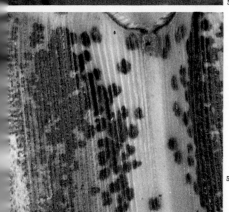

Some fruit that can be grown in the home garden: 55, *Villard Blanc* grapes and 60, *Villard Noir* are used primarily for wine. They also can be eaten at the table. 57, black-sapote is an interesting subtropical fruit. 56, lychee tree provides shade as well as an exotic fruit. It is popular in Florida and Hawaii. 58, a cluster of large-fruited blueberries.

59, straw mulch in a strawberry bed. Besides the usual benefits from mulching, straw keeps berries off the ground and clean.

Examples of insect and disease pests of fruits and nuts: 62, periodical cicada (17-year locust). Adults deposit eggs inside tree twigs and branches, weakening them so they eventually break. Nymphs hatch from the eggs and feed on roots. 63, powdery mildew on pecans damages the developing nuts. 64, scab on apple reduces quality and yield.

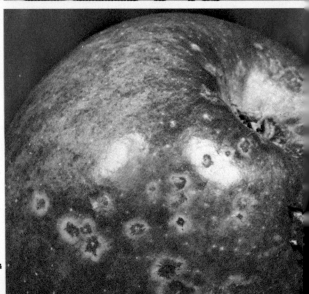

Diseases of peaches include 61, brown rot and 65, scab. A damaging insect is the peach tree borer, 66. Female adult is shown. The larva, which tunnels in the peach tree trunk, is the damaging stage.

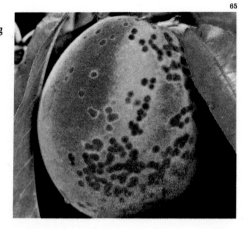

Safe home food preservation methods let you enjoy produce from your garden year-round. 67, jelly-making is a skill easy to learn. 68, to retain top quality, can or freeze sweet corn quickly after picking. Improper canning or storage wastes food. The cloudy green beans, 69, should be discarded without tasting. 70, an assortment of pickles and relishes can add zest to meals throughout the year.

71, peeling tomatoes after they have been scalded for easy removal of skins.

73, taking peach jar from a hot water canner. Jams and preserves should be processed in a hot water bath to retain best quality. 72, canning tomato juice at a community cannery. Such canneries enable you to process large quantities of garden produce in a short time.

Foreword

Bob Bergland
Secretary of Agriculture

GARDENING FOR FOOD AND FUN is a practical book for gardeners of all types—from the beginner to the proficient, from young people to retired persons. Advanced gardeners will find this book helpful as a refresher and as a reference source.

There are four sections in this Yearbook: *Introduction to Gardening, Home Garden Vegetables, Fruits and Nuts,* and *Home Food Preservation.* The last section tells how to preserve and store your garden produce at peak quality for year-round use, and it stresses the need for proper techniques to avoid health hazards.

Gardening is one of America's most popular activities. A U. S. Department of Agriculture study last year found that nearly half the households surveyed either had a garden or intended to have one.

Why do people garden? The survey suggests three main reasons: 1) a preference for the taste of fresh fruits and vegetables, 2) an interest in gardening as a hobby, 3) a desire to save money and cut the food budget.

But besides saving money, a lot of intangible satisfactions come from gardening and home canning. Who knows the value of being able to say, "I raised it myself" or "I prepared it myself"?

We wish you the best of luck in your gardening, and hope this book will be helpful.

Preface

Jack Hayes
Yearbook Editor

Don't start too big. This advice to new gardeners comes from a seed company horticulturist. A smaller garden that is well kept will produce more and better-quality food than a big one that is neglected, she notes, adding: "You want your gardening to be fun."

An Extension specialist says you can have a productive garden in quite a small space—a 10 by 15 foot area, for example. Or your garden may be limited to a balcony or even a windowsill.

Keep your investment in supplies and equipment to a minimum until you find out whether you want to continue gardening. A good time to make a decision is after harvest. You will know about the work and time required, and the expenses involved. And if you want to continue, you will be more of a realist in opting to stay small or to expand.

This Yearbook has been written by knowledgeable specialists throughout the United States and can be your guide for a successful garden. But also read gardening publications prepared by the Cooperative Extension Service in your State, attend gardening meetings that often are held before the growing season, and check your library for useful literature.

Besides the authors, other people with a wide variety of talents and experience contributed to this Yearbook. Personnel of the *Typography and Design Division*, U.S. Government Printing Office, involved in the book include Charles McKeown, Rudie Diamond, Howard Behrens, and Irene Bebber. Denver Browning of the Yearbook staff prepared the index. Other staff members were Mary Vest, Mary McGowan, and Mark Wolkow.

Robert A. Wearne, *Extension Service*, chaired the Yearbook Committee that planned the book. Allan K. Stoner of the *Agricultural Research Service* was assistant chairman.

Yearbook Committee members were:
    William Bailey, *Agricultural Research Service*
    Thomas Barksdale, *Agricultural Research Service*
    Paul Bergman, *Extension Service*

Cecil Blackwell, *American Society for Horticultural Science*
Howard Brooks, *Agricultural Research Service*
Raymond Brush, *American Association of Nurserymen*
William E. Carnahan, *Extension Service*
Robert Falasca, *American Seed Trade Association*
Evelyn Johnson, *Extension Service*
Charles McClurg, *University of Maryland*
Harold Owens, *Extension Service*
David S. Ross, *University of Maryland*
Jane Steffey, *American Horticultural Society*
Fred Westbrook, *Extension Service*

Contents

Foreword xxxiii
  BOB BERGLAND, Secretary of Agriculture

Preface xxxv
  JACK HAYES, Yearbook Editor

Part 1 Introduction to Gardening

Why Folks Garden, and What They Face 2
  CECIL BLACKWELL
Where to Garden—Setting Your Sites 15
  JAMES W. WILSON
Garden Tools and Equipment 24
  JOHN W. BARTOK, JR.
Learning to Make the Best Use of Climate 33
  Preface
    In the Northwest 34
      EARL M. BATES
    In the Northeast 38
      WALTER L. STIRM
    In the Southwest 41
      M. DOUGLAS BRYANT and RICARDO E. GOMEZ
    In the Southeast 43
      PERRY M. SMITH
How Plants Grow—and Let's Hope They Do! 47
  O. B. COMBS
Plant Reproduction 49
  N. CARL HARDIN and BRADFORD C. BEARCE
Plant Pollination 51
  S. E. McGREGOR
Know Your Soil and How to Manage It 54
  LINDO J. BARTELLI, DAVID F. SLUSHER and KELTON L. ANDERSON
Structures—From Trellis to Greenhouse 61
  DAVID S. ROSS
Pest Management Is a Matter of Timing 71
  DAN C. SCHEEL
Organic Gardening—Think Mulch 78
  WESLEY P. JUDKINS

End of One Season Is Start of Another 84
  CHARLES W. MARR

Help! Help! Where You Can Find It 89
  BARBARA H. EMERSON

Gardener's Glossary 93
  Compiled by ROBERT A. WEARNE

Metric Multipliers 100

Part 2 Home Garden Vegetables

Planning Your Vegetable Garden—
Plots, Pyramids, and Planters 102
  GEORGE and KATY ABRAHAM

Growing Vegetable Transplants:
Lights, Containers, Media, Seed 111
  FRANKLIN D. SCHALES

The Complex Art of Planting 119
  CHARLES W. REYNOLDS

Vegetables in Containers Require
Enough Sun, Space, Drainage 126
  KATHRYN L. ARTHURS

Play It Cool With Cole Crops (Cabbage, Etc.);
They Attain Best Quality If Matured in Fall 133
  PHILIP A. MINGES

The Popular, Cultivated Tomato and Kinfolk
Peppers, Eggplant 139
  ALLAN K. STONER and BENIGNO VILLALON

Leafy Salad Vegetables: Lettuce, Celery, Cress,
Endive, Escarole, Chicory 147
  BRUCE JOHNSTONE

Onions Are Finicky as to Growing, Curing;
And Garlic May Not Be a Joy Either 152
  J. S. VANDEMARK

Root Crops More or Less Trouble-free, Produce Lots of
Food in a Small Space 157
  N. S. MANSOUR and J. R. BAGGETT

Greens or "Potherbs"—Chard, Collards, Kale,
Mustard, Spinach, New Zealand Spinach 163
  ALBERT A. BANADYGA

Beans and Peas Are Easy to Grow and
Produce a Wealth of Food 171
  JACK P. MEINERS and JOHN M. KRAFT

Sweet Corn, That Home Garden Favorite, for Good
Nutrition and Eating Pleasure 181
  E. V. WANN

Cucurbit Crops—Cucumbers, Gourds, Melons, Pumpkins,
Squash—Have Uniform Needs 187
    THOMAS W. WHITAKER

Asparagus Starts Up Slow But Goes On and On;
Rhubarb Also Takes Its Own Sweet Time 196

    Asparagus 196
        STEPHEN A. GARRISON and J. HOWARD ELLISON

    Rhubarb 201
        DANIEL TOMPKINS

A Few Rows of Home Garden Potatoes Can
Put Nutritious Food on Your Table 205
    ORRIN C. TURNQUIST

Sweet Potatoes—Buried Treasure 212
    JOHN C. BOUWKAMP

Herbs for Flavor, Fragrances, Fun
In Gardens, Pots, in Shade, in Sun 217
    DORIS THAIN FROST

Okra Is Produced Primarily in the South As
Main Dish Vegetable, and for Gumbos 224
    W. D. KIMBROUGH, L. G. JONES and J. F. FONTENOT

Miscellany, Including Celeriac, Horseradish,
Artichoke, Peanuts, Vegetable Soybeans 228
    HOMER N. METCALF and MILO BURNHAM

Part 3 Fruits and Nuts

Growing Apples, Pears, and Quinces; Pest Control,
Air Drainage Important 246
    ROGER D. WAY

Peaches, Nectarines, Plums, Apricots, Cherries—
Climate Puts Limits on What You Can Raise 253
    JOHN H. WEINBERGER and HAROLD W. FOGLE

Grapes Are Great But You May Have to Wait;
Buying Rooted Vines Can Save You a Year 260
    J. R. McGREW

Strawberries Like Full Sun—and a Good Deal of Attention . 265
    ROBERT G. HILL, JR., JAMES D. UTZINGER, and ELDEN J. STANG

Cane and Bush Fruits Are the Berries;
Often It's Grow Them or Go Without 272
    JOHN P. TOMKINS

Just About Any Home Garden Can Produce Blueberries . . 279
    G. J. GALLETTA and A. D. DRAPER

Nut Crops—Trees for Food, Ornament, Shade, and Wood . . 284
    RICHARD A. JAYNES and HOWARD L. MALSTROM

Subtropical Fruit Choice Wide—From Avocado to Tamarind . 291
    ROBERT J. KNIGHT, JR. and JULIAN W. SAULS

Part 4 Home Food Preservation

The Whys of Food Preservation 298
EDMUND A. ZOTTOLA and ISABEL D. WOLF

How to Minimize Quality Losses 304
GERALD D. KUHN and LOUISE W. HAMILTON

Economics of Home Food Preservation,
or Is Do-It-Yourself Back to Stay? 310
RUTH N. KLIPPSTEIN

Beginner's Guide to Home Canning 313
FRANCES REASONOVER

A Primer on Home Freezing for the Beginner 320
CHARLOTTE M. DUNN

Pressure Canners, Vital for Low-Acid Foods 323
NADINE FORTNA TOPE

Home Canning of Fruits and Vegetables 328
CAROLE DAVIS

Freezing Your Garden's Harvest 334
ANNETTA COOK

Jellies, Jams, Marmalades, Preserves 340
CATHARINE C. SIGMAN and KIRBY HAYES

Pickles, Relishes Add Zip and Zest 345
ISABELLE DOWNEY

Wine Making (with a note on vinegar) 350
PHILIP WAGNER and J. R. McGREW

Home Drying of Fruits and Vegetables 356
DALE E. KIRK and CAROLYN A. RAAB

Storage of Home-Preserved Foods 361
RALPH W. JOHNSTON

Storing Fresh Fruit and Vegetables 365
ANTON S. HORN and ESTHER H. WILSON

Resurgence of Community Canneries 372
F. ALINE COFFEY and ROGER STERNBERG

Questions and Answers on Food Preservation 378
CAROLE DAVIS and ANNETTA COOK

Food Preservation Glossary 383
Compiled by ANNETTA COOK and CAROLE DAVIS

Photography 385

Index 387

PART 1

Introduction to Gardening

Fred S. Witte

Why Folks Garden, and What They Face
by Cecil Blackwell

Interest in home gardening is at its highest level since the Victory Garden era of World War II. Victory Gardens were encouraged then to offset the shortages in commercial production, processing, and transportation of vegetables and fruits during the war years.

The current high interest in gardening is attributed to several factors. Among these is the increasing cost of food which has resulted largely from higher energy and labor costs in producing, processing, and transporting food. Even so, Americans can purchase an adequate food supply with a lower percentage of their take-home pay than can the people of virtually all other countries. But at today's prices, it has been estimated that an average family can save $200 to $300 annually on food costs by growing and processing fruits and vegetables at home. Your savings may be greater or less, depending on the size of your family, the size of your garden, and, more importantly, your skills as a gardener.

But other values derived from gardening are perhaps more important than potential savings on food costs. Some of these can be cited as contributing to the current interest in gardening.

Freshness and quality. Produce harvested at peak maturity from the garden generally has better flavor and higher nutritional value than that harvested at earlier stages of maturity and shipped long distances to the supermarket. This peak quality can also be "captured" and preserved in home-canned and frozen produce.

Therapy, or personal satisfaction. Working with living plants and seeing them respond has therapeutic value. Success from learning and using new skills stirs a sense of pride and achievement. The exercise can be relaxing, even recreational.

A family activity. Gardening, and home preserving of fruits and vegetables, can be a learning experience for the entire family—including the children. What better way to teach "biology" to your children!

The "back-to-nature" trend. This has been cited as one of the reasons why so many young adults (in the 18 to 25 age group) have developed such a keen interest in gardening and food preservation. The desire for organically-grown or "natural" foods is no doubt related to this trend.

Artistic and esthetic values. The vegetable garden and a "home orchard or vineyard" can be an integral part of the home landscaping, enhancing its variety and color at different seasons of the year.

Dwarf fruit trees can be used as "specimen" plants or trained to grow against a fence or wall. Small fruits such as strawberries, blueberries, blackberries and raspberries can be grown in beds or as border plantings. Even the vegetable garden plots can be arranged to add esthetic value to the home landscape.

Neighborliness. In urban areas, next door neighbors often do not really get acquainted with each other. But when a neighbor also gardens you will have much in common to talk about, to share, to compare, to brag on, and to swap in terms of information, ideas, plant materials, products

---

Cecil Blackwell is Executive Director of the American Society for Horticultural Science, Mount Vernon, Va. He formerly was Horticultural Editor of *The Progressive Farmer* magazine, Birmingham, Ala. (1959–65); and Extension Horticulturist at the University of Georgia, Athens (1954–59).

But take courage. The information given here is designed to help you prevent or overcome these problems.

Finally, the work involved in gardening may be a "problem". Sure, there is therapy in planning, planting, and tending a garden, and in harvesting and processing its products. But the work is demanding at times. It may be physically tiring, even exhausting. But, more importantly, things have to be done at certain times during the growing season for best results. This could interfere at times with things like golfing, fishing, or weekend outings.

If you're "addicted" to camping, fishing, golfing and the like, you'd be wise to limit the size and scope of your garden.

Following are some of the more important keys to gardening success.

Selecting Sites

In selecting sites for gardening the most important factors are sunlight, air circulation, protection, competition from trees and large shrubs, drainage, soil structure, and soil fertility.

Most fruit trees and vegetable plants require direct sunlight most of the day for optimum growth and productivity. Some leafy vegetables will tolerate partial shade, but fruit trees and fruit-producing vegetables need direct sunlight. Most plants can tolerate some shade from buildings, fences, and distant trees, particularly during the late afternoon, and some may even benefit from such shade during the hottest part of summer in warmer regions of the country.

Air circulation is important in reducing the likelihood or severity of plant diseases which affect leaves and fruits, particularly in humid areas.

Prolonged dampness of foliage fol-

successes and failures. If your space is inadequate for a home garden and you have a plot in a Community Garden, this comradeship can be broadened beyond the immediate neighborhood.

Beware of Problems

If you are a beginning gardener, have a go at it with open eyes as well as an open mind. Keep in mind that you may face some frustrations and disappointments.

There will be problems with weeds, insects, plant diseases and disorders, and other pests such as rodents and even pets. No doubt about it.

There may well be problems with drought, rain (too much, too little, or at the wrong time), heat, and cold. The weather doesn't always cooperate.

There may be problems from the hot sun, too much shade, competition from tree roots, flooding, poor drainage, and poor soil conditions.

A family garden in the suburbs—great hobby for the kids!

lowing rains or heavy dew favors development and spread of fungus and bacterial diseases that affect plants. Too much shade, especially during the early morning, compounds the problem.

Windbreaks may be beneficial in certain parts of the country, such as the Great Plains, to protect plants from strong prevailing winds. For fruit crops in particular, "frost pockets" should be avoided, since good air drainage helps reduce the hazard of late spring frosts.

Protection from certain animals is often desirable. A close-woven fence around the garden may be needed for protection against rabbits, dogs, and other animals. A fence can also serve as a trellis for crops such as tomatoes, cucumbers, and pole beans. Other types of protection may be needed in some areas against rodents and animals such as mice, moles, ground squirrels, and prairie dogs.

Avoid competition from trees and large shrubs so far as possible. The ill effect of too much shade has already been discussed. Moreover, tree roots often extend considerable distances from trees and will compete with garden crops for moisture and plant nutrients. Cutting tree roots along the edge of the garden will usually provide only temporary relief, since the tree roots will again extend rapidly into the fertile, moist soil of the garden area.

Soil drainage, structure, and fertility can usually be improved. A loamy, fertile soil with a medium clay subsoil is ideal. The structure (and fertility) of heavy clay soils can be improved by adding liberal quantities of organic matter such as compost or manure as well as lime and fertilizer. The fertility and moisture-holding capacity of very sandy soils can be improved in a similar manner, except that less lime should generally be used than on heavy clay soils.

A soil test should be taken as a guide to how much lime and fertilizer are needed.

Good drainage also is important, since plant roots must have air as well as moisture. Adding organic matter improves the moisture-air balance in most soils. However, if your garden site is subject to flooding or has poor surface drainage, you may be able to improve the site by installing drain tile, digging drainage ditches, or plowing deep into the subsoil to break-up an impervious layer called a "hardpan".

If you encounter this problem, seek professional help and advice from your county extension agent, the Soil Conservation Service, or a drainage contractor.

Are street lights near the garden beneficial? Usually not. Are they detrimental? Usually not. However, some vegetables such as spinach, lettuce, cabbage, and radishes are sensitive to daylength or "photoperiodism" in that they tend to "bolt" or go to seed when days are long and temperatures are warm. Thus, street lights can, in effect, create longer days. But most vegetable and fruit crops are not sensitive to daylength. Street lights close to the garden may be beneficial as a security measure against thefts or vandalism.

Preparing Soil for Planting

Soils vary widely in different areas of the country. Ideally, the kind and amount of commercial fertilizer and lime should be applied in accordance with soil test results. Your county Extension office or local garden center can assist you with respect to taking soil samples and sending them to the nearest soil testing laboratory for analysis.

Lime, when needed, has a fourfold effect on most soils:

- It adds calcium, magnesium, and certain other plant nutrients, depending upon the kind of liming material used

- It reduces soil acidity, thereby making some major and minor plant nutrients more readily available (or soluble) for plant growth
- It improves the physical structure of heavy clay soils
- And it improves the environment for certain beneficial micro-organisms in the soil, including those which decompose organic matter, thereby releasing nitrogen and other nutrients for plant growth

Organic matter has a highly beneficial effect on most soils. It makes soil more "mellow" and, when it decays, releases nitrogen, minerals, and other nutrients for use by plants.

There are those who advocate the use of organic matter to the complete exclusion of chemical fertilizers to improve soil fertility and productivity. On the other hand, modern technology makes it possible to supply all the nutrient elements needed for plant growth—in the form of commercial fertilizer. Most organic materials do not contain a "balanced ration" of all the nutrient elements needed by plants. Thus, organic matter will not supply all the nutrient elements needed to correct deficiencies in many soils, the sandier ones in particular. Accordingly, the wiser choice in most cases is to use both organic matter *and* chemical fertilizers in quantities needed for optimum productivity.

Organic materials such as compost and livestock or poultry manures can be mixed directly into the garden soil. However, undecomposed materials such as tree bark, sawdust, leaves, lawn clippings, straw, and refuse from the garden or kitchen should be composted *before* mixing them into the soil—although they can be used as a surface mulch around plants.

Here's why. Humus (decayed organic material) has a carbon/nitrogen ratio of about 10 to 1. Wheat or oat straw has a carbon/nitrogen ratio of about 70 to 1, and the ratio is much higher for woody material such as

sawdust and bark. The decomposition process requires large quantities of nitrogen, and plants will suffer from nitrogen deficiency if substantial amounts of undecomposed organic material is mixed in the soil during the growing season.

When used as a mulch on the surface, however, the organic material can be turned under at the end of the growing season, in which case extra nitrogen fertilizer should be added to speed up decomposition before the next growing season. If, during the next growing season, plants develop nitrogen deficiency (become yellow and stunted in growth), it can usually be corrected by moderate applications of nitrogen fertilizer as needed during the growing season.

What to Grow, How Much

Important considerations in what to grow and how much are nutritional

Spreading thin mulch of sawdust (¼ inch) to protect seed row from drying and crusting.

Garden Planning and Planting Guide—Nutritional Value of Selected Vegetables

Nutritional group	Vegetable	Vitamin content A (I.U.)	Vitamin content C (mg.)	Food energy (calories)
Group I High in Vitamins A and C (plant 3 or more from this group)	Parsley (raw)	8,500	172	44
	Spinach	8,100	28	23
	Collards	7,800	76	33
	Kale	7,400	62	28
	Turnip Greens	6,300	69	20
	Mustard Greens	5,800	48	23
	Cantaloupes	3,400	33	30
	Broccoli	2,500	90	26
Group II High in Vitamin A (plant 2 or more from this group)	Carrots (raw)	11,000	8	42
	Carrots (cooked)	10,500	6	31
	Sweetpotatoes	8,100	22	141
	Swiss Chard	5,400	16	18
	Winter Squash	4,200	13	63
	Green Onions	2,000	32	36
Group III High in Vitamin C (plant 3 or more from this group)	Peppers (mature green)	420	128	22
	Brussels Sprouts	520	87	36
	Cauliflower	60	55	22
	Kohlrabi	20	43	24
	Cabbage	130	33	20
	Chinese Cabbage	150	25	14
	Asparagus	900	26	20
	Rutabagas	550	26	35
	Radishes (raw)	322	26	17
	Tomatoes (ripe, raw)	900	23	22
	Tomatoes (ripe, cooked)	1,000	24	26
Group IV Other Green Vegetables (plant 2 or more from this group)	Green Beans	540	12	25
	Celery	240	9	17
	Lettuce (leaf)	1,900	18	18
	Lettuce (head)	330	6	13
	Okra	490	20	29
	Peas (garden)	540	20	71
Group V Starchy Vegetables (plant 2 or more from this group)	Lima Beans	280	17	111
	Sweet Corn (yellow)	400	9	91
	Onions (dry)	40	10	38
	Peas (field, southern)	350	17	108
	Potatoes (baked in skin)	Trace	20	93
Group VI Other Vegetables (plant from this group for variety in flavor, color, texture, etc.)	Beets	20	6	32
	Cucumbers	250	11	15
	Eggplants	10	3	19
	Pumpkins	1,600	9	26
	Rhubarb	80	6	141
	Summer Squash	440	11	15
	Turnips (roots)	Trace	22	23

Figures are for amounts of vitamins and calories per 100-gram sample for cooked vegetables (unless normally eaten raw). Vitamin A is expressed in International Units (I. U.) per 100-gram sample; Vitamin C is expressed in milligrams per 100-gram sample; and Food Energy is expressed in Bilogram calories per 100-gram sample.

100 grams is equal to about 1/2 cup.

Active adults require daily about 5,000 I.U. of vitamin A for men and 4,000 for women; 45 mg. of vitamin C (men and women); and 2,700 calories for men and 2,000 for women.

Vitamin C values are generally higher if the vegetable is eaten raw. An example is cabbage: 33 mg. cooked; 47 mg. raw.

values, kinds of fruits and vegetables best liked by your family, the kinds that grow best in your area, amount of gardening space available, and what you best like to grow for reasons of pleasure and satisfaction.

Most fruits and the leafy-green and yellow vegetables are among the best sources of vitamins A and C. Some are good sources of other vitamins and minerals, and others provide important bulk in your diet.

The Garden Planning and Planting Guide should help you decide which vegetables to plant—for fresh use and for processing if space permits. Choosing 2 or 3 from each nutritional group will contribute much to a balanced diet for your family.

Selecting Varieties

Choosing the best varieties to plant in your area is highly important. Because of differences in soils and climatic factors such as temperature, length of growing season, and rainfall, varieties of fruits and vegetables that do best in one area may do poorly in other areas. Also, some vegetable varieties may do best in spring plantings and others in late summer or fall plantings, due to differences in temperature, rainfall, and day length.

Plant breeders are constantly developing new varieties. It is usually advisable to try these on a limited scale along with varieties you have been growing, in the same season, for comparison.

Many disease-resistant varieties and strains are now available. These should be selected, if adapted to your area.

It is generally advisable to purchase new seed each year. However, leftover seed of some vegetables can be kept in viable condition several years if stored in a cool, dry place (preferably in an airtight container) and protected from rats, mice, and weevils. Exceptions are sweet corn, leeks, okra, onions, parsley, parsnips, rhubarb, and salsify. These generally keep in a viable condition only 1 or 2 years.

It is usually not advisable to harvest and save your own seed from the garden. Many new varieties are hybrids and will not "come true" from second generation seed.

When to Plant

As a general rule, the most cold-hardy, cool-season vegetables should be planted (or transplanted) 4 to 6 weeks before the last spring frost. The less-hardy, cool-season crops should be planted 2 to 4 weeks before the last spring frost for best results. Refer to table for spring planting dates.

Most cool-season crops thrive best during cool weather and will not tolerate the heat of summer in the lower two-thirds of the country. In this area, early-spring planting is vitally important.

The most cold-hardy of these cool-season crops include broccoli, cabbage, cauliflower, collards, kale, kohlrabi, lettuce, onions, English peas, Irish potatoes, spinach, and turnips. The less-hardy, cool-season crops include beets, carrots, chard, mustard, parsnips, and radishes.

In most areas of the country, the frost-tender, warm-season or heat-tolerant vegetables should not be planted until after the last spring frost and the soil has warmed up. The more cool-tolerant of these can be planted as near the last frost date as possible, and include snap beans, New Zealand spinach, squash, sweet corn, and tomato (transplants). The more heat-tolerant of these should not be planted (or transplanted) until a week or two after the frost-free date, and include lima beans, eggplant, peppers, okra, Southern peas, sweet potatoes, cucumbers, cantaloupes, pumpkins, and watermelons.

Some vegetables have a rather short harvest season from any one planting. To have fresh produce over a longer

Range of Dates for Safe Spring Planting in the Open

Crop	Planting dates for localities in which average date of last freeze [3] is—		
	Jan. 30	Feb. 18	Mar. 10
Asparagus [1]			Jan. 1–Mar. 1
Beans, lima	Feb. 1–Apr. 15	Mar. 1–May 1	Mar. 20–June 1
Beans, snap	Feb. 1–Apr. 1	Mar. 1–May 1	Mar. 15–May 15
Beet	Jan. 1–Mar. 15	Jan. 20–Apr. 1	Feb. 15–June 1
Broccoli, sprouting [1]	Jan. 1–30	Jan. 15–Feb. 15	Feb. 15–Mar. 15
Brussels sprouts [1]	Jan. 1–30	Jan. 15–Feb. 15	Feb. 15–Mar. 15
Cabbage [1]	Jan. 1–15	Jan. 1–Feb. 25	Jan. 25–Mar. 1
Cabbage, Chinese	([2])	([2])	([2])
Carrot	Jan. 1–Mar. 1	Jan. 15–Mar. 1	Feb. 10–Mar. 15
Cauliflower [1]	Jan. 1–Feb. 1	Jan. 10–Feb. 10	Feb. 1–Mar. 1
Celery and celeriac	Jan. 1–Feb. 1	Jan. 20–Feb. 20	Feb. 20–Mar. 20
Chard	Jan. 1–Apr. 1	Jan. 20–Apr. 15	Feb. 15–May 15
Chervil and chives	Jan. 1–Feb. 1	Jan. 1–Feb. 1	Feb. 1–Mar. 1
Chicory, witloof			June 1–July 1
Collards [1]	Jan. 1–Feb. 15	Jan. 1–Mar. 15	Feb. 1–Apr. 1
Corn, sweet	Feb. 1–Mar. 15	Feb. 20–Apr. 15	Mar. 10–Apr. 15
Cress, upland	Jan. 1–Feb. 1	Jan. 15–Feb. 15	Feb. 10–Mar. 15
Cucumber	Feb. 15–Mar. 15	Feb. 15–Apr. 15	Mar. 15–Apr. 15
Eggplant [1]	Feb. 1–Mar. 1	Feb. 20–Apr. 1	Mar. 15–Apr. 15
Endive	Jan. 1–Mar. 1	Jan. 15–Mar. 1	Feb. 15–Mar. 15
Garlic	([2])	([2])	([2])
Horseradish [1]			
Kale	Jan. 1–Feb. 1	Jan. 20–Feb. 10	Feb. 10–Mar. 1
Kohlrabi	Jan. 1–Feb. 1	Jan. 20–Feb. 10	Feb. 10–Mar. 1
Leek	Jan. 1–Feb. 1	Jan. 1–Feb. 15	Jan. 25–Mar. 1
Lettuce, head [1]	Jan. 1–Feb. 1	Jan. 1–Feb. 1	Feb. 1–20
Lettuce, leaf	Jan. 1–Feb. 1	Jan. 1–Mar. 15	Jan. 15–Apr. 1
Muskmelon	Feb. 15–Mar. 15	Feb. 15–Apr. 15	Mar. 15–Apr. 15
Mustard	Jan. 1–Mar. 1	Feb. 15–Apr. 15	Feb. 10–Mar. 15
Okra	Feb. 15–Apr. 1	Mar. 1–June 1	Mar. 20–June 1
Onion [1]	Jan. 1–15	Jan. 1–15	Jan. 15–Feb. 15
Onion, seed	Jan. 1–15	Jan. 1–15	Feb. 1–Mar. 1
Onion, sets	Jan. 1–15	Jan. 1–15	Jan. 15–Mar. 10
Parsley	Jan. 1–30	Jan. 1–30	Feb. 1–Mar. 10
Parsnip		Jan. 1–Feb. 1	Jan. 15–Mar. 1
Peas, garden	Jan. 1–Feb. 15	Jan. 1–Mar. 1	Jan. 15–Mar. 15
Peas, black-eye	Feb. 15–May 1	Mar. 1–June 15	Mar. 15–July 1
Pepper [1]	Feb. 1–Apr. 1	Mar. 1–May 1	Apr. 1–June 1
Potato	Jan. 1–Feb. 15	Jan. 15–Mar. 1	Feb. 1–Mar. 1
Radish	Jan. 1–Apr. 1	Jan. 1–Apr. 1	Jan. 1–Apr. 15
Rhubarb [1]			
Rutabaga			Jan. 15–Feb. 15
Salsify	Jan. 1–Feb. 1	Jan. 15–Feb. 20	Feb. 1–Mar. 1
Shallot	Jan. 1–Feb. 1	Jan. 1–Feb. 20	Jan. 15–Mar. 1
Sorrel	Jan. 1–Mar. 1	Jan. 15–Mar. 1	Feb. 10–Mar. 15
Spinach	Jan. 1–Feb. 15	Jan. 1–Mar. 1	Jan. 15–Mar. 10
Spinach, New Zealand	Feb. 1–Apr. 15	Mar. 1–Apr. 15	Mar. 20–May 15
Squash, summer	Feb. 1–Apr. 15	Mar. 1–Apr. 15	Mar. 15–May 1
Sweetpotato	Feb. 15–May 15	Mar. 20–June 1	Apr. 1–June 1
Tomato	Feb. 1–Apr. 1	Mar. 1–Apr. 20	Mar. 20–May 10
Turnip	Jan. 1–Mar. 1	Jan. 10–Mar. 1	Feb. 1–Mar. 1
Watermelon	Feb. 15–Mar. 15	Feb. 15–Apr. 15	Mar. 15–Apr. 15

See footnotes at end of table.

Range of Dates for Safe Spring Planting in the Open—Continued

Crop	Planting dates for localities in which average date of last freeze is—		
	Mar. 30	Apr. 20	May 10
Asparagus [1]	Feb. 15–Mar. 20	Mar. 15–Apr. 15	Mar. 10–Apr. 30
Beans, lima	Apr. 15–June 20	May 1–June 20	May 25–June 15
Beans, snap	Apr. 1–June 1	Apr. 25–June 30	May 10–June 30
Beet	Mar. 1–June 1	Mar. 20–June 1	Apr. 15–June 15
Broccoli, sprouting [1]	Mar. 1–20	Mar. 25–Apr. 20	Apr. 15–June 1
Brussels sprouts [1]	Mar. 1–20	Mar. 25–Apr. 20	Apr. 15–June 1
Cabbage [1]	Feb. 15–Mar. 10	Mar. 10–Apr. 1	Apr. 1–May 15
Cabbage, Chinese	([2])	([2])	Apr. 1–May 15
Carrot	Mar. 1–Apr. 10	Apr. 1–May 15	Apr. 20–June 15
Cauliflower [1]	Feb. 20–Mar. 20	Mar. 15–Apr. 20	Apr. 15–May 15
Celery and celeriac	Mar. 15–Apr. 15	Apr. 10–May 1	Apr. 20–June 15
Chard	Mar. 1–May 25	Apr. 1–June 15	Apr. 20–June 15
Chervil and chives	Feb. 15–Mar. 15	Mar. 10–Apr. 10	Apr. 1–May 1
Chicory, witloof	June 1–July 1	June 15–July 1	June 1–20
Collards [1]	Mar. 1–June 1	Mar. 10–June 1	Apr. 15–June 1
Corn, sweet	Mar. 25–May 15	Apr. 25–June 15	May 10–June 1
Cress, upland	Mar. 1–Apr. 1	Mar. 20–May 1	Apr. 20–May 20
Cucumber	Apr. 10–May 15	May 1–June 15	May 20–June 15
Eggplant [1]	Apr. 15–May 15	May 10–June 1	May 20–June 15
Endive	Mar. 10–Apr. 10	Mar. 25–Apr. 15	Apr. 15–May 15
Garlic	Feb. 10–Mar. 10	Mar. 10–Apr. 1	Apr. 1–May 1
Horseradish [1]	Mar. 1–Apr. 1	Mar. 20–Apr. 20	Apr. 15–May 15
Kale	Mar. 1–20	Mar. 20–Apr. 10	Apr. 10–May 1
Kohlrabi	Mar. 1–Apr. 1	Mar. 20–May 1	Apr. 10–May 15
Leek	Feb. 15–Mar. 15	Mar. 15–Apr. 15	Apr. 15–May 15
Lettuce, head [1]	Mar. 1–20	Mar. 20–Apr. 15	Apr. 15–May 15
Lettuce, leaf	Feb. 15–Apr. 15	Mar. 20–May 15	Apr. 15–June 15
Muskmelon	Apr. 10–May 15	May 1–June 15	June 1–June 15
Mustard	Mar. 1–Apr. 15	Mar. 20–May 1	Apr. 15–June 1
Okra	Apr. 10–June 15	May 1–June 1	May 20–June 10
Onion [1]	Feb. 15–Mar. 15	Mar. 15–Apr. 10	Apr. 10–May 1
Onion, seed	Feb. 20–Mar. 15	Mar. 15–Apr. 1	Apr. 1–May 1
Onion, sets	Feb. 15–Mar. 20	Mar. 10–Apr. 1	Apr. 10–May 1
Parsley	Mar. 1–Apr. 1	Mar. 20–Apr. 20	Apr. 15–May 15
Parsnip	Mar. 1–Apr. 1	Mar. 20–Apr. 20	Apr. 15–May 15
Peas, garden	Feb. 10–Mar. 20	Mar. 10–Apr. 10	Apr. 1–May 15
Peas, black-eye	Apr. 15–July 1	May 10–June 15	
Pepper [1]	Apr. 15–June 1	May 10–June 1	May 20–June 10
Potato	Feb. 20–Mar. 20	Mar. 15–Apr. 10	Apr. 1–June 1
Radish	Feb. 15–May 1	Mar. 10–May 10	Apr. 1–June 1
Rhubarb [1]		Mar. 10–Apr. 10	Apr. 1–May 1
Rutabaga	Feb. 1–Mar. 1		May 1–June 1
Salsify	Mar. 1–15	Mar. 20–May 1	Apr. 15–June 1
Shallot	Feb. 15–Mar. 15	Mar. 15–Apr. 15	Apr. 10–May 1
Sorrel	Feb. 20–Apr. 1	Mar. 15–May 1	Apr. 15–June 1
Spinach	Feb. 1–Mar. 20	Mar. 1–Apr. 15	Apr. 1–June 15
Spinach, New Zealand	Apr. 10–June 1	May 1–June 15	May 10–June 15
Squash, summer	Apr. 10–June 1	May 1–June 15	May 10–June 10
Sweetpotato	Apr. 20–June 1	May 10–June 10	
Tomato	Apr. 10–June 1	May 5–June 10	May 15–June 10
Turnip	Feb. 20–Mar. 20	Mar. 10–Apr. 1	Apr. 1–June 1
Watermelon	Apr. 10–May 15	May 1–June 15	June 1–June 15

See footnotes at end of table.

Range of Dates for Safe Spring Planting in the Open—*Continued*

Crop	Planting dates for localities in which average date of last freeze is—
	May 30
Asparagus [1]	May 1–June 1
Beans, lima	
Beans, snap	May 25–June 15
Beet	May 1–June 15
Broccoli, sprouting [1]	May 10–June 10
Brussels sprouts [1]	May 10–June 10
Cabbage [1]	May 10–June 15
Cabbage, Chinese	May 10–June 15
Carrot	May 10–June 1
Cauliflower [1]	May 20–June 1
Celery and celeriac	May 20–June 1
Chard	May 20–June 1
Chervil and chives	May 1–June 1
Chicory, witloof	June 1–15
Collards [1]	May 10–June 1
Corn, sweet	May 20–June 1
Cress, upland	May 15–June 1
Cucumber	
Eggplant [1]	
Endive	May 1–30
Garlic	May 1–30
Horseradish [1]	May 1–30
Kale	May 1–30
Kohlrabi	May 1–30
Leek	May 1–15
Lettuce, head [1]	May 10–June 30
Lettuce, leaf	May 10–June 30
Muskmelon	
Mustard	May 10–June 30
Okra	
Onion [1]	May 1–30
Onion, seed	May 1–30
Onion, sets	May 1–30
Parsley	May 10–June 1
Parsnip	May 10–June 1
Peas, garden	May 1–June 15
Peas, black-eye	
Pepper [1]	June 1–15
Potato	May 1–June 15
Radish	May 1–June 15
Rhubarb [1]	May 1–20
Rutabaga	May 10–20
Salsify	May 10–June 1
Shallot	May 1–June 1
Sorrel	May 10–June 10
Spinach	Apr. 20–June 15
Spinach, New Zealand	June 1–15
Squash, summer	June 1–20
Sweetpotato	
Tomato	June 5–20
Turnip	May 1–June 15
Watermelon	

[1] Plants.   [2] Generally fall-planted.   [3] Check county Extension office for date of last freeze.

period, you can make sequential plantings of the following every 2 to 4 weeks as long as the weather is favorable for each: radishes, turnip greens, mustard, leaf lettuce, and spinach among the cool-season crops; and bush snap beans, lima beans, squash, sweet corn, and Southern peas among the warm-season, heat-tolerant crops.

It's better to keep the entire garden plot busy than to let it grow up in weeds, go to seed, and cause trouble the next season. In the small garden, you can easily clear off rows and replant them as soon as the first planting has been harvested.

Some of the cold-hardy, cool-season vegetables do well when planted in late summer or early fall, to mature during cool weather. These should be planted 6 to 8 weeks before the first fall frost (except in the North), and include beets, broccoli, collards, kale, lettuce, mustard, spinach, turnips and rutabagas. Refer to table for fall planting dates.

Cultivating, Mulching

The purpose of cultivation is primarily to control weeds. If the garden is not mulched, cultivate often enough to destroy weeds while they are small. If the soil is hard and crusty, plants may benefit from loosening the soil to provide better aeration—especially while crops are small. After crops become larger, cultivation should be very shallow to avoid damaging roots near the surface.

Avoid cultivating while the soil is wet. This destroys the crumbly structure of the soil.

Mulching is very beneficial in conserving moisture, controlling weeds, and improving soil structure. Organic mulches such as old sawdust, leaves, ground-up tree bark, and partially rotted hay, straw, or lawn clippings will also keep the soil cooler, keep it from packing, and add humus to the soil when turned under at the end of the growing season. Any weeds that push through the mulch can easily be pulled out by hand.

Black plastic sheeting 1½ mils thick

Two cool-season crops in Oregon. Left, green-type kohlrabi, a "stem turnip", in a garden in February. Some leaves have been lost. Right, digging beets in winter.

Range of Dates for Safe Fall Planting in the Open

Crop	Planting dates for localities in which average date of first freeze is—		
	Aug. 30	Sept. 20	Oct. 10
Asparagus [1]			Oct. 20–Nov. 15
Beans, lima			June 1–15
Beans, snap		June 1–July 1	June 15–July 20
Beet	May 15–June 15	June 1–July 1	June 15–July 25
Broccoli, sprouting	May 1–June 1	May 1–June 15	June 15–July 15
Brussels sprouts	May 1–June 1	May 1–June 15	June 15–July 15
Cabbage [1]	May 1–June 1	May 1–June 15	June 1–July 15
Cabbage, Chinese	May 15–June 15	June 1–July 1	June 15–Aug. 1
Carrot	May 15–June 15	June 1–July 1	June 1–July 20
Cauliflower [1]	May 1–June 1	May 1–July 1	June 1–July 25
Celery [1] and celeriac	May 1–June 1	May 15–July 1	June 1–July 15
Chard	May 15–June 15	June 1–July 1	June 1–July 20
Chervil and chives	May 10–June 10	May 15–June 15	([2])
Chicory, witloof	May 15–June 15	May 15–June 15	June 1–July 1
Collards [1]	May 15–June 15	May 15–June 15	July 1–Aug. 1
Corn, sweet		June 1–July 1	June 1–July 10
Cress, upland	May 15–June 15	June 15–Aug. 1	Aug. 15–Sept. 15
Cucumber		June 1–15	June 1–July 1
Eggplant [1]			May 15–June 15
Endive	June 1–July 1	June 15–July 15	July 1–Aug. 15
Garlic	([2])	([2])	([2])
Horseradish [1]	([2])	([2])	([2])
Kale	May 15–June 15	June 1–July 1	July 1–Aug. 1
Kohlrabi	May 15–June 15	June 1–July 15	July 1–Aug. 1
Leek	May 1–June 1	([2])	([2])
Lettuce, head [1]	May 15–July 1	June 1–July 15	July 15–Aug. 15
Lettuce, leaf	May 15–July 15	June 1–Aug. 1	July 15–Sept. 1
Muskmelon		May 1–June 15	June 1–June 15
Mustard	May 15–July 15	June 1–Aug. 1	July 15–Aug. 15
Okra		June 1–20	June 1–July 15
Onion [1]	May 1–June 10	([2])	([2])
Onion, seed	May 1–June 1	([2])	([2])
Onion, sets	May 1–June 1	([2])	([2])
Parsley	May 15–June 15	June 1–July 1	June 15–Aug. 1
Parsnip	May 15–June 1	May 15–June 15	June 1–July 10
Peas, garden	May 10–June 15	June 1–July 15	([2])
Peas, black-eye			June 1–July 1
Pepper [1]		June 1–June 20	June 1–July 1
Potato	May 15–June 1	May 1–June 15	May 15–June 15
Radish	May 1–July 15	June 1–Aug. 15	July 15–Sept. 15
Rhubarb [1]	Sept. 1–Oct. 1	Sept. 15–Nov. 1	Oct. 15–Nov. 15
Rutabaga	May 15–June 15	June 1–July 1	June 15–July 15
Salsify	May 15–June 1	May 20–June 20	June 1–July 1
Shallot	([2])	([2])	([2])
Soybean			June 1–25
Spinach	May 15–July 1	June 1–Aug. 1	Aug. 1–Sept. 1
Spinach, New Zealand			June 1–July 15
Squash, summer	June 10–20	May 15–July 1	June 1–July 15
Squash, winter		May 20–June 10	June 1–July 1
Sweetpotato			May 20–June 10
Tomato	June 20–30	June 1–20	June 1–20
Turnip	May 15–June 15	June 1–July 15	July 1–Aug. 1
Watermelon		May 1–June 15	June 1–June 15

See footnotes at end of table.

Range of Dates for Safe Fall Planting in the Open—Continued

Crop	Planting dates for localities in which average date of first freeze is—		
	Oct. 30	Nov. 20	Dec. 10
Asparagus [1]	Nov. 15–Jan. 1		
Beans, lima	July 1–Aug. 1	July 15–Sept. 1	Sept. 1–30
Beans, snap	July 1–Aug. 15	July 1–Sept. 10	Sept. 1–30
Beet	Aug. 1–Sept. 1	Sept. 1–Dec. 1	Sept. 1–Dec. 31
Broccoli, sprouting	July 1–Aug. 15	Aug. 1–Sept. 15	Aug. 1–Nov. 1
Brussels sprouts	July 1–Aug. 15	Aug. 1–Sept. 15	Aug. 1–Nov. 1
Cabbage [1]	Aug. 1–Sept. 1	Sept. 1–Dec. 1	Sept. 1–Dec. 31
Cabbage, Chinese	Aug. 1–Sept. 15	Sept. 1–Oct. 15	Sept. 1–Nov. 15
Carrot	July 1–Aug. 15	Sept. 1–Nov. 1	Sept. 15–Dec. 1
Cauliflower [1]	July 15–Aug. 15	Aug. 1–Sept. 15	Sept. 1–Oct. 20
Celery [1] and celeriac	June 15–Aug. 15	July 15–Sept. 1	Sept. 1–Dec. 31
Chard	June 1–Sept. 10	June 1–Oct. 1	June 1–Dec. 1
Chervil and chives	([2])	Nov. 1–Dec. 31	Nov. 1–Dec. 31
Chicory, witloof	July 1–Aug. 10	July 20–Sept. 1	Aug. 15–Oct. 15
Collards [1]	Aug. 1–Sept. 15	Aug. 25–Nov. 1	Sept. 1–Dec. 31
Corn, sweet	June 1–Aug. 1	June 1–Sept. 1	
Cress, upland	Sept. 15–Nov. 1	Oct. 1–Dec. 1	Oct. 1–Dec. 31
Cucumber	June 1–Aug. 1	June 1–Aug. 15	Aug. 15–Oct. 1
Eggplant [1]	June 1–July 1	June 1–Aug. 1	Aug. 1–Sept. 30
Endive	July 15–Aug. 15	Sept. 1–Oct. 1	Sept. 1–Dec. 31
Garlic	([2])	Aug. 15–Oct. 1	Sept. 15–Nov. 15
Horseradish [1]	([2])	([2])	([2])
Kale	July 15–Sept. 1	Aug. 15–Oct. 15	Sept. 1–Dec. 31
Kohlrabi	Aug. 1–Sept. 1	Sept. 1–Oct. 15	Sept. 15–Dec. 31
Leek	([2])	Sept. 1–Nov. 1	Sept. 1–Nov. 1
Lettuce, head [1]	Aug. 1–Sept. 15	Sept. 1–Nov. 1	Sept. 15–Dec. 31
Lettuce, leaf	Aug. 15–Oct. 1	Sept. 1–Nov. 1	Sept. 15–Dec. 31
Muskmelon	July 1–July 15		
Mustard	Aug. 15–Oct. 15	Sept. 1–Dec. 1	Sept. 1–Dec. 1
Okra	June 1–Aug. 10	June 1–Sept. 10	Aug. 1–Oct. 1
Onion [1]		Oct. 1–Dec. 31	Oct. 1–Dec. 31
Onion, seed		Sept. 1–Nov. 1	Sept. 1–Nov. 1
Onion, sets		Nov. 1–Dec. 31	Nov. 1–Dec. 31
Parsley	Aug. 1–Sept. 15	Sept. 1–Dec. 31	Sept. 1–Dec. 31
Parsnip	([2])	Aug. 1–Sept. 1	Sept. 1–Dec. 1
Peas, garden	Aug. 1–Sept. 15	Oct. 1–Dec. 1	Oct. 1–Dec. 31
Peas, black-eye	June 1–Aug. 1	July 1–Sept. 1	July 1–Sept. 20
Pepper [1]	June 1–July 20	June 1–Aug. 15	Aug. 15–Oct. 1
Potato	July 20–Aug. 10	Aug. 10–Sept. 15	Aug. 1–Sept. 15
Radish	Aug. 15–Oct. 15	Sept. 1–Dec. 1	Aug. 1–Sept. 15
Rhubarb [1]	Nov. 1–Dec. 1		
Rutabaga	July 15–Aug. 1	Aug. 1–Sept. 1	Oct. 1–Nov. 15
Salsify	June 1–July 10	July 15–Aug. 15	Aug. 15–Oct. 15
Shallot	([2])	Aug. 15–Oct. 1	Sept. 15–Nov. 1
Sorrel	Aug. 1–Sept. 15	Aug. 15–Oct. 15	Sept. 1–Dec. 15
Spinach	Sept. 1–Oct. 1	Oct. 1–Dec. 1	Oct. 1–Dec. 31
Spinach, New Zealand	June 1–Aug. 1	June 1–Aug. 15	
Squash, summer	June 1–Aug. 1	June 1–Aug. 20	June 1–Sept. 15
Squash, winter	June 10–July 10	July 1–Aug. 1	Aug. 1–Sept. 1
Sweetpotato	June 1–15	June 1–July 1	June 1–July 1
Tomato	June 1–July 1	June 1–Aug. 1	Aug. 15–Oct. 1
Turnip	Aug. 1–Sept. 15	Sept. 1–Nov. 15	Oct. 1–Dec. 1
Watermelon	July 1–July 15		

[1] Plants.   [2] Generally spring-planted.   [3] Check county Extension office for date of first freeze.

13

William E. Carnahan

also makes an excellent mulch if used properly. Plastic mulch controls weeds, holds moisture in the soil, and reduces loss of fertilizer by leaching. Since dark colors absorb heat and warm the soil underneath, black plastic is particularly beneficial to early-planted crops. However, soil under black plastic may get too warm during summer unless shaded by plant foliage or covered with a mulching material such as cardboard, sawdust, or fine straw.

Watering will not be needed as often when the garden is mulched. In most areas, garden crops during summer need about 1 to 1½ inches of water (or rain) per week. Sandy soils will not hold as much moisture as will heavier clay soils or those containing ample humus. During prolonged drought, plants on sandy soil (not mulched) may need an inch of water every 5 to 7 days; and those on heavier soils may need about 1¼ to 1½ inches of water every 7 to 10 days.

Thorough waterings at these intervals are more beneficial than light sprinklings at more frequent intervals.

Subsequent chapters cover the prevention and control of insects, plant diseases, and other pests, including the judicious use of pesticides when necessary.

Keeping Records

Don't trust your memory from one year to the next. Keeping records of varieties planted, for example, and taking a few notes on their performance, will help you at decision-making time the next season.

Keeping a record of some of the problems encountered, and what you did about them, may help you avoid the same problems next year.

Did you make some plantings too early or too late this year? If so, a few notes will refresh your memory next time around.

Did you grow too much of some crops and not enough of others? Was one variety of a particular crop better for processing than another? If so, a few records will be beneficial.

How much did your garden cost this year? How much was it worth? What was the return on your investment for seed, plants, fertilizers, pesticides, supplies and equipment? What was your labor worth, and that of your family? Maybe you'd just as soon not figure the value in dollars and cents, but count the enjoyment and personal satisfaction as your reward.

Mulching with grass clippings can save a lot of weeding. This bean planting has been mulched in foreground. Weeds must be pulled from unmulched section.

Where to Garden—Setting Your Sites
by James W. Wilson

You can grow vegetables, fruits and berries successfully in full sun and away from tree roots. But only a few garden sites are far removed from the shade cast by walls, fences, or trees or free from foraging tree roots. Thus, gardening often becomes an exercise in compromise, where you learn to live with site-imposed restrictions and settle for somewhat less than optimum garden and orchard performance and yield.

Home gardeners should look beyond the traditional concept of a single plot as a vegetable and/or fruit garden. More often than not, two or more small plots have advantages over a single garden. Small plots are also easier to dress up with flowers to make them blend into the landscape.

If space permits, a separate orchard or berry plot is to be preferred over a combination garden/orchard. Toxic pest control sprays from fruit trees and berries can drip on vegetables. Also, certain kinds of berries spread aggressively and would invade nearby vegetable rows.

Site selection for fruit or nut trees is more critical than for vegetables, berries and bush fruits because orchards are not "portable". You can't move an orchard around like a vegetable or flower garden. The location of fruit or nut trees, and the form, flower and foliage color of the varieties chosen, can have a significant impact on the home landscape.

Fruit, nut and citrus trees change not only in size but also in form as they mature. Deciduous trees are rather stiff and stark in winter because of pruning, and citrus trees lose color and some of their foliage.

James W. Wilson is Executive Secretary, National Garden Bureau, Los Altos, Calif.

Most professional landscapers prefer not to integrate fruit or nut trees in a landscape plan but would rather set them along the back or side of the property, where they are screened by more graceful trees or large shrubs. Pecan and oriental chestnut trees are an exception; they make excellent shade trees and require no pruning except for shaping and removal of dead branches.

A survey for potential garden and orchard sites on your property might prove disappointing, but you have options today that were not open a few years ago. Gardens on company property or in community plots have experienced a resurgence in popularity. Container gardening permits vegetable and fruit culture where no suitable plots of soil exist.

Let us explore the environmental conditions that largely determine garden performance, to help you assess whether you should attempt to grow vegetables and fruit and, if so, where to begin. Nothing is so demoralizing to a beginning gardener as the emphatic failure that can follow an attempt to grow food crops under difficult-to-impossible environmental conditions. And nothing is so satisfying as success in one's first attempt at gardening and the gradual improvement in results as one learns by experience.

Water Needs

Water availability is an important consideration. In many areas home gardeners rely on rain alone to supply water for vegetable gardens. Even in moist areas, water is needed for transplanting and for bringing vegetables through drought conditions. Therefore, when possible, locate gardens near a faucet where a short hose can reach all parts.

There are no shade-loving vegetables or fruits. All respond to shade by growing slower and taller and maturing later, if at all. All vegetables grow best in full sun, except in extreme desert situations where shade from the afternoon sun can improve growth and prolong the harvest season. All fruit trees and berries prefer full sun and will grow slowly and bear poorly if subjected to shade of even medium density.

Most vegetables can tolerate a certain amount of shade from walls, fences, trees, and other vegetable plants. Plant performance improves in direct relationship to the distance from the shade-casting source. Soil shaded much of the day warms and dries out slowly; bacterial activity and nutrient release are retarded.

As a general rule, it is a waste of time to try to garden within 6 to 8 feet of the northern side of 1-story structures. The south side is an ideal location with its full sun. Gardens close to the west or south sides of low structures grow well, especially warmth-loving species, because the structure radiates heat late in the day. The east side receives full morning sun but plant growth may be slower than on the west side where afternoon sun elevates the soil and air temperature. A short period of shade doesn't retard growth significantly.

If you live in a winter garden area, take special care to locate your garden or orchard away from the long shadows cast by the low-in-the-sky sun. Full sun is essential for satisfactory winter growth, and heat absorbed by the soil during the day promotes root activity and slows the cooling at night.

"High shade" cast by tall trees free of lower branches can often be tolerated by vegetables because the shaded area moves with the sun and covers the garden only a short time. Many trees cast spotty rather than dense shade. With mature trees you can expect widespread roots, often extending to 1½ times the distance from the trunk to the outer branches.

Many homeowners make the mistake of planting their grounds with fast-growing species of shade trees and, within 10 to 15 years, find it impossible even to grow grass, much less fruits or vegetables. At this point consideration should be given to relative values of summer shade versus sunlight on the garden. If the tree is not an attractive or valuable species, trim it to let the sun in. Salvage value, in the form of firewood and ground-up twigs and branches, can reduce the cost of tree removal.

A simple, often overlooked fact about sunlight and shade is its effect not only on photosynthesis (food manufacturing) but also on relative warmth of the soil. Seeds sprout faster and plants grow more rapidly in warm soil.

Areas of your garden exposed to even two to four hours of shade daily will produce slower rates of plant growth than those in full sun. Shaded areas should not be planted with

Plan your garden area so it benefits from the most sun. South and southwest sides of low lying structures are best bet for full sun benefit. Greenhouse is at left.

early maturing vegetables or fruit because their bred-in advantage is defeated.

Wind, Fog and Frost

Certain geographical areas, even within densely settled communities, can be very windy. Some parts of individual gardens may be wind-buffeted while others are tranquil. High or swirling winds are especially destructive to young vines before they are anchored securely to supports, and to mature plants that are heavy with fruit.

In extremely windy areas, such as the Great Plains or in gaps between hills or mountains, gardeners should avoid planting tall-growing varieties because they tend to blow over or break apart. Such difficult sites are improved by planting windbreaks or by erecting walls, fences, or deflecting panels (which also serve to reflect and intensify sunlight). Tree or shrub windbreaks are suitable only for large plots where root competition won't become a problem.

Salt-tolerant shrubs are often planted as windbreaks for shoreline gardens. They trap salt spray before it can damage the foliage of fruits and vegetables.

Corridors between structures are rarely satisfactory for growing vegetables. The wind is concentrated while moving between buildings. Buildings shade the site. The effect of shade can be reduced by painting the adjacent walls white for maximum reflectiveness. In some cases, clear fiberglass windscreens can be erected to deflect the wind without reducing the available sunlight.

Fog causes few problems over much of the United States because it forms only occasionally and for short time periods. Fog can substantially modify the climate and affect plant growth in some low-lying waterside communities and along coastal slopes. In the California and Pacific Northwest fog belts, for example, early maturing vegetable and fruit varieties are advised, and warmth-loving species should be planted near south- or west-facing walls where they can benefit from the heat and reflected light.

Peculiarities in air movement can make certain areas of your garden susceptible to frost damage. Several seasons of observation may be required to confirm the safest site. The garden should be away from fences or other air traps and on a hillside rather than on the hill top or at the bottom of a knoll. Frost occurs in low areas while the hillside allows cold air to drain and warm air to rise past the garden site.

Fortunately, poorly drained soil can be improved by adding organic matter, raising the level of beds above the surrounding ground, or installing drainage tile. However, if you have a choice in sites, select a plot that doesn't flood even after heavy rains. Fruits and vegetables can withstand brief periods of flooding. However, wet, poorly aerated soil promotes the growth of harmful organisms and kills fragile root hairs.

Flood plains of creeks and rivers should be approached with caution as potential garden sites. Inquire of older residents if flooding is a problem. A lot of work can go down the drain or be covered with a layer of mud from flash floods.

Strange and unpredictable changes can occur in drainage patterns when developers build new communities. If you have just moved into a new tract or subdivision, look all around you to determine if your lot is in a swale or low corner. If so, plant small gardens until you can be sure your garden plot will not become a pond or a waterway after heavy rains. If a problem is present, raise the level of your garden to keep it high and dry or select a better site.

Erosion can remove topsoil from

Author watering plants in garden.

the garden. The erosion can occur so gradually that it escapes detection or so suddenly that whole rows of plants are washed out. On sites where the slope is steeper than a gentle grade, vegetable garden or orchard rows should follow the contour to minimize erosion.

Occasionally certain sites make poor gardens because of former uses that compacted the soil, fouled it with weed seeds, or loaded it with feedlot salts. This is especially true of suburban lots that were stripped of topsoil before being developed. With a lot of work, such areas can be improved enough to grow vegetables but they should be avoided if one has other options. Under no circumstances should vegetables be planted on old orchard land where trees were sprayed for decades with lead arsenate.

In urban situations, do not garden on old rights of way where long lasting herbicides were sprayed, or on recently abandoned sanitary landfills. Sanitary landfills generate methane gas that can filter through the usual shallow covering of soil and interfere with plant growth.

Nearby Tree Roots

Roots of nearby trees and shrubs compete for water and soil nutrients. Shallow-rooted annual vegetable plants suffer greatly because the perennial roots quickly invade the fertile soil of the garden. Watering the garden increases soil moisture, which further attracts the tree and shrub roots. Root barriers are generally of little value.

Bamboo and mints are aggressive root-spreading perennials that should never be planted near vegetable or fruit plantings. Be sure to look over the fence or speak to your neighbor before planting near your property line.

Consider the size of existing perennials and their potential for increase over several years.

Container Gardens

The new techniques and materials for growing plants in containers have elevated this method from an expensive hobby to a valid and dependable source of fresh vegetables for apartment and mobile home residents. And, as backyards become increasingly shaded by maturing trees, many suburbanites are moving their vegetables into containers and onto the remaining sunny spots on their patios, sidewalks, balconies, or porches.

Growing berries and bush fruits in small containers is not difficult, but fruit trees are another matter. In areas where winters are extremely cold the pressures caused by freezing and thawing can destroy the large containers necessary to support trees. An alternative is to plant only true dwarf, ultra-compact varieties in portable containers that can be moved into an unheated shelter for protection during winter.

Container gardens of citrus and deciduous fruits, berries and vegetables can be extremely decorative. Wooden barrels or rustic boxes can be treated with copper napthenate which will

not harm plants or animals. Much less expensive, and almost as durable, containers can be made from plastic garbage cans, foot tubs, pickle buckets and plastic laundry baskets.

To reduce watering frequency, containers for vegetables should be of minimum 4-gallon capacity. Tomatoes, 2 large vines per container, can be planted in 20- to 30-gallon plastic garbage cans. Berry and fruit tree plantings call for sturdier containers to support the weight of the root ball. Half-barrels of 20-gallon capacity work well. Avoid small pots; they dry out quickly, are a nuisance to water, and will blow over in a stiff breeze.

Window boxes occasionally are used to grow strawberries and small varieties of leaf or root vegetables. These boxes present a large evaporative area and dry out quickly. Spring and fall crops are the most successful as rain and cool weather reduce the stress on plants. Recently introduced midget varieties of cucumbers, tomatoes and melons can be grown in window boxes that provide about 1½ cubic feet of soil for each plant.

Commercial growers cannot afford crop failures and many have switched almost entirely to artificial soil mixes for growing container crops. Some garden soils are dense and extremely heavy, slow to drain, apt to shrink and pull away from container walls, full of weed seeds and disease-carrying bacteria and fungi, and occasionally contaminated with herbicides. The two advantages of garden soil are its buffering effect that minimizes damage from overfeeding, and its natural component of micronutrients which can be partially duplicated in artificial soil mixes by using additives.

Lightweight, near-sterile artificial soil is made by mixing slow-to-decay organic matter and inorganic particles graded to uniform size to promote drainage and to add a little weight.

Such mixtures contain little or no plant nutrients or lime, and require either frequent applications of liquid plant food or feeding with the recently perfected long lasting controlled release fertilizers. Artificial soil mixes drain rapidly; plants grown in them need to be watered daily during dry or windy weather. The slowly decomposing woody matter creates substances that retard growth of organisms which cause plant diseases.

Excellent soil mixes can be made

Atop New York building, gardener grows vegetables and flowers by using containers.

by combining either coarse sphagnum peat moss and graded fine sand or pulverized pine or fir bark, composted sawdust, graded fine sand, and steam-sterilized manure.

Garden soil can be substituted for the fine sand, but it should be pasteurized at 140° F for 30 minutes. Allow an additional 30 minutes for warmup and for cool down. (This can be a smelly, messy procedure.)

Your county Extension office may have bulletins on preparing and maintaining artificial soil mixes for container crops.

Front Yard Gardens

Front yard vegetable gardens, once frowned on in the suburbs, are becoming commonplace. Often the most desirable garden sites are in front yards where sunlight and air drainage are ideal. Some turfgrass areas may have to be sacrificed and, for cosmetic purposes, garden plots need to be mulched or planted in a cover crop during winter.

Front yard vegetable gardening tends to become "social." It is difficult to garden on the customary solitary plane when neighbors and passersby are stopping and inquiring about how you manage to grow such early sweet corn or those large tomatoes. Yet gardening "out front" has its advantages. When surpluses of leaf lettuce or zucchini develop, you can sell them quickly.

If you live in a conservative neighborhood, start with a few decorative vegetables in the front yard. Enlarge your plot gradually and dress it up with flower borders. If you have to dust or spray, wait until dusk or dawn. Certain people still look on food gardening as a rather vulgar, sweaty activity that should be restricted to the backyard. They will blow the whistle on you if local ordinances give them support.

Vandalism and theft in front yard gardens can be minimized by letting neighborhood kids watch as you plant and water, and by letting them sample fresh peas, carrots, and tomatoes. Melons, however, are much too attractive to plant out front.

Hedges of large annual flowering or foliage plants, such as cleome, four o'clock or kochia, can screen the garden from small children and obscure low temporary fences for keeping out dogs. A few cities have ordinances against permanent fences in front yards, so use a low temporary barrier such as chicken wire or snow fence.

Front yards, unless tightly fenced, are a poor location for fruit, nut or citrus trees or bush fruits. Passersby not only steal the crop but often climb trees and break limbs. The required spraying does not sit well with neighbors when it drifts on their property or automobile.

Place salad or herb gardens near an entrance where leaves, roots, buds, and seeds can be gathered quickly and conveniently with a minimum of walking. Beds of staples such as potatoes, sweet potatoes, beets, and carrots can be grown some distance from the house because they are harvested less often.

Community Gardens

Community gardening almost died out in the years following World War II, before inflation began to be felt and the price of fresh and processed vegetables started to climb. In the early 1970's groups of young people, often in university towns, took up gardening as part of the ecological movement. They located the pressure points and resource areas in local, State and national bureaus and established many thriving gardens that have helped to feed people in low income areas.

At present, no organized Federal or State programs exist to fund and supervise community gardens except in impacted areas where minority groups are involved. Where funding

can be arranged, the compensation for leaders is usually small and on a short term basis.

Community gardens where you can grow as you please have begun to spring up. Most are in medium to large sized towns since small towns offer spacious home grounds and vacant lots for individual gardens.

Private entrepreneurs have developed large gardens in a few towns and rent plots of 500 to 1,000 square feet. In other areas, institutions such as savings and loan companies have converted large acreages to garden plots as a public relations gesture. State governments, through their Extension Service, have launched a few pilot programs, mostly aimed at low income areas.

By and large, the most successful programs have been organized by volunteer committees including experienced gardeners and citizens who know how to twist arms to procure land and startup funds.

A number of corporations are making company land available to employees for gardening. Some small gardens contain no more than a dozen plots, others as many as 400. Some are merely strips of land too awkwardly situated for buildings or parking lots; others are special areas carved out of lawns. Prosperous corporations have hauled in organic matter, piped the gardens for water, added paved access paths, and fenced in the garden area for security. Most are convinced that the benefits derived from happy employees more than repay costs of the programs.

Every community or corporation garden needs someone in charge to assign plots, collect fees, charge out tools from the supply shed, provide advice, and arbitrate disputes. Fees are usually set high enough to cover water bills, spring tillage, maintenance around roads and fence lines, and a stipend for the supervisor. Fees can be slightly higher if there are

Photos by William E. Carnahan

Top, youngster plants lettuce, and bottom, another youngster holds some produce he grew, at Children's Garden of Brooklyn Botanic Garden, New York.

21

Scenes at 50-plot community garden in Silver Spring, Md. Top left, garden coordinator grinds up corn stalks in power shredder. They will be added to compost pile in background. Next to compost is pile of stable manure. Top right, boy holds pumpkin grown in his plot. Bottom, some participants in garden.

Photos by William E. Carnahan

This inner city garden project of North Chicago was once a stony, rubble-covered vacant lot. Top, youngster cultivates his plot. Bottom, woman checks corn.

Photos by Fred S. Witte

special services such as hauling in manure or other organic material.

Experienced organizers and supervisors of community gardens often follow seemingly unorthodox priorities in selecting garden sites. Quality of soil is not first. Guidelines include:

—Accessibility and parking facilities. Better two or three small garden sites no more than a mile or two from users than one large garden site that attempts to serve people from 5 to 10 miles away.

—Availability after working hours and on weekends.

—Security. A high, stout fence and a locked gate controlled by the supervisor is, unfortunately, a "must."

—Water. More squabbles occur over water than any other service. The best arrangement is at least one hose faucet for every four plots.

—Long term guarantee. No one likes to work hard to improve a garden that can be taken away in a year or two. It is not smart to plant fruit or nut trees in community gardens because they require from three years (almonds) to seven (pecans) to begin bearing significant crops.

—Roads. A wide bisecting or perimeter road is ideal.

—Sheds. The ideal arrangement is a large shed that provides secure lockers for each individual's tools and supplies, and large cabinets for holding tools and community supplies purchased in bulk.

—Peripheral space. A resourceful supervisor can promote all sorts of organic waste material for composting, if he has a place to pile it.

—Fertile soil. A determined gardener with access to lots of compost can grow pretty good vegetables on soil that was once compacted for a parking lot, or on the subsoil and rubble remaining after a building demolition. With built-up beds of compost, the fertility and structure of the base material becomes less important.

Garden Tools and Equipment

by John W. Bartok, Jr.

Size of your garden, the jobs to be done, and the money you wish to spend are important matters to consider when you purchase garden tools and equipment.

Basic tools needed for the small or beginning garden are a spading fork ($4 to $8), steel rake ($4 to $7), and garden hoe ($4 to $7). With these you can turn the soil, smooth and remove the stones and sod, form the rows, cultivate, and dig the root crops.

Garden centers, hardware stores and catalog houses carry a large selection of hand and power tools. It pays to purchase good quality equipment, as it will give better service, stay sharp longer, and when properly cared for may last a lifetime.

In selecting hand tools, look for handles with straight grain and a sturdy attachment between the tool and its handle. To get good service and long life from a tool, use it for its intended purpose. For example, shovels, spades and hoes should not be used as crowbars—don't pry with them.

Clean tools after use and before returning them to storage. Wash or brush off all soil and grass. Protect stored tools from rust by coating them with oil, grease, or other rust inhibitors. Keep cutting tools sharp by filing or grinding the cutting edges. Roughened handles should be sanded smooth and a coat of raw linseed oil applied. Also, keep the handles tight in their tools. This can sometimes be done by soaking the tool head in raw linseed oil or motor oil. The oil swells the wood fibers and excludes any moisture.

John W. Bartok, Jr., is Extension Agricultural Engineer, University of Connecticut, Storrs.

Store hand tools in an orderly way. Attach a pegboard or plywood rack to the wall of a basement, garage or utility shed. Nails or special hooks can be used to support the tools. With a location for each tool, you can then tell if one has been left outside. Also, you limit the possibility of someone getting injured from stepping on a loose tool lying on the floor.

As your interest and enthusiasm increase and your garden gets larger, you may need additional hand tools. Multi-use tools or equipment that will be used throughout the gardening season should be purchased first.

The round point shovel ($5 to $10) with either "D" or long handle fits the above classification. It is useful for turning sod, digging holes for fruit trees and bushes, moving soil, or edging a bed. For transplanting, a trowel ($1 to $2) is handy.

A hand-held or wheel-supported cultivator ($5 to $40) can often save time and labor in removing weeds from a large garden. This implement with curved tines is dragged or pushed between the rows to pull out weeds. It is also used to loosen the soil so rainwater may penetrate more easily. Hand forks or claws are miniature cultivators ($1 to $2) and often all that's needed in a very small garden.

In areas where stones are numerous, a pick ($6 to $10) will aid in their removal. It is also useful for starting postholes and removing tree roots.

Many types of seeders ($8 to $40) are available. These take the backache out of planting. Most are adaptable to a variety of seed sizes. Some seeders are designed to open the furrow, drop and cover the seed, and firm the soil in one operation.

Lime and fertilizer are usually ap-

William E. Carnahan

William Aplin

plied to the soil before planting. On small gardens this can be done by hand. On larger gardens a precision or broadcast spreader ($15 to $30) will reduce the labor and give a more uniform application. The spreader can also be used to fertilize and lime the lawn.

Several pieces of equipment will aid you in maintaining the berry patch and fruit orchard. Hand pruning shears ($3 to $6) are used to prune branches and twigs up to about a half inch in diameter. Lopping shears with handles 18 to 24 inches long will cut

Above, some basic tools for small garden. Right, wheel-supported cultivator removes weeds and loosens dirt around plants while saving a lot of labor in a larger garden.

branches up to 1½ inches in diameter. For larger branches a pruning saw ($4 to $6) can be used. These tools will need to be sharpened from time to time. A fine-tooth file or sharpening stone can be used. Maintain the same cutting angle.

A wheelbarrow or cart ($15 to $40) can make the work of moving soil, stones, peat moss, and tools much easier. Select one with large diameter wheels for easier pushing. Also keep pneumatic tires properly inflated and bearings lubricated.

Many other tools are available that have been developed for either a specialized task or a single crop. In certain circumstances these tools will save considerable labor.

Several pieces of power equipment have been developed to aid the gardener planting a large area. Most are powered by single cylinder air-cooled engines. A few are powered by small electric motors.

Rotary tillers with steel tines on a powered horizontal shaft can prepare a seedbed in one operation. The most common type uses the rotating tiller shaft to also provide forward motion. Weight of the engine helps provide penetration into the soil. Digging depth is regulated by a set of gage wheels or depth control bar.

A second type uses the engine to power both the tines and the wheels. The speed of forward motion can be regulated. This type is not affected as much when stones are hit.

The size of engine required varies with the width of cut, but in general 4 to 6 horsepower is needed. Some tillers are equipped with a reverse

Left, broadcast spreader gives uniform application of lime or fertilizer. Below, compact tractor and cart are handy for moving fertilizer, lime, peat and other bulky garden supplies.

William E. Carnahan

gear that is helpful in getting out of tight corners. Most tillers can be adjusted to cultivate between rows as close as 16 inches apart.

Daily maintenance consists of keeping the tines free of weeds and roots, checking the oil and gas levels in the engine, and lubricating moving parts on the tiller.

Compact tractors, or garden tractors as they are often called, have become increasingly popular as an aid in garden and yard work. This type tractor normally is equipped with an air-cooled engine of 7 to 16 horsepower. Several battery-powered electric tractors also are available. These are recharged after use by plugging into a 115-volt convenience outlet.

Many options are available if you buy a compact tractor. These affect ease of operation, the jobs that can be done, and the tractor's cost.

Four types of starting systems are used on compact tractors, but three of them seem to be phasing out. Rope starter, rope rewind, and impulse which uses a handle to wind up a spring are now being replaced with an electrical starting system. The electrical system, which requires a battery, has the advantage of being easier to operate.

The clutch is the link in the power train between the engine and the transmission. It allows the operator to shift gears or to stop the tractor. Most manufacturers use a system of V-belts that are tightened to engage the engine and transmit the power. A few tractors use plate clutches.

V-belts usually need to be adjusted every few months. Some are underdesigned and can cause troublesome repairs and expense.

The transmission reduces the high speed of the motor to drive the wheels. Some tractors use a transmission similar to a standard shift car, others have a hydrostatic transmission which compares to the automatic shift.

Hydrostatic transmissions have been gaining in popularity even though they add $100 to $125 to the tractor's cost. Advantages are in providing an infinite range of speeds, and easier operation.

In general, the larger the tire diameter the better. The tire is less likely to become stuck. For lawn work, a wide flotation tire is desirable. For most garden work, a lug-type tire is preferred for its greater traction. A universal design tire is also available where both conditions exist. Additional traction can be obtained by adding wheel or frame weights.

The power take-off (PTO) supplies power to attached implements such as tillers and lawn mowers. PTO's should be examined for adequacy of drive and ease of attaching imple-

Garden soil preparation can be made easier with a rotary tiller that can be adjusted to various depths. Speed can be regulated on some tillers.

ments. Guards that cover the drive shaft and belts should always be used to prevent clothes from becoming tangled and causing injury to the operator. Wear good shoes to protect your feet while you operate the equipment.

Compact tractors can be equipped with several implements that will make your gardening easier. A moldboard plow which turns the sod and soil over can do a good job when mounted on a tractor of 10 or more horsepower. It can be maneuvered in small plots.

To prepare the seedbed you can use a disk harrow, spike tooth harrow, or spring tooth harrow. The harrows are pulled over the plowed ground to cut, pulverize, and level the soil before planting.

Other pull-behind attachments include carts, fertilizer spreaders, seeders, and sprayers.

A rotary tiller with its powered tines cuts the sod and pulverizes the soil in one operation. Depth of tillage is easily adjusted but should be at least 5 inches deep. A minimum of 10 horsepower is required. The rotary tiller does not work well on very stony soil or a high clay soil.

When considering purchasing a particular model tractor, talk with persons who have owned one for several seasons. They can tell you how it handles and what it is capable of doing. Also, operate the tractor yourself on your own property if possible. Finally, buy the tractor from a dealer who has facilities to service and repair the equipment he sells.

Water is essential for germination and growth of plants. When rainfall fails to provide a constant supply of moisture, artificial watering may be required. Water needs vary with the kind and maturity of the plant. For example, lettuce with its large leaf area requires more water than carrots.

How often to water depends on the amount of rainfall your garden gets,

water-holding capacity of the soil, and how fast the moisture evaporates. A quick test is to ball a handful of soil from the root zone of the plants. If the soil crumbles when the pressure is released, the garden needs water. When watering, soak the soil to a depth of 6 to 8 inches. This will be sufficient moisture to last from 7 to 10 days under normal conditions.

Frequent light waterings will cause roots to grow near the soil surface. The roots are easily damaged during hot dry days or when the garden is cultivated.

A watering can or garden hose is sufficient for the small garden. Buy a hose long enough to reach all points in the garden. Hoses are made of vinyl plastic, rubber, or a combination of the two. Better grade hoses are usually reinforced with nylon. A double reinforcing gives added life.

Continuous shallow watering leads to drying out of deeper soil. Soak soil to good depth. Plants then will be able to sink deep, healthy roots.

Norman A. Plate

Purchase a hose that is guaranteed for several years. Also, get one with ½-inch or ⅝-inch inside diameter, as it reduces your watering time considerably compared to a ⅜-inch diameter.

For larger gardens a lawn sprinkler saves time. A sprinkler should be positioned so the garden gets uniformly watered. Do not water on a hot sunny day or when it is dry and windy, as much of the water may evaporate before it reaches the soil.

Trickle irrigation can be used by the home gardener. This new system has small diameter (1 to 2 inches) plastic tubing with tiny holes punched every 4 to 8 inches along its length. The tube is placed along the row of plants. Water is supplied from a garden hose.

As the system operates at very low pressure (3 to 5 pounds per square inch), a pressure-reducing valve is needed. In addition a fine strainer is

Top, trickle irrigation tubing is set close to base of plants. Below, water oozes from tubing.

inserted in the line to remove foreign matter so the holes are not plugged. Because of the slow rate of flow, several rows or the whole garden can usually be watered at one time. This system needs to remain on for a longer time than when a sprinkler is used.

The advantage to this type of watering system is that only the soil near the plants gets wet, and the saving in the amount of water used can be 50 percent or more. Trickle irrigation equipment is available from some garden centers and greenhouse suppliers.

The successful use of pesticides to control insects and other garden pests depends largely upon three factors:
* Selection and correct dosage of the proper chemicals
* Proper timing of the application
* Use of spraying equipment that is properly adjusted, calibrated and operated

For most home garden pests a general purpose spray or dust will do a good job. These can be bought at a local garden center. For the small garden, purchasing the chemical in an aerosol can or self-contained duster is the most economical and the most convenient to use.

Where special chemicals are needed or for the large home garden or orchard, some type of pesticide application equipment is needed. The simplest of these is the hand atomizer or sprayer, available in capacities of from ½ pint to 2 quarts. This sprayer is inexpensive so several can be purchased, one for each type of spray material used. Separate sprayers are advised for weed killers and insecticides to prevent plant damage that could occur.

The plunger type duster is commonly used to apply insecticides and fungicides in powder form. A morning with a slight dew on the leaves and no wind is best for applying dusts.

The compressed-air sprayer provides better atomization and spray coverage, especially to the underside of leaves. It is available in capacities from 1 to 5 gallons. Its spray will reach to the top of most dwarf fruit trees.

Since these sprayers are not equipped with an agitator, they must be shaken frequently when wettable powders are sprayed. Use should be limited to calm days to avoid spray drift to nearby plants or a neighbor's yard.

The compressed-air sprayer is the most popular type of home garden sprayer.

A knapsack sprayer contains a 4- to 5-gallon tank carried on the back and shoulders of the operator. A pump located at the bottom of the tank and operated by a handle maintains an air pressure. When the valve on the nozzle is open, the air in the tank forces the spray out.

Power sprayers may be operated with either a small gasoline engine or an electric motor. In most power sprayers pressure is developed by a pump action directly on the liquid spray material. Generally they are capable of spraying the tops of trees 30 to 40 feet high.

Jobs for which power sprayers are used include spraying garden crops, fruit bushes, and orchard trees. Most are mounted on two wheels and equipped with handles for moving manually. Some are available to pull behind a garden tractor. A pressure regulator, a relief valve, and a pressure gage should be on all power sprayers.

Safety is important when using spray equipment and applying pesticides. *Read the label carefully before applying any pesticide.* The label will tell you the crops that can be sprayed, the amount to use and how it should be applied.

Some pesticides require protective clothing and a face shield. Store this clothing in a separate place and wash it separately from other clothes. Store all pesticides in a locked cabinet or box to keep children and pets from being poisoned. Wear gloves while applying a pesticide or wash your hands immediately after you are done.

Small animals may attack a home garden. Woodchucks, rabbits, raccoons and household pets will feed on and damage the plants. Hexagonal wire netting or welded mesh wire can be fastened to wooden or steel posts to fence out the animals. Box traps or smoke bombs may be more effective on some animals. All these materials are available from garden centers or department stores.

Good quality seeds and plants are important in obtaining high yields and quality vegetables. Information on the varieties best adapted to your area and the recommended cultural practices are available from your county agent at the nearest Cooperative Extension Service office. Seeds can be purchased from garden centers, department stores or through seed house catalogs. Advertisements of seed companies can usually be found in the garden section of your local newspaper in early spring.

Some vegetables are better grown from transplants or bedding plants.

Seeds are started in early spring in greenhouses and then marketed after the last frost by garden centers and roadside markets. Transplants will give you vegetables much earlier than plants grown from seed in your own garden.

Fruit tree stock and small fruit bushes are also available from garden centers in spring and fall. They are available either as bare root stock or as container plants. Both will start to grow easily if given the proper soil conditions.

The home gardener wishing to start seed for early plants will need pots, starting soil, and fertilizer. Peat pots or peat pellets are available for this use. These items and other gardening aids such as plant stakes, bean poles, and tomato cage wire are available at local garden centers.

Plants need nutrients to grow. Some are provided naturally from the soil, but others must be added either as organic or synthetic fertilizers. Bone meal, fish meal, and animal manures are examples of organic fertilizers. These are available from garden centers by the bag.

If you live in a rural area, you may be able to get cattle or horse manure from a local farmer. Because farm manures are often diluted with straw or other bedding materials, they help lighten the soil and allow it to hold more moisture.

Murray Lemmon

Left, working with peat pots which come as wafer-like disks and swell when placed in water. Top, discarded food cartons are among many types of containers that can be used to start plants.

Commercial fertilizers can be added to the soil to increase the amounts of nutrients. They work fast. Commercial fertilizers are always labeled to show the amount of nitrogen, phosphorus and potassium. These are available in two ways:
- A liquid or a water-soluble form which is mixed with water and applied with a watering can, or
- A granular form that is spread by hand or applied with a fertilizer spreader

Mulches are materials that are used between the rows and around the plants to keep the weeds down, conserve moisture, and improve the garden's appearance. Organic mulches, those that can be later incorporated in the soil, include lawn clippings, leaves, peat moss, wood chips, bark, and straw. Aluminum foil and black polyethylene plastic are both inorganic and can usually be rolled up and reused the following year. These materials are available from most garden centers.

Often your garden will produce more than you can use at the time. Most vegetables can be preserved by freezing or by canning. The major expense associated with freezing is purchasing a freezer. This will cost several hundred dollars and is an investment that will take a number of years to pay off. It will also increase your monthly electricity bill by several dollars. A self-defrosting freezer costs more to operate than the nonself-defrosting type and should not be purchased. For freezing you will need some containers—either plastic or paper.

Some specialized canning equipment is essential for preserving safe and attractive products. This equipment includes a canner—either water bath or steam pressure, pint or quart jars, and several kettles for precooking or blanching the foods to be canned. This equipment is usually found in most department stores.

Before using any of the equipment discussed, read and follow the instruction manual. The manual will tell you how to safely operate the equipment and how to maintain and store it.

Learning to Make the Best Use of Climate

PREFACE

For the reader's convenience this chapter is divided into four parts, representing the major climatic regions of the 48 adjoining states. Each part was written by an author within the region and familiar with its climatic variations. The regions are the Northwest, Northeast, Southwest, and Southeast.

Gardening generally requires a mild, sunny climate with adequate moisture. However, if you consider terrain features which favorably influence the immediate climate, you may be able to garden locally in a seemingly unlikely area for this activity. An attempt is made here to explain climate in the United States and its use to gardeners.

Temperature and rainfall are probably the two climatic features that gardeners first worry about. Sunshine, average cloud cover, and wind also are important. Adequate moisture is absolutely necessary, but if all other climatic features are suitable for gardening in an area, water may be supplied by some irrigation method. Thus, natural rainfall is not necessarily among the most critical points when looking for a gardening location.

A gardener needs to thoroughly understand the term "growing season." The true growing season is determined not by climate alone but by climate plus the climatic tolerance of a specific crop. The number of frost-free days does not really describe a growing season length because most vegetables and fruits have a certain amount of tolerance for a temperature of 32°F. Some species even have a temperature tolerance lower than the freezing point of water.

Soils are a reflection of climate because they develop and improve naturally as climate acts upon the parent material. Where climate is suitable

Four large, general gardening climate regions exist in 48 adjoining States. Texas, with its great breadth and elevation variations, lies in 2 regions.

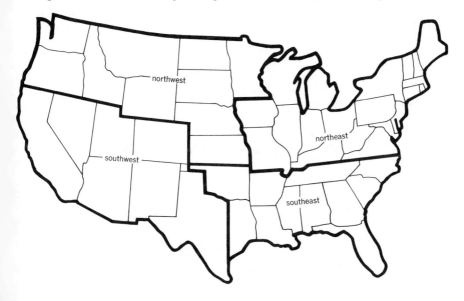

for successful gardening, soils generally are acceptable.

Climate is involved in crop quality. High temperature hastens ripening and reduces quality. Sugar content increases in maturing crops during sunny, cool, dry weather. Coloring is also increased under sunny, cool days and clear cool nights. Rapid ripening rates of sweet corn, under extremely hot weather, may reduce the harvest period to half the usual length. Extended warm, wet weather produces fruit that is soft, watery, has low sugar content, and needs special attention in storage and processing.

Climate influences disease and insect development. Warm, wet periods activate leaf blights, scab, bacterial disease, and fruit rots. Insect build-up occurs under hot, dry conditions. Cool-dry, and cool-moist periods will suppress disease and insect problems respectively.

Understanding the local climate, and learning to plan with the weather, are vital preliminaries to successful gardening.

IN THE NORTHWEST
by Earl M. Bates

The Northwest region has widely varied terrain. Although much of the region is mountainous and lies as far north as 49° latitude, gardening can be profitable and pleasurable.

Two principal climate types are a marine climate west of the Cascade Mountains of Washington and Oregon and a continental climate east of the Cascades, through the northern Rocky Mountains and into the northern plains.

The marine climate is characterized by a mild, wet winter and only a moderately warm, dry summer. A considerable amount of summer cloudiness prevails, which reduces the sunshine.

This area has a long frost-free season, but spring is frequently too cool and wet for early planting. Therefore, the effective growing season is relatively short. Gardens are practical here, but the vegetables selected should generally be the "cool season" type. Peas, carrots and cole crops do well, but such species as corn and tomatoes are practical only in certain warmer valleys.

Crops grown in the area may be either those requiring much or little water because even though the summer is dry, gardeners can feel confident of enough irrigation water for all crops.

In the continental area from the Cascade Mountains eastward to the northern prairie States, climate varies considerably due to elevation.

In the mountain valleys from Washington and Oregon eastward through the Rockies, gardening is usually successful at elevations at or below 1,500 feet above sea level.

Across the northern prairie States gardening is done at even higher elevations because large amounts of sunshine and stirring of the air by the plains winds have a favorable influence on this spring climate. However, late frosts occur in the northernmost tier of States, and the frost-free season may be as little as 130 to 140 days in places.

East of the Rockies, gardens of the hardier vegetables can be grown. Such species as corn, tomatoes and melons require special care and only the experienced gardener is likely to have success.

From the Cascades eastward through the Rockies and into the northern plains, rainfall is generally less than adequate for a garden. Irrigation water seems to be no problem at today's population level and probably will not be for at least a few decades. Sunshine is abundant across

---

Earl M. Bates is Advisory Agricultural Meteorologist, National Weather Service, Corvallis, Oreg.

Climatic Data for Representative Areas of the Northwest

City		Apr	May	Jun	Jul	Aug	Sep	Oct	Annual *	Mean number of frost-free days
Boise	Temp.	49.9	57.7	64.8	74.7	72.1	63.2	52.6	50.8	171
	Precip.	1.16	1.29	0.89	0.21	0.16	0.39	0.84	11.43	
Bismarck	Temp.	43.0	55.4	64.5	72.2	69.8	58.7	46.2	41.8	136
	Precip.	1.22	1.97	3.40	2.19	1.73	1.19	0.85	15.15	
Cheyenne	Temp.	42.6	52.9	63.0	70.0	67.7	58.6	47.5	45.9	141
	Precip.	1.57	2.52	2.41	1.82	1.45	1.03	0.95	14.65	
Corvallis	Temp.	51.7	57.0	61.6	66.6	66.4	62.7	54.4	53.0	210
	Precip.	2.05	1.77	1.15	0.33	0.55	1.31	3.78	39.70	
Denver	Temp.	47.5	56.3	66.4	72.8	71.3	62.7	51.5	49.8	165
	Precip.	2.05	2.20	1.64	1.36	1.43	1.08	1.01	14.20	
Grand Island	Temp.	49.6	60.6	71.0	77.2	75.3	65.2	53.4	50.1	160
	Precip.	2.47	3.78	4.40	3.00	2.54	2.51	1.08	23.41	
Helena	Temp.	43.3	52.9	59.5	68.4	66.2	56.0	45.6	43.4	134
	Precip.	0.83	1.56	2.23	1.03	0.89	0.95	0.66	10.85	
Huron	Temp.	46.0	58.0	68.1	75.5	73.3	62.4	49.8	45.5	149
	Precip.	1.84	2.36	3.14	1.81	2.07	1.53	1.15	17.33	
Miles City	Temp.	45.7	57.3	65.6	75.3	72.6	61.0	49.0	45.8	150
	Precip.	1.06	1.73	2.71	1.34	1.24	0.96	0.87	12.17	
Minneapolis	Temp.	44.7	57.6	67.3	73.0	70.5	60.5	48.3	43.8	166
	Precip.	1.85	3.19	4.00	3.27	3.18	2.43	1.59	24.78	
Pendleton	Temp.	52.0	59.6	65.8	73.6	71.9	64.2	53.7	52.8	196
	Precip.	1.09	1.12	1.17	0.22	0.28	0.63	1.18	12.38	
Salt Lake City	Temp.	50.1	58.9	67.1	76.6	74.4	64.2	52.9	51.3	202
	Precip.	1.76	1.56	.91	.61	.97	.74	1.34	14.74	
Seattle	Temp.	51.3	57.4	62.2	66.7	65.6	60.5	52.6	52.2	233
	Precip.	2.15	1.58	1.43	0.66	0.81	1.83	3.50	36.11	
Spokane	Temp.	47.3	56.2	61.9	70.5	68.0	60.9	49.1	47.8	169
	Precip.	0.91	1.21	1.49	0.38	0.41	0.75	1.57	17.19	
Topeka	Temp.	54.5	63.8	73.8	79.2	77.6	69.0	57.9	54.6	200
	Precip.	3.50	4.32	4.54	3.25	4.66	3.44	2.56	33.09	
Yakima	Temp.	50.5	58.5	64.4	71.0	68.6	61.3	50.5	47.8	177
	Precip.	0.47	0.54	0.81	0.13	0.20	0.35	0.60	7.86	

Temperature and precipitation. Monthly and annual

* Average annual temperature, and total annual precipitation.

this region and provides for fast growth.

Data in the table show a large variation in rainfall and temperature from east to west in the Northwest region. The marine West Coast climate can be distinguished from the continental by the small temperature variation from summer to winter at Seattle and Corvallis as opposed to large variations at Spokane or Miles City. Also, the marine climate has much cooler summer months with a longer frost-free period than many of the continental climate localities.

The frost-free period in this table is the number of days from the last average date of 32°F (0°C) in spring to the average first occurrence of 32° in autumn.

You need to study the climate in deciding what crops to grow. Or, if you are interested in certain crops, consider their climate needs.

Limited climatic information can be obtained from a National Weather Service Office nearest the desired area, or some may be obtained through the Weather Service Forecast Office for the State. In States with a State Climatologist office, more detailed climatic information is available. Many libraries have the publication called *Climates of the States* for their particular State and it contains valuable climate information.

The number of "frost-free days" officially reported may not hold true for your community. Temperature is measured at an official climatological station, and the number of frost-free days describes the climate at that location. But due to temperature variation with distance, the number of frost-free days in an adjoining valley may be different.

Growing season length is influenced by elevation and slope of terrain. Higher elevations such as foothills—or higher plains areas as in western Nebraska and Kansas or eastern Colorado—have shorter growing seasons than lower valley regions like Boise or Yakima.

For the coastal area west of the Cascade Mountains, the number of frost-free days does not describe the length of growing season. The marine influence causes large numbers of frost-free days but many spring days are only a few degrees above freezing which is too cool for crop growth. This shortens the true growing season.

The growing season has natural controls, but is also partly what the gardener makes it. If you take full advantage of the local climate, garden crops may be grown even though the climatic record for the region may indicate the season is too short.

Tender crops may be started earlier in spring if a gardening area slopes so as to cause light night drainage winds which are a frost preventative. South-sloping terrain generally warms more quickly in spring and may be less susceptible to frost; these same conditions, and sunny locations, will give the warmest soil in spring which aids seed germination and contributes toward early garden development. Natural large water bodies of the high plains and mountain sections have close-by climates which may be mild. Gardening can be carried on in these areas which may not be possible a few miles away.

Some gardener modification of the local climate may be practiced to prevent frost and to make water available to plants. Sprinkling crops in the spring is an effective way to prevent night frosts. Using hot caps or other cover, or down-the-row irrigation, can give some protective warmth. If a

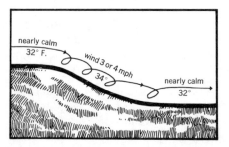

Left, air at night moves down a slope, in general. Frost is most likely to occur at hilltop and valley floor away from foot of slope. Turbulent mixing of air in down-slope breeze helps prevent frost. Above, clouds both absorb and reflect incoming radiation from sun and thus reduce surface temperatures.

Photos by Terence O'Driscoll

season is known to be a little too short for a crop, the plant may be started in a greenhouse, then put in the garden. These cultural practices for extending the growing season can be effective for several different fruits and vegetables.

Crops requiring more moisture than supplied naturally in the region may be grown by careful mulching and fine cultivation of the soil to conserve available soil moisture. Plants started at the earliest possible date in spring may produce a crop with natural moisture before the full strength of the summer's drought. Irrigation is, of course, the best way to produce a crop in a dry area. In most parts of the Northwest which have little summer precipitation, streams provide a good water source, or wells of adequate capacity can be dug.

The Pacific northwest area has grayish-brown soils that are leached by the abundant winter rains. These require some additives at times to give the best garden production.

Eastern Oregon, southern Idaho and Utah have extensive areas of grayish soils of the arid West with some alluvial soils. This soil calls for much irrigating. Northern Idaho, western Montana, and much of Wyoming and Colorado have variations of the gray soils of the arid West and the brown soils of the semiarid grasslands. Soils of the mountain slopes may be thin and also leached by abundant rainfall. Valleys and creek bottoms are usually most productive.

The high tablelands and foothills of eastern Montana, eastern Wyoming and eastern Colorado, as well as western North and South Dakota, western Nebraska and western Kansas, have brown soils of semiarid grasslands. There are some poorly formed and underdeveloped soils. The brown soils are quite productive with the application of water and with good cultural practices.

The Great Plains area of the eastern Dakotas, Minnesota, eastern Nebraska and Kansas have dark prairie soils. These are generally fine soils and some are nearly black. Drainage must be assured in the black soils for good production. This area produces well with natural precipitation, but some irrigating may help.

Transplanted seedlings of frost-sensitive crops must be kept warm during chillier weeks of planting season. One way to do this is by using hot caps, which are warmed by sun. They can be commercial types, as in top photo, or home-made, as seen by use of paper bags in lower photo.

IN THE NORTHEAST
by Walter L. Stirm

Gardening can be successful in every State of the Northeast region but is dependent on the climate. The region consists of Iowa and Missouri, and States east of the Mississippi and from the Canadian border southward to and including Kentucky and Virginia. Two basic climatic areas in this humid region are one with small temperature changes and another with moderate changes.

The small temperature change area comprises primarily all the Northeast region north of 40° latitude. This is a cold climate varying from hot summers (average above 71.6°F) and cool long winters as in the New England States. The area is forested and has frozen soil and snow cover for several months in winter. Rainfall is adequate in all seasons.

The moderate temperature change area comprises most of Illinois, Indiana, Kentucky, Ohio and the mid-Atlantic States. The climate is warm and rainy. There is no distinct dry season. Winters are cool and short, with frozen soil and snow cover a month or less in duration. Weather in the summer growing season varies from hot in much of the area to cool long summers in the Appalachian Highlands. Rainfall is adequate and sometimes more than needed during the growing period. A marine-like climate exists at times along the Atlantic Seaboard States.

Geographic features are favorable for gardening throughout the Northeast. Numerous streams provide a good drainage system. Swamps and high terrain areas are minimal in extent. Elevations range from 200 to 800 feet above sea level in the eastern Great Plains and Great Lakes area, to 0 to 200 feet above sea level in the middle and north Atlantic States.

Appalachian Highlands range from 2,000 to 4,000 feet above sea level and are 10° to 12°F cooler in summer because of the elevation. Mountain valleys and high plateaus are usually problem areas in the highlands due to cold air drainage and strong winds. Western slopes of the Appalachians are also cooler and cloudier during spring.

Areas within 20 to 30 miles of the south and east shores of the major Great Lakes have extensive moderation in temperature and precipitation conditions during all seasons. Numerous lakes in this area provide some moderation effect. Spring and summer rainfall is diminished, but rainfall increases in fall and winter. Temperatures are cooler for a longer time in spring but warmth is extended in autumn.

Large fertile flood plains are characteristic of river drainage systems throughout the Northeast. These areas have more water than needed in late winter and spring.

Temperature and rainfall are important to success in gardening through all of the Northeast. Average temperature and precipitation data are given in the table for representative cities in each State of the region. Average monthly temperatures are used as a guide in planting and for determining length of the growing season. The beginning and end of the growing season for cool season crops closely follows the time when average temperatures warm to 40° F in spring and cool to 40° in fall. For warm season crops an average temperature near 50° is used.

As a general rule, temperatures reach the level for safe planting of cool season crops about March 15 in southernmost States of the Northeast region. The northward progression of safe planting dates is about 5 to 7 days later for each 100 miles. In sou-

---

Walter L. Stirm is Advisory Agricultural Meteorologist, National Weather Service, West Lafayette, Ind.

Climatic Data for Representative Areas of the Northeast

City		Apr	May	June	July	Aug	Sept	Oct	Annual *	Mean number of frost-free days
Madison	Temp.	44.4	56.1	66.1	71.1	69.5	61.0	49.9	45.0	177
	Precip.	2.57	3.34	3.95	3.58	3.37	3.32	2.21	30.16	
Springfield	Temp.	53.4	63.9	74.0	78.0	75.2	67.6	56.7	53.6	205
	Precip.	3.59	3.88	4.45	3.49	2.74	2.93	2.91	34.83	
Lansing	Temp.	45.7	57.1	67.4	71.7	70.2	62.0	51.3	47.6	185
	Precip.	2.87	3.73	3.34	2.58	3.05	2.60	2.50	31.18	
Indianapolis	Temp.	50.8	61.4	71.1	75.2	73.7	66.5	55.4	52.1	193
	Precip.	3.74	3.99	4.62	3.50	3.03	3.24	2.62	39.25	
Lexington	Temp.	54.4	64.5	73.6	77.4	76.0	69.3	58.1	55.6	198
	Precip.	4.04	3.85	4.72	3.98	3.21	2.80	2.28	44.73	
Columbus	Temp.	50.8	61.5	70.8	74.8	73.2	65.9	54.2	52.0	196
	Precip.	3.49	4.00	4.16	3.93	2.86	2.65	2.11	36.67	
Harrisburg	Temp.	51.8	62.7	71.3	76.2	74.1	66.9	55.7	53.3	201
	Precip.	3.02	3.90	3.42	3.51	3.65	2.82	2.97	37.65	
Charleston	Temp.	55.3	64.8	72.0	74.9	73.8	68.2	57.3	55.6	193
	Precip.	3.68	3.71	3.69	5.67	3.95	2.92	2.58	44.43	
Richmond	Temp.	58.1	67.0	75.1	78.1	76.0	70.2	58.7	58.1	220
	Precip.	3.15	3.72	3.75	5.61	5.54	3.65	3.00	44.21	
Baltimore	Temp.	54.2	64.4	72.5	76.8	75.0	68.1	57.0	55.2	234
	Precip.	3.60	3.98	3.29	4.22	5.19	3.33	3.18	43.05	
Trenton	Temp.	51.7	62.3	71.0	76.0	73.9	67.1	56.8	53.9	211
	Precip.	3.21	3.62	3.60	4.18	4.77	3.50	2.84	41.28	
Albany	Temp.	46.2	57.9	67.3	72.1	70.0	61.6	50.8	47.6	169
	Precip.	2.77	3.47	3.25	3.49	3.07	3.58	2.77	35.08	
Hartford	Temp.	48.5	59.9	68.7	73.4	71.2	63.3	53.0	49.8	180
	Precip.	3.73	3.41	3.70	3.61	4.01	3.65	3.18	42.92	
Boston	Temp.	47.9	58.8	67.8	73.7	71.7	65.3	55.0	51.4	192
	Precip.	3.77	3.34	3.48	2.88	3.66	3.46	3.14	42.77	
Burlington	Temp.	41.2	53.8	64.2	69.0	66.7	58.4	47.6	43.2	148
	Precip.	2.63	2.99	3.49	3.85	3.37	3.31	2.97	33.21	
Portland	Temp.	42.5	53.0	62.1	68.1	66.8	58.7	48.6	45.0	169
	Precip.	3.73	3.41	3.18	2.86	2.42	3.52	3.20	42.85	

* Average annual temperature, and total annual precipitation.

thern mountainous areas, cool season crop plantings can begin April 1. The planting season progresses northward 5 to 7 days later per 100 miles.

Warm season crops can safely be planted April 15 in southernmost States of the region. The season progresses northward with a delay of 5 to 7 days per 100 miles. Mountain areas need to delay an additional 5 to 10 days, starting after April 20 in the South.

Cool season crops begin biological growth around 40° F and warm season crops begin growth around 50°. Cool season crops include asparagus, broccoli, brussels sprouts, cabbage, chives, collards, kale, kohlrabi, rutabaga, spinach, turnip, beet, carrot, cauliflower, celery, endive, lettuce, parsnips, potato, and salsify. Warm season crops include snapbean, sweet corn, tomato, most vine crops, eggplant, lima bean, pepper, melons, and sweet potato. Tree fruit and small fruit are included in cool season crops.

Average length of the frost-free growing season ranges from 180 to 234 days in the south and from 90 to 120 days in the north and mountainous areas.

The growing season rainfall March through October varies from 20 inches in the west to 40 inches in the south and east. Rainfall is the least in September through October and the greatest in June. Summer rainfall comes mainly from thunderstorm activity and often is poorly distributed during July and August. Dry, hot periods lasting 2 to 3 weeks are common during midsummer through the Northeast. In these periods warm season crops, double crop plantings, and tree fruit often suffer from stress and require supplemental irrigation to survive.

Extensive wet periods often occur in late May and June. Cool wet periods in April delay planting and establishing gardens with a frequency of 2 or 3 years in 10 years. These wet periods produce seed rot and disease problems.

Evaporation and transpiration may exceed precipitation during July through September, and in July may reach a 2 to 1 ratio.

Climate effects must be considered in handling garden soils. Garden crops have a limited root system and require a continuous supply of nutrients and moisture. Excessive rainfall leaches out soluble nutrients and increases acidity. Wetness causes aeration problems and poor root growth.

Climate and soil texture are important for soil warm-up in spring. Sandy soils warm earliest and have the widest daily variation. Heavy soils are cool in spring, warm slowly, and hold heat longer in autumn. Soil temperature at planting depth usually exceeds the air temperature by 5° to 10° F during the afternoon, and cools to several degrees lower than the air temperature at night. Soil temperature variations are largest when the soil is dry and smallest when it is wet.

Plant growth limitation occurs at various temperature thresholds. Temperatures above 86° F stop the growth of warm season plants, while temperatures as low as 77° halt the growth of cool season crops. Such temperatures occur as frequently as 10 days per month during summer in southern areas, and 2 to 3 days per month in the north. Warm season crops have lower growth limits around 50° and leaf tissue is killed at 30° to 31°. Cool season crops have a wider low temperature tolerance and some may withstand temperatures down to 20° before the leaf tissue dies.

Effects of the small scale climate close to the soil surface or near the crop canopy are important in gardening. Temperatures at the soil surface may be 20° to 40° F higher than the air temperature on hot, dry, sunny days and 5° to 10° cooler than air temperatures at night. Leaf temperature variations are 2° to 3° above air temperature on sunny afternoons and 1° to 2° cooler at night.

During low temperature periods, differences between forecast and surface level temperatures are important. Local forecasts predict what the temperature is expected to be at eye-level height of about 5 feet. Surface temperatures can be 2° to 8° F cooler on calm clear nights. Freezing temperatures down at crop level are possible with forecast temperatures of 35° to 40° F.

Direction of slope of the garden area is important. Southeast, south and southwest sloping surfaces are warmer, receive more solar radiation and dry out faster This may permit planting a week or more early. Gardens in low depression areas surrounded by high terrain are subject

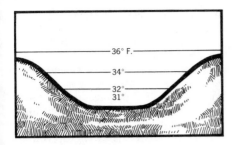

to cold air drainage and more frequent frost-freeze problems.

Freeze protection for garden plants is a necessity in the Northeast. In winter, perennial garden crops need protection from freeze-thaw action of the soil and protection from extremely cold soil temperatures. Straw mulches are effective.

Soil freeze depths in the Northeast range from 2 inches to over 24 inches, depending on snow cover. Root tissue damage develops at soil temperatures of 15° to 20° F.

During spring, protection by covering or by sprinkling on potential freeze nights pays dividends. Sprinkling is highly beneficial, particularly on dry cool nights when evaporative cooling causes leaves to supercool. Sprinkling must begin a few degrees above freezing, and continue until all ice has melted and the temperature is several degrees above freezing.

Sprinkling prior to covering is effective during dry cold periods. This reduces radiation losses through plastic covers.

Many crop management techniques can modify the local climate to reduce weather stress. Some are mulching, minimum tillage, plant population adjustment, shading, increasing organic matter, and adjusting fertility.

Coolest night air is usually found in lowest portion of a valley or in a bowl-shaped depression. Such places are generally frost-prone.

IN THE SOUTHWEST
by M. Douglas Bryant and Ricardo E. Gomez

The Southwest region has many plant growth zones because of temperature and precipitation differences due to great variation in elevation. Deciduous tree fruits with high chill requirements—such as apples, pears, peaches and cherries—are grown in the mountain valleys of Colorado, New Mexico, Arizona and California. Citrus is grown in selected low areas of southern California, Arizona and Texas. Vegetables and fruit crops are grown from below sea level to above 8,500 feet where frost-free days range from 365 to less than 60.

Carefully choose types and varieties of fruits and vegetables adapted to your specific locality.

The Southwest has low rainfall, extremely low relative humidity (often as low as 4 percent), sunny days, high solar radiation, temperatures that are high in daytime but drop as much as 50° F at night, and periods of drying winds that are often hot during summer. Warm, sunny days in mid-winter and late winter followed by sub-freezing temperatures at night make fruit production hazardous in many localities.

Rainfall is usually low and erratic, requiring at least supplemental irrigation. Water for irrigation is limited in most areas and even critically low in places. Water is impounded in reservoirs and dispersed through canals to fields, orchards and gardens, turning barren deserts into lush production.

Areas without access to reservoir water must rely on pumping from underground supplies. However, much

M. Douglas Bryant and Ricardo E. Gomez are Extension Horticulturists, New Mexico State University, Las Cruces.

of the underground water is high in salts, possibly toxic to plants.

With the rising demand for water by agriculture, industry, and domestic use, most States have passed water use regulations that allocate given amounts of water to each user. Many water/irrigation districts have been closed to additional expansion because of limited water supply.

Some growers and gardeners have installed "drip" or "trickle" irrigation systems which meter a very small amount of water to plants daily. These systems use only a fourth to a tenth the total amount of water applied by the conventional furrow and flood irrigation systems, but are more expensive to install. Mulching with plastic, straw, or grass clippings is a common technique practiced by gardeners to conserve moisture in the arid Southwest.

Prevailing wind east of the Sierra Nevada Mountains is from the southwest and is very dry and strong—especially during spring. Low rainfall combined with strong, dry winds makes gardening difficult in spring. Soil moisture is lost very rapidly, requiring frequent irrigation for young plants. Blowing sand is a problem from early March to mid-May, cutting off seedlings at ground level. Hail is a hazard to gardens during summer.

In vegetable gardening, give special care to variety selection since Southwestern areas have various types of climates. For example, some varieties of tomatoes do not set fruit well in the hot part of summer (probably due

Climatic Data for Representative Areas of the Southwest

City		Apr	May	Jun	Jul	Aug	Sep	Oct	Annual *	Mean number of frost-free days
Albuquerque	Temp.	55.8	65.3	74.6	78.7	76.6	70.1	58.2	56.8	198
	Precip.	.68	.68	.68	1.46	1.25	.94	.73	8.40	
Denver	Temp.	47.5	56.3	66.4	72.8	71.3	62.7	51.5	49.8	165
	Precip.	2.05	2.20	1.64	1.36	1.43	1.08	1.01	14.20	
Flagstaff	Temp.	42.1	50.1	58.2	65.6	63.6	57.5	47.0	45.2	118
	Precip.	1.28	.74	.52	3.11	3.03	1.69	1.48	20.92	
Las Cruces	Temp.	60.0	68.0	76.9	80.0	78.1	71.7	61.2	60.5	208
	Precip.	.25	.34	.58	1.58	1.79	1.21	.70	8.68	
Los Angeles	Temp.	58.8	61.9	64.5	68.5	69.5	68.7	65.2	61.7	359
	Precip.	.97	.38	.09	.00	.01	.20	.54	14.77	
Lubbock	Temp.	60.0	68.5	77.1	79.7	76.4	71.0	60.0	59.7	205
	Precip.	1.44	2.37	2.55	2.05	1.92	2.91	2.35	18.82	
Phoenix	Temp.	67.7	76.3	84.6	91.2	89.1	83.8	72.2	70.3	304
	Precip.	.43	.12	.08	1.00	.93	.71	.44	7.62	
Reno	Temp.	46.8	54.6	61.5	69.3	66.9	60.2	50.3	49.3	155
	Precip.	.53	.50	.39	.29	.33	.27	.44	7.73	
Sacramento	Temp.	59.6	65.3	71.3	75.9	74.9	72.5	64.5	61.4	307
	Precip.	1.04	.54	.16	.00	.00	.30	.80	15.88	
Salt Lake City	Temp.	50.1	58.9	67.1	76.6	74.4	64.2	52.9	51.3	202
	Precip.	1.76	1.56	.91	.61	.97	.74	1.34	14.74	
San Antonio	Temp.	69.6	76.0	82.2	84.7	83.7	79.3	70.5	68.8	282
	Precip.	3.25	3.50	2.58	2.30	1.64	2.86	2.33	26.79	

* Average annual temperature, and total annual precipitation.

to low humidity and high temperatures), but do set in spring and early fall. Some tomatoes require too long a growing season to be of any value at high altitudes.

Soils vary in texture, and for the most part are alkaline. This alkalinity may present problems, especially to gardeners from humid regions. Most gardening books suggest adding lime or such materials as wood ashes to improve soil characteristics. Their addition would at best be of no value, and might even deteriorate the soil.

Some manures—depending on source—have a high salt content and may be harmful rather than helpful.

Yet organic matter plays an important role in soil management and this is especially important in the Southwest where water is scarce and relative humidity usually low. Organic matter not only makes sandy soils retain more moisture, but also loosens up heavy or clayey soils.

Since soils and/or water for irrigation are usually salty, fertilizer practices—as well as other cultural practices—are somewhat different. Very light rains cause some problems because salt is brought to or near the soil surface where seedlings are more apt to be damaged.

Because some high elevation areas of the Southwest have short growing seasons, mulching with plastic films is advantageous. About one or two weeks may be gained on the weather. If mulching is combined with transplanting, you can expect another three- to four-day jump on the weather.

Mulching with organic materials is also very useful. It helps fight weeds and conserves water. This is a practical way of adding organic matter to the soil.

Cool season vegetables of excellent quality may be grown as summer crops in the cooler high altitude areas. In other parts, three distinct gardens may be grown on the same plot of land: spring, summer and fall gardens.

Spring crops will probably be of lower quality than fall crops because temperatures are getting higher when the crop is maturing, exactly opposite to what occurs in fall.

Excellent quality produce may be grown in the Southwest by even a new arrival if factors such as moisture relations, plant varieties, soils and climate are taken into consideration.

IN THE SOUTHEAST
by Perry M. Smith

Climate of the Southeastern States is primarily mild and humid. Occasionally there are short-lived extreme temperatures. The Southeastern States are generally considered to be North Carolina, Tennessee, South Carolina, Georgia, Florida, Alabama, Mississippi, Louisiana, Arkansas, Oklahoma and Texas. However, West Texas and West Oklahoma do not fit into this climate pattern—and should be considered in the Southwest region.

Climatic conditions in the Southeast allow a wide variety of fruits and vegetables to be grown. Fruit ranges from citrus in Florida, Southern Louisiana, and Texas to apples in North Carolina. In the Lower South, vegetable gardening is a year-round job. During recent years, home food production has increased substantially.

Americans are continually moving. Those gardeners who move south should become aware of the southern climate as it relates to gardening. Times for planting vegetables vary greatly from those of the North. The best advice one can give a new or inexperienced gardener is, "Go to your

---

Perry M. Smith is Extension Horticulturist-Vegetables, Auburn University, Auburn, Ala.

Climatic Data for Representative Areas of the Southeast

City		Apr	May	Jun	Jul	Aug	Sep	Oct	Annual *	Mean number of frost-free days
Atlanta	Temp.	61.1	69.1	75.6	78.0	77.5	72.3	62.4	60.8	244
	Precip.	4.61	3.71	3.67	4.90	3.54	3.15	2.50	48.34	
Charlotte	Temp.	60.8	68.8	75.9	78.5	77.7	72.0	61.7	60.5	239
	Precip.	3.40	2.90	3.70	4.57	3.96	3.46	2.69	42.72	
Columbia	Temp.	64.1	72.1	78.8	81.2	80.2	74.5	64.2	63.5	252
	Precip.	3.51	3.35	3.82	5.65	5.63	4.32	2.58	46.36	
Dallas	Temp.	66.4	73.8	81.6	85.7	85.8	78.2	68.0	66.2	249
	Precip.	4.72	4.85	3.27	1.80	2.36	3.25	3.18	35.94	
Houston	Temp.	69.4	75.8	81.1	83.3	83.4	79.2	70.9	68.9	309
	Precip.	3.54	5.10	4.52	4.12	4.35	4.65	4.05	48.19	
Jacksonville	Temp.	68.1	74.3	79.2	81.0	81.0	78.2	70.5	68.4	313
	Precip.	3.07	3.22	6.27	7.35	7.89	7.83	4.54	54.47	
Lexington	Temp.	55.3	64.7	73.0	76.2	75.0	68.6	57.8	55.2	198
	Precip.	3.87	4.16	4.31	4.83	3.40	2.65	2.12	44.49	
Little Rock	Temp.	61.7	69.8	78.1	81.4	80.6	73.3	62.4	61.0	244
	Precip.	5.25	5.30	3.50	3.38	3.01	3.55	2.99	48.52	
Miami	Temp.	75.0	78.0	81.0	82.3	82.9	81.7	77.8	75.5	†
	Precip.	3.60	6.12	9.00	6.91	6.72	8.74	8.18	59.80	
Mobile	Temp.	67.9	74.8	80.3	81.6	81.5	77.5	68.9	67.4	298
	Precip.	5.59	4.52	6.09	8.86	6.93	6.59	2.55	66.98	
Nashville	Temp.	60.1	68.5	76.6	79.6	78.5	72.0	60.9	59.4	224
	Precip.	4.11	4.10	3.38	3.83	3.24	3.09	2.16	46.00	
New Orleans	Temp.	68.6	75.1	80.4	81.9	81.9	78.2	69.8	68.3	302
	Precip.	4.15	4.20	4.74	6.72	5.27	5.58	2.26	56.77	
Richmond	Temp.	57.8	66.5	74.2	77.9	76.3	70.0	59.3	57.8	220
	Precip.	2.77	3.42	3.52	5.63	5.06	3.58	2.94	42.59	
Tulsa	Temp.	60.8	68.8	77.3	82.1	81.4	73.3	62.9	60.2	216
	Precip.	4.17	5.11	4.69	3.51	2.95	4.07	3.22	36.90	

† Freeze occurs in less than 1 year in 10.
* Average annual temperature, and total annual precipitation.

local county Extension office for garden information."

Compared with many areas of the country, the Southeast has a long growing season. With the exception of high elevation areas in North Carolina and Tennessee, there are at least 180 days without frost. This ranges to well over 300 days in South Florida and Texas.

Often there is confusion regarding freezing temperatures and frost. Frost is the deposit of ice crystals on the surface of plants or other ground objects. It is caused by freezing of water vapor at or below 32° F.

Frequently, because of dry air, frost does not occur with the subfreezing temperatures that injure plants. But it can occur on low growing plants when the reported temperature, which is taken at an eye-level height of about 5 feet, is above freezing. In other words, when frost occurs the temperature in the area of the frost is 32° F or below, while at the 5-foot level it may be several degrees higher.

Most areas of the Southeast receive about 50 inches of rain. Some areas in the mountains report as much as 80 inches.

Even though this much rain is enough to grow several crops, irrigation is often needed for optimum gardening. The heaviest rainfall comes in late winter and early spring. In most years prolonged periods without rain occur during the growing season.

High rainfall and humidity, although needed for plant growth, also create gardening problems. The moist conditions are often ideal for many diseases of plant foliage which are not a problem in more arid regions. Because of this it is not generally recommended that gardeners save their own seed. Most vegetable seed should be grown in arid areas of the West.

The Southeast is known for its mild temperature. Temperatures well below zero have been recorded, but these are usually of short duration. the biggest damage to gardens comes with late freezes in spring. For example, a freeze in late March 1955 virtually wiped out the peach crop all over the Southeast. It also severely damaged many vegetables. Fortunately such widespread damage has not occurred since.

In the Lower South, gardening comes to a virtual standstill in midsummer due to high temperature. During most summers temperatures may hit the upper 90's. Readings beyond 100° F occur but not frequently. At these times vegetable crops, such as tomatoes and beans, do not set fruit.

Altitude

The southeastern rim of these States is bordered by the Atlantic Ocean or the Gulf of Mexico. The elevation starts at sea level and rises as you go inland. In Alabama, for example, the elevation at Mobile is 11 feet; Montgomery, 169 feet; and Birmingham, 598 feet. In the mountains of North Carolina and Tennessee many areas are above 2,000 feet.

For each 400-foot increase in altitude, flowering of plants of the same species is retarded by 4 days. Also, for each degree of latitude north of the Equator, flowering is retarded 4 days.

Soils of the Southeast vary from lakeland sand to the heaviest prairie soils. With the abundant rainfall, they are subject to erosion year-round. They are rarely frozen for any length of time during winter, when much of the rain occurs. The light sandy soils are also subject to heavy leaching. Organic matter should be added frequently as it is dissipated rapidly by rainfall and high temperatures.

Pests

While the Southeast's mild and humid climate is ideal for a wide variety of crops, it is also great for a number of pests. Garden insects are more prevalent; more generations appear during a long growing season.

Some insects are problems throughout the year. However, the insect population is normally very high for late summer and fall crops. For example, early cucumbers and cantaloupes miss the pickle worm; control of this insect is a must for growing a late crop.

Another common garden pest in the region is nematodes. Here, again, the population increases during the season. New gardeners in the region should have their soil tested for nematodes through the county Extension office. A high percentage of old garden sites are infested.

The climate is ideal for development of many plant diseases. Using resistant varieties (when available), crop rotation, sanitation and fungicides should be considered to reduce losses from disease.

The climate is good for growing weeds and grasses, as well as for gardening. Rarely is there a garden meeting without someone asking how to control nut grass. Using chemicals to control weeds in a garden is somewhat limited. The best recommendation is the "Santa Claus method"—hoe! hoe! hoe!

Varieties

With both fruits and vegetables, gardeners should use varieties tested and found adapted to the Southeast. For example, Everbearing strawberries in some areas of the country do well; however, in the Southeast they might be called never-bearers. Factors that are most important are disease and nematode resistance and the ability to withstand high temperature.

The importance of using recommended varieties applies not only to the region as a whole but for specific areas within it. A good example of this is peaches. Most older varieties require 900 to 1,200 chilling hours (hours below 45° F) to break dormancy. Generally, these are used in the Piedmont areas of North Carolina, South Carolina, Georgia and Alabama. But in more southern areas, varieties with less chilling requirements must be used. Some varieties with a chilling requirement of 650 hours and some with even less have been developed in Florida.

Gardeners should obtain a list of recommended varieties for their area from the county Extension office.

Fruits and vegetables with specific adaptation to local conditions include:

Okra—This crop is closely associated with cotton. It requires hot weather and a long growing season. Though a favorite in the South, many people in other areas have never eaten it.

Sweetpotatoes—Another crop that requires warm weather and a long growing season.

Collards—This vegetable is found in most southern gardens. Extreme cold—the low teens—early in winter may kill the plants. Collards have contributed much nutritionally for Southern gardeners. They are very versatile as far as planting and growing are concerned.

Rhubarb requires cooler weather than is found in most of the South. It should not be grown except in mountain areas. Broccoli, cauliflower, brussels sprouts, garden peas, lettuce, onions, beets and carrots should be planted in winter or early spring because hot weather may damage these crops. Some gardeners plant them in early fall and harvest before extreme winter cold.

Many Southern gardeners grow Irish potatoes. These are planted in late winter or early spring and production is good. However, the big question often raised is why Southern-grown potatoes do not store like Maine and Idaho potatoes. Southern potatoes are harvested in extremely hot weather when not fully mature and are easily skinned. Also, this crop requires a cool storage area.

A final thought: The best way for gardeners to cope with the weather is to follow recommendations of the county Extension office.

How Plants Grow—and Let's Hope They Do!
by O. B. Combs

Fruit and vegetable plants are made up of tiny cells. They grow and reproduce by increasing the number, size and nature of these cells. As growth occurs, several important processes take place. Seeds germinate and become young plants that develop roots, stems, leaves and flowers. When flowers are pollinated, seeds and fruits begin to form.

These processes involve intake of carbon dioxide and oxygen, largely from the air and water, and minerals from the soil. Foods manufactured from these raw materials and stored in fruits, seeds, stems, tubers, leaves and flowers provide nourishment, directly or indirectly, for people and animals.

The most important single process carried on by green plants is called photosynthesis. Energy from the sun is trapped by the green plant, simple sugars are formed, and oxygen is released into the atmosphere. Complex carbohydrates, fats, proteins and vitamins are formed by the incorporation of sugars with mineral elements from the soil.

As plants multiply, enlarge their cells and build new tissues and organs, they break down foods with the aid of oxygen to release energy. This process is known as respiration; carbon dioxide, energy and water are released.

Most vegetables are reproduced sexually from seeds. Most fruits and a few vegetables are reproduced asexually or vegetatively from plants or plant parts. Seeds contain the embryo (live plant) and stored foods. For rapid germination, seeds must be alive and strong enough to break the seed coat and emerge from the soil. They need water, oxygen, and a suitable temperature.

The temperature best suited for rapid germination varies with different plants. Seeds of cool season vegetables such as carrots germinate rapidly at soil temperatures between 50° and 60° F, and seeds of warm season vegetables such as beans between 70° and 80°. Sweet corn seeds, which may require ten days to germinate at 50°, may germinate in two to three days at 80°.

Roots anchor and support plants and absorb water and minerals from the soil. These materials are taken in through root hairs and then moved through the stem to the leaves. Roots are the major storage organs of such vegetables as beets, carrots, and sweet potatoes.

Roots have no chlorophyll, thus they depend upon the leaves for their food supply. Roots must be healthy, unrestricted and undamaged by diseases, insects, or deep cultivation if they are to perform their functions. They must have adequate oxygen.

Too much water from a high water table or over-irrigation on heavy soils may result in insufficient oxygen and serious interference with proper root growth and function.

Root systems differ markedly in their form, spread and depth. The tap roots of carrots grow almost directly downward to considerable depths, and give off many lateral branch roots. The fibrous roots of onions grow an extensive, relatively shallow network of small roots. Some plants, such as apple and other tree fruits, have several large roots for anchorage, with many smaller branch roots.

Stems make up the above-ground framework of plants. They provide support for the plant and contain the food storage and transport tissues.

O. B. Combs is Professor of Horticulture, University of Wisconsin, Madison.

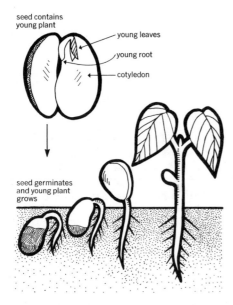

Water and nutrients from the soil move to the leaves, where sunlight and leaf chlorophyll—through the process called photosynthesis—manufacture food for plant growth and production of the edible portions. Water, nutrient, and manufactured food movement within the plant occurs in the conductive tissue, known as the xylem and phloem.

Leaves originate from stems and are the principal food-manufacturing organs of green plants. They take in carbon dioxide from the air and allow oxygen and water vapor to escape through small openings—mainly on their lower surfaces—called stomata. Plants are able to regulate these stomatal openings and thus partially control the loss of water from their leaves.

Flowers are necessary for production of seeds. Beans, peas, and sweet corn are annuals; they produce seeds each year and die when the seeds have matured. Cabbage, beets and carrots are biennials and need some part of two seasons to produce seeds; they die when their seeds have matured. Asparagus and rhubarb are perennials which—once established—may produce and mature seeds each year, but the plants continue to grow in the same location for several years.

Light is the source of energy for photosynthesis, the food manufacturing process carried on only by green plants. It also influences the movement and position of plant organs and the size, form and structure of plants. The amount of food manufactured depends upon the duration, intensity and quality of light. Whether a plant flowers or not is often determined by the relative length of day and night.

Formation of the green coloring matter (chlorophyll) in plants goes on only in light. Pure white cauliflower and creamy white, blanched celery, endive hearts and asparagus spears are produced by excluding light from these plant parts.

Plants arrange their leaves to insure suitable exposure to light. Excess exposure causes stunting; insufficient light causes tall, weak, light green plants.

Plants such as spinach and chinese cabbage go to seed if growing when days are longer than 15 hours.

The above-ground parts of plants generally are positively phototropic, they grow toward light; roots of most plants are negatively phototropic, they grow away from light. Tops of plants are negatively geotropic, they grow upward away from the force of gravity; plant roots are positively geotropic, they grow downward toward the force of gravity.

Temperature markedly influences plant growth. All important functions of plants—respiration, photosynthesis, absorption, digestion and reproduction—are influenced by temperature. Each plant has a temperature range in which it grows best, other factors being equal.

How bean plant develops from a seed.

Cool season vegetables such as carrots and spinach grow best at lower temperatures than those preferred by warm season vegetables such as beans and melons. Likewise, cool season vegetables are less susceptible to injury from frost.

Plants started indoors frequently are "hardened" by gradually exposing them to somewhat lower temperatures before they are set outdoors. Certain vegetables such as asparagus, horseradish, Jerusalem artichoke, parsnip, rhubarb and salsify will withstand very low temperatures and may be left in the soil over winter even in areas with severe climate.

Water is the most frequent factor limiting plant growth. It is the principal constituent of plants, and an essential raw material in the manufacture of food. Mineral salts must be dissolved in water before they can move into plants through the root hairs. Oxygen and carbon dioxide enter and leave plants in water solution. Mineral salts and manufactured foods move throughout the plant in water solution.

Water keeps plant cells turgid so that they can carry on their functions. It also helps to keep plant surfaces cool through evaporation from the leaves and stems.

At least 15 chemical elements are needed for the growth of fruit and vegetable plants. These include carbon, hydrogen, oxygen, phosphorus, potassium, nitrogen, sulphur, calcium, iron, magnesium, boron, manganese, zinc, copper and molybdenum. Some of these—such as boron, zinc, manganese, iron, copper and molybdenum —are referred to as minor or trace elements, since they are needed by plants in very small amounts.

The successful gardener must know something about how plants grow. He must also know the essential needs of plants and strive diligently to fulfill these needs with care and at the proper time.

PLANT REPRODUCTION
by N. Carl Hardin and Bradford C. Bearce

All plants reproduce themselves somehow, either sexually by seed or spores or asexually by splitting off parts such as bulblets, roots, buds and leaves from which new plants are formed.

To produce seed, plants must first produce flowers which come in varied forms among the many plant species. They vary greatly in size and brightness of color. A complete flower has each of the four kinds of parts: sepals, petals, stamens and ovaries. An incomplete flower lacks one or more of these.

The sepals are the outermost parts and together form the calyx, usually green. The calyx is the outer protective cover on a bud with the petals, stamens and ovaries inside. The often brightly hued petals are for attracting pollinating insects.

Next inside are the stamens which furnish pollen, and in the center of the flower are one or more ovaries. The ovaries contain ovules which develop into seeds when fertilized with pollen.

Some species of plants possess individuals with flowers lacking either stamens or ovaries. For such species to form seeds, pollen must be transferred from the stamens of one individual to the ovaries of another. Birds, insects, wind and water are important pollen carriers for these plants.

Once a single pollen grain has been transferred from stamen to the often sticky stigma, top surface of the ovary, it germinates to form a tube which grows through the stigmatic surface and down through an often elongated column of tissue called a

---

N. Carl Hardin and Bradford C. Bearce are Professors of Horticulture at West Virginia University, Morgantown.

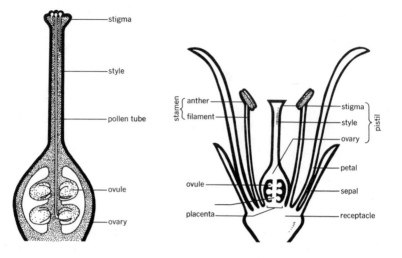

style into the ovary. Here the pollen tube penetrates an ovule.

A sperm nucleus from the pollen tube fuses with the egg nucleus of the ovule to form a zygote, the seed embryo. Another sperm nucleus fuses with other nuclei of the ovule to develop the seed endosperm. The tissue of the ovary surrounding the developing seed becomes the flesh of a fruit or the hardened shell of a nut, depending on the species of plant.

Parts of the Seed

Every seed has at least an embryo and one or more seed coats, while many seeds also possess an endosperm. The embryo in most seeds has a plumule or rudimentary shoot; one or more cotyledons or seed leaves; a radicle or rudimentary root; and a hypocotyl, the part of the embryo between radicle and cotyledons.

The seed coats protect the other parts enclosed within them. Endosperm and cotyledons serve as sources of stored food for the embryo as it develops within the seed and as it germinates. The cotyledons may also act as leaves to carry on photosynthesis, manufacture of food, until the first true leaves of the seedling have expanded.

Many seeds, even though alive, can not germinate right after maturity even under the proper conditions. This delayed germination is called dormancy and serves to prevent the seed from germinating during unfavorable times of the year when adverse climatic conditions might kill the young seedling plant.

Dormancy can be caused by hard seed coats which prevent entrance of water and air to the embryo, by immature embryos, or by presence of natural chemical inhibitory substances. Dormancy can be removed to allow seed germination by acids in the soil dissolving the seed coat, a cold period (winter, for example), or light which can cause breakdown of chemical inhibitors in the seed.

Asexual or vegetative reproduction occurs through cell division which is a normal process of growth and regeneration. In asexual reproduction a cell can split the chromosomes and divide into two daughter cells. As a result, the complete chromosome sys-

Left, germination of pollen grains on stigma, and growth of a pollen tube to ovule. Right, diagram of "perfect" flower with both male (stamen) and female (pistil) parts.

tem of an individual cell is duplicated in each of its daughter cells. The chromosomes produced will be the same as in the cell from which they came.

Asexual reproduction or vegetative reproduction is important in horticulture because the unique characteristics of an individual plant can be maintained. For instance, the millions of Golden Delicious apple trees throughout the world can all trace their ancestry to a chance seedling found in Clay County, West Virginia.

Cuttings and Rootings

The technique of removing a portion of the parent plant and placing it in a favorable condition to produce new roots and shoots is called cutting and rooting. Stems, leaves, and roots may be used.

The advantages of cutting and rooting with those plants that root easily are that many new plants can be made from a few stock plants. The method is simple, inexpensive, rapid, and does not require special skills.

It is important to have nearly ideal conditions to get the cutting to root. Generally, a temperature of 65° to 70° F, 100 percent humidity, and sterile soil produce satisfactory results.

Grafting is the art of joining parts of plants together so that they will unite and continue their growth as one plant. The part of the graft that is to become the lower part is termed the rootstock. All methods of joining plants are properly called grafting, but when the scion part is a small piece of bark containing a single bud the operation is called budding.

The reasons for budding and grafting are (1) to perpetuate a plant that cannot be reproduced by cuttings, layers, division, or other asexual methods; (2) to obtain special forms of plant growth; or (3) to obtain the benefits of certain root stocks. Root stocks may have disease resistance or growth controlling characteristics. Grafting or budding are also done to change the variety of an established plant, to increase production, or to grow a more popular variety.

Tubers (potatoes), bulbs (onions), and tuberous roots (sweet potato) differ botanically but have one thing in common. Each is an enlarged underground portion of the plant. The gardener reproduces these plants by cutting up or pulling apart the thickened structures into pieces from which new plants grow.

Many plants can be reproduced either sexually or asexually and often both methods are used. In other instances a combination of both methods is used.

Apple and other fruit trees are produced in great numbers by planting the seeds to grow into root stocks, then grafting or budding the desired scion wood or variety to the root stock.

PLANT POLLINATION
by S. E. McGregor

Many plants can be propagated vegetatively by cuttings or underground parts. This is not always as practical as growing the plant from seed. But for seed to be produced the plant must flower and the flower needs to be pollinated.

There must be a union of the sperm of the pollen with the ovule, or developing seed, within the flower. This union has to take place at precisely the right time and in a manner carefully prescribed by the flower, as we shall see. First we need to get acquainted with the flower.

Within every flower there is a sexual column, usually surrounded by petals. The female part, the pistil, consists of the ovary, style and stig-

---

S. E. McGregor is a Collaborator at the Bee Research Laboratory, Agricultural Research Service, Tucson, Ariz.

ma. In the ovary the seed or fruit develops. There may be only one seed, as in the peach, or hundreds of seeds, as in the melon. The stem-like style extends beyond the ovary. At its outer tip is the stigma, the area on which pollen must land if seed is to be produced.

The male part of the flower usually consists of numerous hair-like filaments, the stamens, bearing on the outer ends the pollen-producing anthers. When an anther matures it splits and disgorges the microscopic pollen grains.

Transfer of pollen from anther to stigma is termed pollination.

When a pollen grain lands on the stigma of a receptive flower it sprouts a pollen tube. This tube grows down the inside of the style to the ovary. The sperm nuclei of the pollen grain in the tube contact an ovule in the ovary and seed development is initiated. This union is referred to as fertilization.

Some plants are not receptive to their own pollen. These are referred to as self-sterile. Most commercial apple varieties are self-sterile. They produce fruit only when pollinated by another apple variety, and in some cases only by a specific variety.

If the plant is receptive to its own pollen it is referred to as self-fertile. If fruit or seed can be produced by the plant without the aid of any outside agency it is self-pollinating and self-fertilizing. When two varieties will pollinate each other they are compatible or cross-fertile.

There is an important difference, which is often overlooked, between self-fertility and self-pollination. Most peaches and muskmelons, for example, are self-fertile but they are not self-pollinating. An outside agent is necessary to transfer the pollen from anther to stigma. At least one pollen grain must land on the stigma at the right time for each seed that develops.

William E. Carnahan

Only one viable pollen grain is necessary to produce a peach. About 10 seeds must develop within an apple if the fruit is to be uniform in shape. This means that at least 10 viable pollen grains, of a compatible variety, must land at precisely the right time on the stigma. Hundreds of grains must land on the melon stigma, sometimes within only a few minutes, if a perfect melon is to be produced.

Agents of Pollination

There are numerous pollinating agents: wind, insects, birds, bats, raindrops, and to a degree gravity. Wind and insects are the primary cross-pollinating agents of cultivated crops.

Plants that are insect-pollinated usually have colorful flowers which produce nectar and pollen attractive to insects. The pollen grains are coated with a sticky material that tends to hold them together. Nectar is secreted in nectaries that are usually located within the flower near the base of the sexual column.

Flowers of wind-pollinated plants are usually inconspicuous. They have small petals, or none at all. They produce pollen in great abundance that, when dry, is easily carried by wind. The stigmas are often relatively large, complex, and exposed so as to in-

Honey bee is most effective of all pollinating insects.

Some Fruits and Vegetables That Benefit From Insect Pollination

Almond	Mango
Apple	Muskmelons
Apricot	Pawpaw
Avocado	Passion fruit
Blackberry	Peach and nectarine
Blueberry	Pear*
Broadbean	Peppers
Chayote	Persian melon
Cherry	Plums and prunes
Chinese gooseberry	Pumpkin
(kiwi apple)	Quince
Cucumber	Raspberry
Currant	Scarlet runner-bean
Eggplant	Squash
Gooseberry	Strawberry*
Honeydew melon	Tangelo
Lima bean	Tangerine
Macadamia	Watermelon

*Some varieties

crease the likelihood that wind-carried pollen will contact them.

Corn is an example of a wind-pollinated plant. Pollen is produced in the anthers of the tassel at the top of the plant. The silks on the ear are the styles leading to the ovules or grains of the ear. Both wind and gravity aid in pollinating corn.

Most fruits and many vegetables are insect-pollinated.

The honey bee is the best of the pollinating insects, because it visits flowers of many different plants, is widespread, and can be manipulated by man. It collects large quantities of nectar and pollen for maintenance of the colony. In the process of collecting this food it accidentally transfers pollen from anthers to stigma of the flower.

Each bee usually visits many plants but only one plant species on a foraging trip. Therefore it effectively pollinates many flowers.

There are other pollinating insects, including "wild bees", ants, beetles, butterflies, moths, and wasps. Only the wild bees provision their nests with nectar and pollen; this makes them, like the honey bee, more efficient pollinators.

On some crops, such as apples, intensive bee activity on the flowers for one day is enough to produce an excellent crop of fruit. Their activity may be required for three or four weeks on cucumbers.

A few plants will produce fruit without any form of pollination. Such fruit is seedless, and referred to as parthenocarpic. The seedless oranges, seedless raisin grapes, certain cucumbers, certain pears, pineapples, some figs, and bananas are examples of parthenocarpic fruit.

Home gardeners sometimes keep a colony of honey bees to insure pollination.

Insecticides should not be applied to flowers in such a way that pollinating insects are killed.

Know Your Soil and How to Manage It
by Lindo J. Bartelli, David F. Slusher and Kelton L. Anderson

Essential elements of successful fruit and vegetable gardens are suitable soil, an adequate water supply, enough sunlight, and climatically adapted plants. These, of course, are supplemented by the gardener's know-how and good management. Water, sunlight, and suitable plants are available to most home gardeners. Well suited soils may not be.

The soil must permit adequate root growth to support the plant and must supply water and oxygen. It should be free of toxic elements. An ideal garden soil is nearly level or gently sloping and has favorable air and water movement. It is medium acid to neutral, pH 5.5 to 7.0, and has a good supply of organic matter in the surface layer.

How then does the home gardener, except by trial and error, determine how suitable the soil is or what properties need special attention to make it as productive as possible? A good guide is a soil survey.

Soil surveys have been made and published for more than a third of the counties in the United States by the U.S. Department of Agriculture in cooperation with State and other Federal agencies. Each year about a hundred additional soil surveys are made and published. Copies for a county are available in most public libraries. They include maps on which the gardener can locate his garden plot and determine the name of the soil.

Text of the soil survey describes the soil in terms of properties important to plant growth, such as depth of root zone, available water capacity, soil texture, wetness or soil drainage, acidity or alkalinity, and the rate of air and water movement. General statements of the organic matter content, fertility level, and the need for lime and fertilizer are often given. Plants suitable for certain soils are also listed.

If the land area contains two or more soils, you can select the most desirable for your garden. The soil survey can be very useful in selecting the most desirable sites if such alternatives exist.

The major soil type in each area outlined on the soil map is given a name. The soil description gives the properties of each soil layer, usually to a depth of several feet. From the

---

Lindo J. Bartelli retired in December 1976 as Director, Soil Survey Interpretations Division, Soil Conservation Service. David F. Slusher is Assistant Director of the Division. Kelton L. Anderson is Leader, Extension Agronomy, Mississippi State University.

Section of soil survey map showing kinds of soil by symbols within each outlined soil area. County soil survey report describes characteristics of soils. Besides soil areas, this map section includes a town name (Gladys), a railroad, road intersection, and powerline.

Gordon S. Smith

Examine soil to depth of two or three feet.

description you can learn about the soil properties important to plant growth and some of the practices that may be helpful in good soil management.

Many soils in urban areas have been disturbed to some degree during excavation for utilities and foundations for buildings and during construction of roads and walks. In areas of naturally steep slopes, cuts and fills may have altered the soils drastically. However, the soil survey can be helpful because it states the properties of underlying material in cuts or can be used to predict the composition of fill or soil moved only a short distance as a result of site preparation.

Most grading operations around homesites result in a greater degree of soil compaction than that in natural soils, and water or root penetration may be restricted.

If you have an opportunity to supervise grading during site preparations, you should see that the natural topsoil is stockpiled and used as top dressing in the final grading and smoothing operations.

Soil is a mixture of mineral particles, organic matter, air, and water. It typically occurs as a series of horizontal layers from deposition of geologic material or the result of soil-forming processes. The surface layer of an undisturbed soil normally is a more desirable medium for plant growth than underlying layers.

Relative composition of desirable soils is about 45% mineral matter, 5% organic matter, 25% air, and 25% water. Also important are depth to water table, permeability to air and water, and available water capacity. These are all given in the soil survey.

If no soil survey is available, you should examine the soil to a depth of 2 or 3 feet to evaluate its properties. It's best to do this when the soil is moist but any time is O.K. Use an auger or spade, dig down a few inches at a time, and examine each layer for thickness, color, stickiness, texture, and content of stones. If the soil is dry, moisten a small amount with water and rub it between your fingers to note the sand content and stickiness. The importance of these and other properties in gardening are discussed separately.

Texture—A soil contains different sizes of particles called sand, silt, or clay. Clay is the smallest and sand the largest. A loamy soil has a desirable amount of each. For example, a mixture containing about 20% clay, 40% silt, and 40% sand is a loam—a desirable texture for a garden soil.

Slightly different composition results in textures of silt loam, sandy loam, or clay loam, all of which are desirable for garden soils. Texture of soil layers of each soil is given in the soil survey.

Too many clay particles result in a soil that feels smooth and plastic when moist and retains its shape when molded. Clays have very small pores, and air and water movement is too slow. They are sticky and plastic when wet, and hard when dry.

Too many sand particles make a texture that is loose and feels gritty between the fingers. Sands are easy to work but do not have the capacity to store water for plants. Some soils contain rocks and stones that interfere with cultivation unless removed.

Surface texture can be changed by adding silt, sand or clay and working it into the surface layer. This is quite difficult and seldom practical on a good-sized garden. A better alternative is to cover the garden with 4 to 6 inches of desirable topsoil from some other source.

Five cubic yards (a common truck load) will cover a 400-square-foot garden (20x20) with 4 inches of soil. The same amount will cover a 270-square-foot garden (16x17) with 6 inches of soil.

Your soil survey also will rate the suitability of the different soils in the area as a source of topsoil.

Some effects of unfavorable surface soil texture can be overcome by adding organic material. That and other important functions of organic matter will be discussed later.

Soil drainage, permeability—Garden plants grow best in soils with good drainage and aeration. The drainage property is given in the soil survey in terms such as well drained or somewhat poorly drained. Permeability is the rate of water movement through the soil. The rate is given in the soil survey as inches per hour or in terms such as moderate permeability or slow permeability.

Another way to find out about drainage is to dig a hole a few inches across and about 3 feet deep. Observe the hole for presence of water, especially during the wettest part of the season. If the water table (free water level) is near the surface during the growing season, most plants can't be expected to do well unless a drainage system is installed.

For technical assistance, contact your soil conservation district office, Extension office, or a drainage contractor.

Another way to check for the rate of water movement is to dig holes 2 to 3 feet deep and fill them with water. Allow the water to drain away and fill them again. After the second filling, the water level should fall at least 1 inch every 45 minutes and disappear entirely within about 24 hours. If it takes longer than this, the soil has slow permeability and you can expect only very shallow-rooted crops to do well.

Wet conditions in gardens can result from a low garden location or from water flowing onto the garden from higher elevations. Leveling the garden to eliminate low places, sloping the surface toward a drain, or raising the garden's center to permit drainage away from the garden may be helpful. In some places each row is planted on a slight ridge rather than in the furrow because drainage is improved and the top of the row warms up sooner in spring.

Organic matter—The importance of soil organic matter cannot be overemphasized. Almost any soil can be made fit for growing plants if enough organic matter in the form of decayed leaves, compost, peat moss, decayed sawdust, ground bark, or animal manure is added.

Dark brown to black colors indicate a high organic matter content. Natural topsoil contains the largest amount. That is why it's important to locate the garden on undisturbed soil if at all possible This also is the reason that top soil should be stockpiled during grading for construction and then used as the final top dressing.

Organic matter has several functions. The most important is as a source of plant nutrients and as a soil conditioner. Organic matter adds body to sandy soils. It increases the capacity to hold moisture and nutrients. It promotes granular structure in clay soils, which aids in root growth and the entrance of water and air into the soil. It makes the soil easier to till, results in better seedbeds, and reduces crusting that affects emergence of seedlings.

Raw organic material is decomposed by soil micro-organisms to produce humus and release nutrients that are used by plants.

Animal manure has a fairly good balance of nutrients and decomposes rather rapidly. On the other hand, sawdust has a low level of nutrients (especially Nitrogen) and decomposes rather slowly. However, decomposition can be speeded up quite markedly by adding generous amounts of fertilizer (particularly Nitrogen) to sawdust compost or to soil into which sawdust has been incorporated. The same applies, to a slightly lesser extent, to other "woody" organic materials such as ground bark, leaves, and grain straw.

Organic materials added should be worked into the soil's upper 3 to 6 inches. Such matter needs to be added each year on most soils. Don't use organic wastes contaminated with weed seed or pesticide residue.

Soil moisture and watering—If a soil has been saturated by rainfall or irrigation and then is drained by gravity until the larger pores are free

Contoured garden saves water and reduces soil erosion.

of water, the soil moisture content is at field capacity. This takes a few hours. When plants have used up the available moisture and permanently wilt, the soil moisture content is at the wilting point.

The amount of water in the soil between field capacity and the wilting point is the available water capacity. Compaction decreases the water-holding capacity and additions of organic matter increase it slightly.

Plants vary in their water needs. Water requirements also vary with the soil conditions and the climatic conditions.

Most garden plants require about 1 inch of water per week. A soil with the capacity to hold .1 inch of water per inch of soil can supply only a week's growth with roots in the top 10 inches of soil. If rainfall is less than 1 inch per week, supplemental watering is required.

Know your soil. For example, the water-holding capacity of the soils of the District of Columbia ranges from .04 to .26 inches of water per inch of soil. The soil with the lowest rate will need watering every 3 days during the growing season, while the soil with the highest rate can support

A. S. Harvey

plant growth for 2½ weeks without watering.

Plan a watering schedule that fits your soil and climate. Start watering when about a third to half of the available water has been removed. Do not apply more water than the top 10 inches of soil can hold. Most garden sprinklers apply about a quarter of an inch of water per hour.

Slope and erosion control—Some slope is desirable on most gardens to eliminate wet spots. About 1 foot per 100 is usually enough. On strong slopes water may run off too rapidly, eroding the topsoil and carrying away organic matter and plant nutrients. Contouring or terracing permits the water to move across the slope instead of down. Thus the soil absorbs more rain and is less susceptible to drought.

To contour a garden, make rows across or perpendicular to the slope of the land. This does not require special skill or equipment. An efficient device can be made from an ordinary carpenter's level mounted on a long 2 by 4. Lay the 2 by 4 across the slope and move one end up or down until the bubble on the level is centered; mark the spot with a stake. Repeat the process across the slope to establish the contour guide line.

Plant the rows of vegetables parallel to this line. As you cultivate the garden, leave small channels between the rows to collect and hold the moisture so that it soaks into the soil.

Surface drainage may be required on some soils. You may need to give a slight slope to the contoured rows to provide for drainage.

Plant, Soil, Fertilizer Relations

Plants must have chemical nutrients to live, grow, and reproduce. The soil is the storehouse for plant nutrients. If the soil's natural content is low in any required nutrient, the nutrient can be added through chemical fertilizers or soil amendments.

Plants get nutrients through their roots. The plants contain complex uptake and distribution systems that move essential materials to every living cell. Growth and nutrition of all plants depend on an adequate supply of nutrients distributed where needed.

Plant nutrients can also be absorbed through openings in leaves and stems, and sometimes it is practical to apply them in this manner.

Nutrients dissolved in water enter plant roots and circulate throughout the plant. A moist but unsaturated soil is desirable. For plants to get nutrients, the nutrients must be in the root zone and must be in a soluble form.

Nitrogen deficiency is usually the most limiting to plant growth. It is an essential part of proteins and amino acids in protoplasm and nuclei of all living plant cells.

Nitrogen is usually concentrated in fast-growing areas such as tips of shoots, buds, and new leaves. It is often called the growth element. Adequate nitrogen encourages a dark green color, rapid growth, and profuse fruiting. Excess nitrogen causes excessive vegetation and reduces fruiting.

Pale greenish yellow leaves and stunted plants indicate a nitrogen deficiency. Older leaves are affected first, and tissue along the leaf midribs will eventually die.

Phosphorus provides the "energy currency" of all living cells. Adequate phosphorus encourages root development, growth, and fruiting. It is essen-

To lay out contour line, use carpenter's level mounted on 2 by 4.

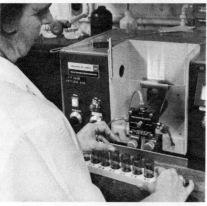

tial in activation and regulation of certain enzyme systems. Phosphorus deficiency causes slow growth and stunting, and the leaves turn dark green or blue green.

Phosphorus moves slowly in the soil. Highly water-soluble or weak acid soluble sources are needed for quick plant response.

Potassium increases plant vigor and strength, reduces lodging, encourages early root formation, and increases resistance to certain diseases. Inadequate potash results in spindly, poorly developed plants that lodge easily and are more susceptible to leaf and stem diseases. Severe deficiency causes leaf burn along the margins from the tip, beginning with the lower leaves and moving up the plant.

The pH of a soil indicates the degree of acidity or alkalinity, not the amount. For example, two sandy soils with the same pH require about the same amount of lime to raise the pH one unit. A clay soil, however, with the same pH may require several times as much lime as the sandy soil to raise the pH one unit. In addition, soil testing laboratories run special tests to measure reserve acidity in order to make accurate recommendations for lime or other soil amendments.

The pH of the soil solution greatly influences nutrient availability to plant roots. Although extremely acid soil solutions may injure living plant tissue, soil acidity in nature is seldom toxic to plants.

Certain elements such as aluminum and manganese found in large quantities in some soils may, however, become soluble to levels toxic to plant growth at a pH below 5.5. In the

Top, filling in address on soil sample bag, which is sent to Pennsylvania State University lab for analysis, center, and results returned to gardener in print-out form, bottom.

Southeast, the subsoils may be leached to the point that subsoil acidity actually reduces the rooting depth of certain plants. Soils having these properties are identified in soil surveys.

Plant species differ in the pH range at which they grow best. Most plants are at their best in a pH range of 6.0 to 6.5 if most nutrients are available in sufficient but not toxic quantities. Examples of plants that do well in acid soils are blueberries and Irish potatoes.

Soil pH can be changed by adding lime or sulfur to meet the pH requirements of the plant to be grown. Adding a liming material increases pH by reducing soil acidity. Adding sulfur increases soil acidity by reacting with water in the soil to form sulfuric acid, thus reducing soil pH.

Test for Fertilizer Needs—One way to determine lime and fertilizer needs is to have the soil tested. Reliable soil testing services are available from both private and public laboratories throughout the Nation. Many land grant universities provide this service through the cooperative Extension service. Extension offices are located in each county and some cities.

Soil samples should be representative of the area they are taken from. Use the soil map as a guide for locating sample sites. Sample different soil types separately. Small garden areas should be taken as one sample and the crops to be planted listed. Soil testing helps take the guesswork out of lime and fertilizer programs.

Kind, rate, frequency, and method of fertilizer application are everyday questions of gardeners. The kind of fertilizer, type of plant, and soil and moisture conditions all affect these decisions.

The nature of the major plant nutrients and their movement within the soil is important. If placed on the soil surface, nitrogen moves rapidly into the soil with water, phosphorus moves in very little, and potassium moves fairly well into porous soils. Movement is mostly downward, but there is some upward and lateral movement in some soils as wetting and drying occurs. These features are discussed by kinds of soil in many soil survey reports.

Nitrogen is easily lost from most soils, so frequent small applications to the surface or shallow incorporation is advised. Watering will help speed response.

Phosphorus should always be placed in the soil before or at planting if practical. Potassium applied as a topdress or sidedress should be incorporated in the upper few inches. One application of phosphorus or potassium per crop or year usually is enough for most vegetable or orchard plants.

Many long-season vegetables and most orchard plants will need several applications of nitrogen.

Special liquid, mixed fertilizers for spraying the leaves of orchard and vegetable crops are available. They must be used in dilute solution, which requires several applications to provide enough plant nutrients for one crop. This is a good way to apply several of the minor nutrients such as boron, zinc, copper, iron, and manganese.

A soil survey and a soil test are a good introduction to your garden. They help you select the most suitable site. Through knowledge of soil properties you can learn to manipulate your soil for a wide selection of plants.

Structures—From Trellis to Greenhouse
by David S. Ross

Structures can add to your gardening enjoyment in a number of ways. The beginning gardener may get along with a minimum, just a place to keep tools and supplies. Others will find additional structures advantageous and reasonable.

Coldframes, hotbeds, and greenhouses provide a place to start seedlings or to grow an early crop of lettuce. A storage shed or area is needed for rakes, hoes, power tillers, fertilizers, and numerous other items. Gardening space can be expanded vertically by using a fence or trellis to support a growing vine or even a fruit tree. And when the harvest is over, a place must be available to store the surplus produce until it can be consumed.

The environment in plant growing and food storage structures is quite important. Environmental control equipment such as heaters, coolers, ventilation fans, and thermostats are necessary to maintain the appropriate temperature conditions. Light and water are essential for plant growing and must be supplied in a suitable way.

Plans and information releases from the U.S. Department of Agriculture are available from your county Extension office. These offices have additional information and will be able to answer questions on specific gardening subjects.

A hotbed or coldframe is well suited for starting seedlings early for transplanting into the home garden. It can double as a place to grow cool-season vegetables in the spring for table use or to extend the fall growing season.

The coldframe is generally a wooden frame box about 3 by 6 feet in size with the back (north side) higher than the front so as to slope the top to capture the most sunlight, and for rain runoff. The sloped top, attached by hinges, is made of window sash, storm windows, or a frame covered with polyethylene film. The hotbed differs by having a source of heat to warm the soil and in colder climates to warm the air.

A sunny well-drained location with wind protection is ideal. A location near the south side of the house is good because it is close to water and electricity, and the young plants can be given the frequent attention they need.

Good construction conserves the soil heat at night. Make the joints as airtight as possible. Soil can be banked up around the sides to keep it warmer. The wood parts should be painted with a primer and one or two coats of white paint to reflect the light.

A wood preservative, such as 2 percent copper naphthenate, is safe near plants and can be used to give additional protection before painting. Creosote and pentachlorophenol are

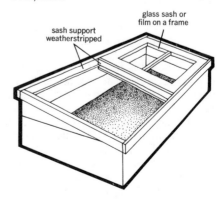

Coldframe.

David S. Ross is Extension Agricultural Engineer, University of Maryland, College Park.

toxic to plants and should not be used.

A convenient size may depend on the building materials available, but a 3- by 6-foot coldframe or hotbed gives plenty of room for young tomato, pepper, cabbage, onion, and other plants for a big home garden. Height of the frame can be 12 inches in front (south side) and 18 inches in the back (north side). However, these dimensions can be reduced; the top may be open most of the time after the last frost when the plants are tall. The length can be in multiples equal to the width of the window sash or other material used for the cover.

Heat for the coldframe comes from the sun which warms the soil and air. At night the heat is slowly lost through the cover. The temperature must be controlled during the day so it does not get too high for the plants (maximum of 100° F). The cover must be raised to permit ventilation, or you can install a small ventilation fan controlled by a thermostat.

Cool-season crops, such as cabbage, cauliflower, and lettuce, can stand a day temperature of 60° to 65° F. Warm-season crops, such as eggplant, peppers, tomatoes, and melons, do better at 65° to 75° day temperature. The night air temperature can be 5° to 10° lower with good results. Coldframe temperature may not be easy to control, depending on weather conditions and the frequency with which someone can check the temperature. However, with experience a satisfactory operating procedure should be found.

The hotbed is heated by an electrical cable placed under the soil. In northern areas of the United States, 12 to 16 watts of heating cable per square foot of bed area is needed. In southern areas 10 watts per square foot should be adequate.

Heating cables vary in length and wattage rating. Select the number of cables needed for the total wattage.

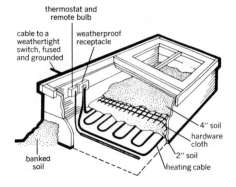

Be sure adequate electrical service with ground fault interrupter (GFI) is available to meet requirements of the National Electrical Code and any local codes.

Lay the heating cable on the ground 6 inches below the intended planting surface. Carefully spread the cable so it is uniformly spaced on the bottom. Avoid kinks. Do not lay cable across another cable as overheating damage may result.

Cover the cable with 2 inches of soil or sand. Then place ½-inch mesh hardware cloth across the surface of the sand to protect the cable from gardening tools. Add a 4-inch layer of soil or soil mix on top of the wire mesh for growing plants. Sometimes 2 inches of sand or vermiculite is placed below the cable to conserve heat.

Early hotbeds had 12 to 24 inches of animal bedding and manure in them to provide heat as the manure composted. These beds generally grew better plants because of carbon dioxide released by the manure.

A thermostat, normally built into the heating cable, is used to control soil temperature. A separate thermostat with remote sensing bulb works well because the sensing bulb can be placed in the root zone. A temperature of 70° to 75° F is ideal for ger-

Layout of heating cable in hotbed. Cable is covered with soil.

minating most seeds. A good thermometer should be used to check operation of the thermostat at seed depth and to measure air temperature.

Cost of operation depends on the weather, location, and construction. A 3- by 6-foot bed will use 1 to 2 kilowatt-hours of electricity per day.

Operating costs can be reduced by adding insulation to the sides, covering the bed at night with straw or other insulating material, and keeping the frame in good repair and airtight.

Watering is important. Try to have the bed moist at all times but not soaked. Apply water in the morning so plant foliage can dry by evening.

Greenhouses

A greenhouse can open a new world of enjoyment for the seasonal gardener who otherwise has to put away his gardening tools each fall. It can also increase the pleasure of the indoor plant grower who needs better light and better control of temperature and humidity.

Rising fuel costs, the relative high cost of the structure, and the needed environmental equipment limit the number of people who can invest in a hobby greenhouse. Although food can be grown in the off-season, the expense generally will be greater than buying the product in the supermarket.

A small greenhouse can be designed and built by an individual or can be purchased from one of the many manufacturers. Some shapes and styles look quite different from the conventional and show that one's imagination is the limit. They can be attached to the house or freestanding, away from any building.

The home greenhouse can be attached to the home where space is limited or where it blends well into the architecture and landscaping. The attached greenhouse is conveniently accessible and water, electricity, and heating facilities may be shared with the house. Heat loss is less since one side is shared with the house and not exposed to the weather. The amount of sunlight may be less because of shading if the greenhouse is on the east or west wall of the house.

Choice of location for your greenhouse may be limited but there are several things to consider. Of primary importance is available sunlight. A southern or southeastern exposure would be first choice for good winter lighting when young seedlings are growing. The east side is second choice because it gives exposure to the morning sun. The north side of a building is the least desirable location unless only shade-loving plants are grown.

The greenhouse can be partially shaded in summer by deciduous trees, but should not be shaded in winter. Keep in mind that the winter sun is lower in the sky and shadows of evergreens and buildings are much longer.

The building site should be well-drained with the greenhouse floor built up and the surrounding ground sloped from the site to carry rainwater away. Water, fuel for heating, and electricity should be available nearby so utility connections can be made. Check local electrical and plumbing codes, and get qualified people to do the work. Shelter from wind helps reduce heat loss.

Size of the greenhouse will often be limited by cost and available space. If cost is a major factor, consider a coldframe or hotbed instead. The temperature is more difficult to control in a very small greenhouse and heat loss is high compared to a larger house. A 6- by 6-foot greenhouse could better be replaced by one or two 3- by 6-foot hotbeds.

Greenhouse size depends on the amount of growing space you feel you need and can maintain. Also bear in mind that many people soon find they run out of space. Space may be needed in the greenhouse for working and for storing soil mix, chemicals, tools, and other items. However, heating the work and storage space will be expensive. Estimate the amount of use throughout the year in justifying an investment.

Width of the greenhouse is fixed once construction is completed. But with prior planning, the length can be increased in the future to give more growing space. Benches generally are 2 to 4 feet wide for easy working from one side, and 2½ to 3 feet high. Aisles would be a minimum of 2 feet wide, or 2½ to 3 feet wide if a cart, wheelchair, or wheelbarrow is used.

Leave at least 6 inches between each sidewall and the bench to permit good air circulation. Do not crowd yourself on headroom either. Gable roof houses should have at least 5½ feet eave heights and a minimum of 8 feet at the ridge.

A minimum greenhouse size is about 8 by 10 feet for a freestanding house. A lean-to greenhouse is limited to widths of 7 to 12 feet by the roof slope. Roof slope for glass greenhouses should be 6 inches of rise per foot of width, and 7 to 8½ inches of rise per foot for fiberglass and polyethylene (film plastic) houses.

The steep slope is to permit moisture condensing on the inside of the roof to run off instead of dripping onto the plants, promoting disease. It also allows snow to slide off to let light in.

Ideally, a greenhouse has a minimum framework needed to support a translucent cover so the maximum amount of light is received. The supporting frame can be made of wood, aluminum, plastic, and black or galvanized iron. The cover may be glass, rigid plastic or fiberglass, or plastic film. Each material has advantages and disadvantages which should be considered for the type greenhouse planned.

Glass greenhouses are long lasting; require a good foundation and strong,

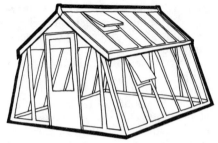

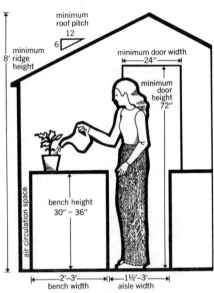

rigid frame; and have high initial cost. Maintenance is reduced but heat loss is increased if a metal frame is used instead of wood. Fiberglass has a variable service life depending on the quality product used.

Film plastic has a service life of one or two years and can be used on quite temporary frames. Annually or every second year, commercial people replace polyethylene on good frames which last many years. This spreads

Top, freestanding, slant-leg greenhouse. Bottom, suggested minimum dimensions for a greenhouse. Benches should be placed far enough from side wall to allow adequate ventilation.

out the cost of the cover over the life of the greenhouse.

Regardless of the cover used, be sure it is a high quality material made for greenhouses.

A recent development is the use of two layers of film plastic spaced ¾ of an inch to 4 inches apart to reduce heat loss by 30 percent. The two layers are often installed together and then separated by a small fan.

Polyethylene houses lose heat by radiation more rapidly than glass and fiberglass and therefore require more heat. Polyethylene houses are tighter than glass houses so less heat is lost by air infiltration or leakage.

The small greenhouse requires heating and ventilating equipment. Part of the winter heat is received from the sun, and on mild sunny days you may have to ventilate to reduce the temperatures. However, at night no heat is received from the sun and heating units are required to maintain the temperature.

Heat can be distributed in the greenhouse in three ways: forced hot air; natural convection from small space heaters, hot water or steam pipes; and direct radiation. Any of the common fuels may be used.

A good quality heater has a thermostat and is vented to the outside, except for electrical units. Unvented heaters can discharge gases harmful to plants and deadly to humans into the greenhouse, and should never be used.

The amount of heat required for a greenhouse depends on the maximum temperature difference between outside and inside, the surface area of the greenhouse, the quality or tightness of construction, and the wind.

A simple formula for determining heat loss is (surface area of transparent greenhouse cover) times (maximum temperature difference to be maintained) times (heat loss factor for the covering and wind) equals (heat loss, in BTU/hr.) To convert to kilowatts for electric heaters, divide by 3,413. Heater output should equal the heat loss calculated.

Heat loss factors are given in the table. Maximum heat loss is normally about an hour before sunrise.

Ventilation is an exchange of air inside the greenhouse with outside air. Ventilation is needed for cooling, to reduce high humidity, and to replenish carbon dioxide. The air exchange can be achieved by opening vents and doors or by using fans.

Exhaust fans in combination with inlet louvers are a good ventilation

Heat Loss Factors
Typical Small Greenhouse

	Heat loss factor	
Greenhouse covering	Calm area	Windy area
Polyethylene or Fiberglass	1.2	1.4
Glass	1.5	1.8
Double layer plastic or glass	0.8	1.0

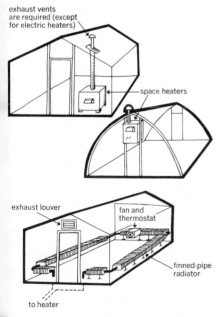

Some heating systems for greenhouses.

65

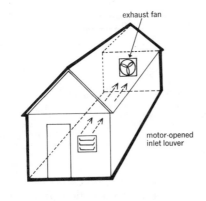

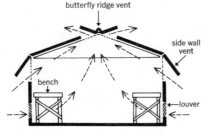

system used by many greenhouse operators. Fans can be wired to a thermostat for automatic operation, and sized to seasonal requirements of the greenhouse. A two-speed fan is desirable to permit spring and fall ventilation at half the summer rate. The exhaust fan and inlet louver should be placed at opposite ends of the greenhouse.

Small fans can be used in the greenhouse to circulate the air continuously to maintain a more uniform temperature during the heating period.

The ventilation rate is calculated by multiplying the greenhouse floor area, in square feet, times 12 to give the fan capacity, in cubic feet per minute (CFM). Since there are pressure losses in moving air through the greenhouse, the fan should have the required capacity at a pressure rating of ⅛ inch of water static pressure (s.p.). Check your catalog for the pressure rating or ask the salesperson for assistance.

The number 12 is cubic feet per minute per square foot of floor area. The number is 1.5 to 2.0 times the average greenhouse height so the fan capacity is, in reality, 1.5 to 2.0 times the volume of the greenhouse. This volume is necessary to remove the summer heat load.

Vents at the ridge, side walls, and door can be used for natural convective air exchange. Using the principle that warm air rises, the vents are opened and closed manually or by motorized unit in response to the inside temperature. Cooler outside air enters to replace escaping warm inside air.

Shading devices over the greenhouse can reduce the light intensity and heat load in summer. Evaporative coolers can be used effectively where the temperature exceeds 90° F for more than 10 to 15 days a year. In dry areas of the country the inside temperature can be reduced 20° or more. In humid areas the temperature reduction is less. The evaporative cooler fan pulls air through a wet pad. Heat is removed from the air to evaporate the water, thus reducing the air temperature.

Big Investment

The greenhouse requires the greatest commitment of money. Shortcuts in equipping the greenhouse with proper heating and ventilating equipment can lead to disappointments. Consult local greenhouse owners before making your investment. A good manual for hotbeds, coldframes, and greenhouses is available from the Northeast Regional Agricultural Engineering Service for $2.00. Ask for *Hobby Greenhouses and Other Structures*, NRAES-2, from NRAES, Riley Robb Hall, Cornell University, Ithaca N.Y. 14853.

A propagation unit is used to star

Ventilation systems for greenhouses Top, mechanical, and bottom, convec tive.

plants from seeds or cuttings by providing the environmental factors which encourage good growth. The unit typically has lights and may be equipped with a water misting system and heating cable or other means of providing bottom heat to the soil. A ventilation fan is often needed to remove lamp heat.

Good controls are needed to do the job right. Time clocks can control the lighting and misting cycles. Thermostats can be used to control the air and soil temperatures.

Typically, people start vegetable seed in their homes under fluorescent lamps. The plant containers may be wooden or plastic boxes. To maintain a high humidity, the propagating containers are enclosed by a plastic cover. Heat may be supplied by the room air temperature or by locating the propagating unit over a radiator or other heat source. This simple arrangement works well for the once-a-year spring project—if you exercise care.

Young vegetable plants grow best with a lot of light, and a minimum of 1,000 foot-candles is desirable. Good light can be supplied by 40-watt cool white fluorescent lamps spaced 6 inches apart and hung a few inches above the plants. A height equal to the spacing between lamps will give fairly uniform lighting.

At a 6-inch height the plants receive about 500 foot-candles from a standard 40-watt, 2-lamp fixture. Two 1,500 milliampere fluorescent lamps give 1,000 foot-candles at 6 inches. The 2 types of lamps cannot be interchanged. Reflectors improve the lighting at plant level.

Major disadvantage of the fluorescent lamp is that the light decreases as the lamps age, so a lamp lighted for 8,000 hours may give less than half the light of a new one. For

Left, corrugated fiberglass 8- by 12-foot portable greenhouse. Right, plants being started in greenhouse. This type greenhouse was built and sold by Future Farmers of America chapter as part of school vo-ag project in Burlington, Wash. Plans are available from Northwestern Vocational Curriculum Management Center, Commission for Vocational Education, Building 17, Airdustrial Park, Olympia, Wash. 98504, for $1.25.

growing plants they should be replaced before that time.

Incandescent lamps are the ones commonly used around the home. They are easy to install and the least expensive. However, they have a low efficiency and a short operating life (1,000 hours).

Fluorescent light is broad spectrum, operates cooler, and has greater light intensity so it is better suited for plant lighting. Incandescent light is in the red and far red portion of the spectrum which triggers certain plant processes; it can be used to supplement fluorescent lamps.

Continuous light can be used for most seedlings for the first 2 and 3 weeks, but then the young plants should have a dark period of 4 to 8 hours per day. The best way to control the lighting period is with a 24-hour time clock.

Air and soil temperatures are important factors in germination. A heating cable with thermostat is one way to maintain the soil temperature at a desired 70° to 75° F. The presence of lamps helps keep the soil warm.

As the plants get older, the temperature should gradually be reduced so the temperature change at transplanting outdoors will not be too great. Young seedlings will grow tall and spindly in high air temperatures and poor light.

Rapid growth is encouraged by high humidity and good soil moisture. Watering or misting frequently can meet these requirements.

Most garden supply houses have fairly simple and inexpensive watering systems. Typically, a time clock is used to control a small solenoid valve which turns the water on and off automatically. A misting system requires a time clock which permits 6 seconds of mist each 6 minutes, as an example. Lamps must be protected from the water.

Carbon dioxide is another impor-

tant factor for good growth. If a small fan is used to move the lamp heat away, it will also keep fresh carbon dioxide moving in contact with the plant leaves. Direct a gentle breeze of air over the plants.

While these environmental factors are needed to achieve the best growth, success also depends on use of high-quality seed and stock plants, a suitable growing medium, and simultaneous control of all environmental factors. The growing medium must provide good moisture conditions, a high nutrient level, and proper aeration.

Your imagination is the limit on using fences, trellises, and other structures to support plants. One person grew squash in a small 8- by 15-foot garden by getting the vines to grow up a 6-foot wooden fence. The vines were partially supported by string. The string was first tied tightly around vertical boards of the fence, and then the ends were tied loosely around the squash vines.

Tomatoes can be tied up to stakes. Peas and beans can grow on a fence made of several strands of string attached to closely spaced stakes. Cucumbers and other vines can grow on a trellis made by nailing wood strips on stakes driven into the ground. Varieties of fruit trees are adaptable to trellis arrangements and can be used as a hedgerow in the landscape.

Trellis provides support for vine plants as well as decorative touch.

The plant support structure can be functional to create additional garden space in a crowded area, or it can be decorative and express creativeness in the landscape design if done properly.

In windy areas, take steps to protect the vegetable or fruit plantings from damage. Wind can cause structural damage, and windborne soil particles can injure leaves. An 80 percent solid fence near the planting will give good wind protection. Buildings, hedges, and trees afford wind protection if the garden can be located to take advantage of it.

For long range protection, plant short and tall trees, hedges, and shrubs on the windward side. A good tree windbreak reduces wind velocities for 10 to 30 tree heights downwind and for 5 to 10 tree heights upwind. Give consideration to both summer and winter wind directions and to the resultant winter snow drifts.

Equipment Storage

Gardening requires an investment in tools, fertilizers, pesticides, sprayers, seeds, stakes, and numerous other items. Amount of the investment depends on how much you wish to spend and the size of your garden. Hand-operated equipment will be less expensive than powered units.

A clean, dry place is needed to keep tools, treated seeds, fertilizer, and pesticides out of reach of children and pets. A safe, lockable cabinet in a dry, well-ventilated location would be ideal for keeping harmful tools and chemicals away from children. This is a simple safety requirement but is often disregarded.

Tools and leftover supplies should be prepared for storage in fall unless a label indicates a chemical should not be stored for another season. Hand tools can be washed to remove dirt particles. After they dry, apply a light coat of oil with a cloth to protect against rusting.

Clean up your power equipment according to the manufacturer's directions. Gasoline engines should have the gas drained, the oil drained and replaced with new oil, and a small amount of oil put into the cylinder through the sparkplug hole. With the sparkplug removed, turn the engine over by hand so the cylinder walls are lubricated.

A light coat of oil can be put on parts which contact the soil when in use; other bare metal should be painted if the original paint has been damaged. Properly maintained equipment will give many years of service.

The modern home differs from homes of earlier generations when people produced much of their own food and stored it for winter use in root or storm cellars, attics, closets, cellar shelves, and even in the garden. Today we use a minimum of a refrigerator/freezer and kitchen shelves to store our food supply. Of course, the "old ways" are not gone and are being revived as more people grow their own food and seek to preserve it.

Specific details on harvesting maturity and storage requirements can be found in other chapters. In general, note that well matured, good-quality vegetables or fruits are the first requirement. The storage space must have proper moisture and temperature conditions. Frequent sorting and removal of decayed produce is essential.

Providing the proper conditions for all items may not be easy, and storage planning must be done before the harvest season. As an example of the varied conditions required, here are three groups and their needs: *warm dry* for squash and pumpkin; *cool dry* for onions and dried peas and beans; and *cool moist* for root crops, potatoes, cabbage, and apples.

Warm dry storage conditions are 50 to 70 percent relative humidity and

40° to 50° F temperatures. These conditions can be provided in a room not heated for family use but having minimum heat to maintain the cool environment in areas of extreme temperature.

Cool dry storage means 25° to 32° F and 70 to 75 percent humidity. Unheated attics or closets or even a garage may be used. An old freezer or refrigerator may be adapted to provide these conditions in warmer areas.

Cool moist storage calls for temperatures of 32° to 40° F and high humidity. This is difficult to obtain in a home. An insulated room in a basement can be vented to the outside for cooling. The insulation protects stored produce from basement heat and outside frost. Ventilation is used to drop the temperature in fall and to hold the temperature in winter. Outside air is brought in when it is cooler than inside air. A fan and thermostat arrangement can be used to automate the temperature control. Humidity is maintained by sprinkling the floor frequently with water.

Carrots, beets, rutabagas, turnips, and parsnips, for example, need high humidity to reduce shriveling. These vegetables can be stored in containers or buried in sand and covered with moist burlap or cloth to keep the air moist. Parsnips can be left in the ground over winter and dug in spring before they start to grow.

Excess fresh fruit and vegetables can be stored, but unless you have sufficient quantities to justify setting aside a room, building a special room, or buying a used refrigerator or storage unit, the expenses may be greater than the benefit. Be careful to evaluate the savings in the supermarket versus the cost of the investment in facilities and their operating cost at home.

The U.S. Department of Agriculture has made up plans for various types of garden structures. Some of these can be obtained through your county Extension office. The office has additional publications available and can answer specific questions.

Plans are available from the Extension Agricultural Engineer at your State land grant university if the county Extension office does not have them. The county Extension agent can request the plans for you.

For Further Reading:

Building Hobby Greenhouses, U.S. Department of Agriculture Inf. Bul. 357, on sale by Superintendent of Documents, U.S. Government Printing Office, Washington, D.C. 20402. 30¢.

Electric Heating of Hotbeds, U.S. Department of Agriculture L 445, on sale by Superintendent of Documents, U.S. Government Printing Office, Washington, D.C. 20402. 35¢.

Hobby Greenhouses and Other Gardening Structures, NRAES-2, 1976, NRAES, Riley-Robb Hall, Cornell University, Ithaca, N.Y. 14853. $2.00.

Mini-Hotbed and Propagating Frame, U.S. Department of Agriculture M 1184, on sale by Superintendent of Documents, U.S. Government Printing Office, Washington, D.C. 20402. 5¢.

Storage Building for Garden Tools, U.S. Department of Agriculture M 1192, on sale by Superintendent of Documents, U.S. Government Printing Office, Washington, D.C. 20402. 25¢.

Storing Vegetables and Fruit in Basements, Cellars, Outbuildings, and Pits, U.S. Department of Agriculture H&G Bul. 119, on sale by Superintendent of Documents, U.S. Government Printing Office, Washington, D.C. 20402. 40¢.

Utility Shed, U.S. Department of Agriculture M 1196, on sale by Superintendent of Documents, U.S. Government Printing Office, Washington, D.C. 20402. 25¢

Pest Management Is a Matter of Timing
by Dan C. Scheel

Total pest control may not be necessary nor realistic, as the loss of a few leaves or flowers does little real harm to a plant. If a garden is generally healthy, a few chewed or discolored leaves, and some less than perfect fruits, are neither economically nor ecologically harmful. There are various ways to control pests, including chemical pesticides, physical methods, cultural practices, and natural and biological methods.

Pest management should be your goal. This means that you want to prevent disease and insect problems during the growing season, and to control those pests you cannot prevent. A combination of prevention and cure, natural controls, and chemical controls working together make up pest management.

Learn what to expect by getting literature on the various vegetables and fruits you grow. Informative pamphlets and bulletins can be obtained from Federal and State Agriculture Departments, as well as Cooperative Extension Service offices. These Extension offices are usually located at county seats, and are affiliated with State colleges and universities where agricultural research is done. Newspapers with a gardening section generally offer timely local pest information as the seasons progress. Your local garden center, or the place where you purchase plants, fertilizer, pesticides, and other garden supplies, is often a good source for literature and information.

You must learn what to do in your own garden, as pests are selective. Some insects and diseases attack only certain plants. These pests are usually not a problem during the entire season. They attack at specific stages of a plant's growth, or when favorable weather conditions exist for their own growth and development. To develop your plan for pest control, consider (1) What you are growing, (2) Which pests are most likely to attack, and (3) When these pests will generally be a problem.

Timing is the major key to success

Pickle worm, a melon pest, is shown in larva, pupa and adult stages. This pest also damages cucumbers and squash.

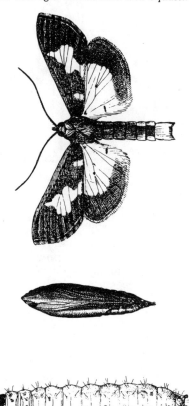

Dan C. Scheel is the owner of Strawberry Orchard Fruit Farm and president of The Elegant Farmer, Inc., Mukwonago, Wis.

duce or eliminate the need for chemical control. Specific chemical recommendations are not given because these vary from State to State, and because the Environmental Protection Agency may cause pesticide labels to be changed from time to time.

This "what to do" is a basic guide only. You must follow local recommendations on rates, restrictions, and alternatives. Never use a chemical unless the label says you can use it on the plant to be sprayed. Follow all label instructions and precautions.

You can make up your own pest control chart, using the one with this chapter as a model. Or, you could list the month first, and then note all the pest control measures you are likely to use in, say, April.

The critical times listed are for chemical prevention of the most likely problems at the most likely time. A regular check of your plants is necessary to detect and then control these and other problems as they may occur.

Keep dated records that include the varieties planted, the steps taken for prevention, and successes or failures in control. Note the weather conditions, as they are a major factor controlling pest development. With a little effort and a plan, it won't be long until you will know precisely what to do for your garden. You will have developed your own "pest management program" that maintains a balance between chemical and other pest control methods.

Pests can be controlled using one or more of several techniques. Your pest management program will need to consider and use the best combination of methods available. A good

once you have determined what measures to apply. There are certain critical times when you have to take some steps in order to prevent pest problems or to control insects and diseases. The chart will aid you in this timing. If you can't or don't do anything else all growing season, plan to take the necessary steps at the critical times listed. You should be able to prevent most pest problems. But throughout the season, be on the lookout for the early stages of insect and disease attack.

The "what to do" section of this listing refers to chemical measures only. Other types of control may re-

A bikini is fine for beach but not for spraying garden chemicals, which can easily get on skin and may cause harm. When spraying, wear full-length clothing, gloves, and wide-brim hat.

Pest Control Program

Crop	Critical time(s)	Pest and what to do
Fruits	As a general rule, critical times for preventing the most likely fruit pests are from the start of growth in spring through fruit set.	
Apples and pears	Early spring, when buds have ¼-inch green tip	Apple scab—Spray with fungicide, every 7 days through bloom
	Early spring, when buds have pink tips	Overwintering insects on tree—Spray with dormant oil spray
	Late spring at petal fall	Codling moth, leaf roller, plum curculio, apple scab—Spray with insecticide and fungicide
	Normal or wet summer	Apple maggot fly (railroad worms)—Spray every 10-14 days with insecticide
Brambles raspberries, blackberries, gooseberries, currants, etc.	Spring when leaf buds are ⅛ inch long	Anthracnose, spur and cane blight—Spray with fungicide
	Spring—just before blossoms open	Sawfly, borers, anthracnose, spur & cane blight—Spray with fruit combination spray containing fungicide and insecticide. Repeat in 3 days and then in 10 days
Grapes	Spring, when new shoots are 6-8 inches long	Black rot—Spray with fungicide
	Spring, just before blossoms open and after petal fall	Black rot, leafhoppers, rose chafer, moths—Spray with fungicide and insecticide
	Spring, just after blossoms fall	Powdery mildew & above insects—Spray with fungicide and insecticide
Stone Fruits cherries, peaches, plums, apricots, etc.	Dormant, before growth starts in spring	Peach leaf curl, black knot—Spray with fungicide
	Petal fall	Brown rot, leaf roller moth, curculio—Spray with fungicide and insecticide

Pest Control Program—*Continued*

Crop	Critical time(s)	Pest and what to do
	7-10 days after petal fall	Brown rot, leaf spot, curculio, fruitworm—Spray with fungicide and insecticide
Strawberries	Early spring, right after uncovering or just before buds appear	Leaf spot, leaf scorch, aphids, plant bugs, leaf roller—Spray with fungicide and insecticide
	While in blossom, every 10 days until 3 days before first picking	Leaf spot, leaf scorch, berry rot, berry mold—Spray with fungicide
Vegetables	Most vegetables can be put into one of two categories for critical time(s): 1. As seedlings, 2. At blossom and fruit set. Use seeds treated with a fungicide to assure good seed growth.	
Beans	Blossom through fruit set	Leafhoppers, aphids, lygus bugs, beetles—Spray with insecticide
Beets and Chard	Before planting	Cutworms—Spray ground with insecticide
Cabbage and related plants as broccoli, brussels sprouts, kohlrabi, cauliflower, collards, kale	As seedlings and mature plants	Aphid, looper, webworm, diamondback, cabbage worm, harlequin bug—Spray with insecticide
Carrots	As seedlings	Aster leafhopper—Spray with insecticide
Celery	As seedlings	Aster leafhopper, blight—Spray with insecticide and fungicide
Cucumber and related vine crops as squash, pumpkins	As seedlings	Striped cucumber beetle—Spray with insecticide
Melons	At vining out and/or at blossoming	Striped cucumber beetle, pickle-worm, squash vine borer—Spray with insecticide
Lettuce	As seedlings	Aster leafhopper—Spray with insecticide

Pest Control Program—*Continued*

Crop	Critical time(s)	Pest and what to do
Okra	During fruiting	Corn earworm—Spray with insecticide
Peas	At blossom and fruit set	Cowpea curculio, pea weevil—Spray with insecticide
Peppers	At blossom and fruit set	Pepper weevil, aphid, European corn borer—Spray with insecticide
Potatoes	Before planting	Grubs, wireworms—Spray ground with insecticide
	Seedlings to blossom	Beetles, aphids, leafhoppers, blight—Spray with insecticide and fungicide
Radish	Before planting	Root maggots—Spray ground with insecticide
Sweetcorn	At silking, and every 5 days for 2 more sprays	Earworms—Spray with insecticide
Tomatoes	At blossom, and possibly at weekly intervals thereafter	Leaf spot, blight, flea beetles, hornworms, leaf miners—Spray with fungicide and insecticide
Turnip and Mustard	First true leaves	Aphid—Spray with insecticide

rule of thumb is not to apply a chemical unless it is needed to prevent damage or to control a pest. In some instances, natural pest predators and parasites will help control a problem while chemicals may kill the good guys as well as the pests.

Good cultural practices and physical controls also provide a garden environment conducive to reducing needs for chemical control.

Chemical Controls

Chemicals are classified according to their primary use as follows:

Pesticide—General term meaning all chemical control materials as a group

Insecticide—Chemicals to kill insects

Fungicide—Those used to prevent or cure diseases caused by fungi attacking plants

Bacteriocide—Those used to prevent, or cure, bacterial diseases of plants

Herbicide—Weed preventers or killers

Nematicide—To kill nematodes, tiny microscopic round worms, that live mostly in the soil

Rodenticide—To control rodents

Pesticides can be purchased in dry or liquid form, and are usually applied as dusts, sprays, or granules. They are formulated to contain a percentage of the active chemical ingredient. This percentage is shown on the label, such as 50WP. The amount of a chemical recommendation always refers to a particular formulation of the chemical, and it is as important to have the correct formulation as the correct chemical.

Dusts and sprays are practical for small applications. Dusts are applied in the form as purchased, while spray material is diluted with water before applying. You should make an even, thorough application, forcing the material into foliage so that it coats all sides of the leaves, stems, and fruits.

Pesticides used improperly can be injurious to man, animals, and plants. Follow all label directions, and read the cautions for handling. Do not use the same sprayer for herbicides (weed killers) and insecticides and fungicides, as minute traces of weed killers can kill garden plants.

Pesticides should be your "ace in the hole" when other methods may not work or have not worked. They will control a large number of pests, but will not eradicate all insects or eliminate all diseases. Timely application is critical for reasonable insect control. Diseases can rarely be cured, they must be prevented. The following measures can help create conditions in and around your garden which will control many pest problems and reduce the need for chemicals.

Physical Controls

You can keep your garden relatively healthy by taking these two steps:
- Keep the soil in good shape by adding soil amendments and nutrients
- Remove any plant that is severely damaged by pests, rotten, wilted or moldy.

Once you have a healthy garden, keep it that way by raking up any debris, and by digging in fresh mulches each year. Sanitation is a useful pest control practice.

Pick off plant eaters from the plants on a regular basis. Remember that pests are a problem only at certain times under certain conditions. So crunching a few bugs before they lay eggs may solve a major problem. Many books have been written on organic gardening and they contain a wealth of information for the non-chemical control of pests.

Cultural Controls

Methods you are already applying probably fit into the area of cultural controls. Common garden procedures are effective in pest management.

Many plant troubles blamed on pests can be caused by poor soil drainage. Use fertile, well-drained soil. Rotate annual crops, alternate between leaf and root crops. Do not crowd plants, let them breathe. Good aeration will let leaves dry rapidly after rain or dew. This in turn will lessen leaf spot diseases.

Plant crops that are suited to your soil, exposure, location and climate. Keep weeds under control as they are the nesting and breeding grounds for many pests. Keep all surrounding fences and lot lines as free and clean as your garden. Buy certified seed of disease-resistant varieties. Purchase virus-free, disease-free plants of resistant varieties. You will have to do some homework on varieties. Most labels on seed packets and descriptions in seed and nursery catalogs will indicate which varieties are resistant, and whether nursery stock is certified virus-free.

The cultural control methods are numerous and you can probably come up with your own. They are closely related to, and assist in, promoting biological and natural controls.

Biological Controls

The balance of nature contains predators and parasites working for you to naturally control pest problems. Primarily, natural controls refer to beneficial insects controlling pests. But the natural controls range from birds and toads (they eat insects) to earthworms (they work the soil).

Learn what the beneficial insects are and how to develop them in your garden. They include lady bugs, lacewing, praying mantis, certain mites, spiders, some wasps, ground beetles, pirate bugs, assassin bugs, and syrphid flies. In addition, some natural (biological) insect controls are available commercially, such as *Bacillus thuringiensis*. This bacterial product will control many caterpillars including the cabbage looper.

Prevention itself is a pest control, and your pest management program should aim at preventing or reducing pest problems no matter what combination of controls you use. Again, *timing* is the key to everything.

A certain amount of pest damage is inevitable, so do not demand or even expect total eradication of pests. Control to an acceptable degree is what you should seek.

Pests other than insects and diseases can cause considerable damage. Different parts of the country may have problems unique to that particular area.

Rodents may burrow and eat roots and tubers, or eat plant tops and fruits. Control is limited to traps or chemicals. Chemical baits are available for mice in fruit orchards.

Rabbits can be repelled by sprinkling-dried blood meal on the ground throughout the garden. It must be re-applied after a heavy dew, rain, or after a few days. Buried fencing around small gardens can be effective.

Deer can be repelled with electric fences, dried blood meal as above, moth balls scattered around, creosote rags in various spots, dogs, noisemakers or fencing.

Birds can be kept off fruit trees with nets available for this purpose, scarecrows, and noisemakers.

Your pest management program is the result of many variables, applied to your location and unique conditions. Season-to-season adjustments will be necessary, as the weather changes.

Successful "pest control" is primarily dependent upon prevention, and prevention is a direct result of the timing of control measures. A diary is a necessity for keeping a record of all problems, measures taken, results, variable factors, and weather.

You will develop a balance between all methods of prevention and cure—natural, physical, cultural, or chemical. This balance you develop is pest management.

Left, adult and larva of lady bug feeding on aphids. Right, a bit of creativity, as well as fun, goes into making scarecrow to ward off birds.

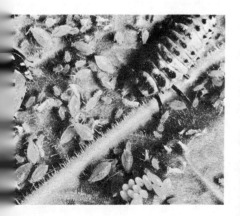

Organic Gardening—Think Mulch
by Wesley P. Judkins

Organic gardening is the production of crops without the use of inorganic chemical fertilizers or pesticides. This means that only organic fertilizers such as manure, sewage sludge, cottonseed meal, bone meal, or dried blood are used. Also, diseases, insects, and weeds are controlled by natural resistance, birds, predator insects, or mechanical means, rather than by using pesticides or herbicides.

Organic methods may be followed with success in the home vegetable garden. This is especially true if mulches are used during the summer, and the crop refuse is returned to the soil each year to replenish and increase the organic matter content.

Regardless of the original source, fertilizer in the soil must break down into its ionic form before it can be used by plants. The ions which are then absorbed are identical, whether derived from an organic or inorganic source. Therefore, in terms of benefits to plants, when similar quantities of nutrients are available there is no advantage for either organic or inorganic fertilizer.

Organic materials are less caustic than inorganic fertilizers and, except for poultry manure, may be used with little possibility of damage to plants. The nutrients from organic fertilizers are rather slowly made available to plants. This may be an advantage where delayed release is desired to promote plant growth over an extended period.

Manure is commonly salvaged as a low cost fertilizer by farmers who produce livestock or poultry. It is usually not readily available to home gardeners. Dried manure with an analysis of about 1-1-1 may be used.

Other materials such as cottonseed meal, bone meal, and dried blood are available as organic fertilizers. These, and dried manure, are much more expensive than inorganic fertilizers. Also, from a practical conservation point of view, cottonseed meal, bone meal, and dried blood should be recycled as feed for livestock or poultry, and their manure then salvaged as fertilizer.

Most inorganic chemical fertilizers are much cheaper than organic types per pound of nutrient element. These elements become available for plants quite rapidly when applied in the inorganic form. Because of their concentration and solubility, inorganic commercial fertilizers are somewhat caustic and must be used with care to avoid damage to roots or foliage.

Sewage Sludge

Sewage sludge may serve as a desirable source of nutrients for gardens if a properly processed product is used correctly.

Raw sewage sludge, which is sometimes called primary or settled sludge, is a solid, lumpy material with an offensive odor. It is not recommended for use in the production of any type of crop, because it is a potential carrier of harmful pathogenic organisms.

Digested sludge is settled sludge which has been anaerobically decomposed. During this process, much of the organic matter is converted to gases and soluble material, and most of the pathogenic organisms are destroyed. The nutrient content of digested sludge is low and variable with an average of about 2% nitrogen, 2% phosphoric acid, and .5% potash. This type of sludge may be

---

Wesley P. Judkins is Emeritus Professor of Horticulture, Virginia Polytechnic Institute and State University, Blacksburg.

William E. Carnahan

used on the home garden if applied in fall before the soil is prepared for planting the next spring.

Dried activated sludge is settled sludge which has been treated with large amounts of air to cause aerobic decomposition. The resulting product is then heat dried and ground. The aeration and heat treatments inactivate any harmful pathogenic organisms. Activated sludge has a composition of about 6% nitrogen, 4% phosphoric acid, and .5% potash. This product may be used for gardens.

Benefits of a Mulch

One of the most important uses of organic matter by the home gardener is as a mulch. When so used it reduces the damaging impact of rain or irrigation on the soil. This is very important because it increases the infiltration of water and reduces erosion. An organic mulch conserves soil moisture by reducing evaporation, and helps to suppress weed growth.

Some of the best organic materials to use as mulch in the garden are leaves, lawn clippings, fresh sawdust, fine wood shavings, pine needles, chopped straw, ground corn cobs, shredded tobacco or cane stems, peanut hulls, or cottonseed hulls. They will not add important amounts of nutrients, or have a significant effect on the pH of the soil.

The dead vegetable and flower plants in your garden should be chopped down and left on the ground as a protective mulch during winter. This trash mulch will help reduce erosion, and improve the organic matter content of the soil when the garden is prepared for planting in spring.

Cut corn stalks, tomato vines, and other tall plants into 8-inch pieces with a sickle or pruning shears. Chop up low plants like beans and bushy flowers by running along the row with a rotary lawn mower.

Cover Crops

The planting of rye or wheat in fall is sometimes recommended as a means of controlling erosion during winter, and providing organic matter to improve the soil. Most home gardens are small and on relatively level land, so erosion is not a serious problem. Also, the use of mulch, leaves, lawn clippings, and plant refuse from the garden and flower bed usually will provide more organic matter than would be secured from the cover crop.

If the garden is planted in early spring, and a fall garden of hardy vegetables is maintained until late fall, there will not be enough time to plant an overwintering cover crop and secure a beneficial growth to plow into the soil the following year. Therefore, the use of overwintering cover crops is usually not recommended for the home garden.

Checking soil moisture under organic mulch of lawn clippings.

For gardeners who have a relatively poor soil and cannot secure adequate supplies of mulching material, the use of an overwintering cover crop may be desirable. Sow rye or wheat broadcast at the rate of 3 to 4 pounds per 100 square feet, and scratch into the top inch of soil with a rake. This should be done several weeks before the average date of the first fall frost.

If crops are still growing in the garden, the rye or wheat may be planted between the rows.

Plow or rototill the cover crop into the soil in spring when the garden is prepared for planting, about a month before the average date of the last frost.

Mulching with fresh sawdust, lawn clippings, or other organic material is an ideal way to raise vegetables. Such a mulch will conserve soil moisture, help control weeds, and improve the infiltration of rain or irrigation water into the soil.

Plant your vegetable seeds at the recommended depth and cover with soil in the usual way. Then spread a band of sawdust or vermiculite about 4 inches wide and ½ inch thick on top of the row. This mulch helps to conserve moisture, reduces crusting of the soil, and allows the young seedlings to emerge easily.

When the vegetable seedlings are about 6 inches tall, apply between the rows a 1-inch mulch of fine organic material such as fresh sawdust, lawn clippings, or peanut hulls, or a 2- or 3-inch layer of loose unshredded leaves.

If weeds are present which are 2 inches tall or more, kill them by cultivating or hoeing before the mulch is applied. Mulch will smother weeds less than 1 inch tall.

Be sure the sawdust or fine shredded bark you use as mulch is fresh, or well aerated if old. Sawdust or shredded bark from the inside of a pile may go through anaerobic decomposition and become very acid, with a pH of about 3.0, and have a pungent odor. Such material is very toxic to plants.

Some weeds may continue to come up through the mulch. Pull these out by hand or carefully cut them off with a sharp hoe. Weeds are easy to pull when the ground is moist after a rain. Do not cultivate because this will mix the mulch with the soil and reduce its effectiveness.

The mulch should be worked into the soil by plowing, rototilling, or spading when the garden is prepared for planting in spring. This will add organic matter to the soil and improve soil structure and workability. It gives all the advantages of adding compost, plus the benefit of a full season with mulch.

Apply 4 pounds of 5-10-5 fertilizer per 100 square feet when the garden is plowed. This will counteract any tendency toward nitrogen deficiency, and provide adequate nutrients for the vegetable crops in the garden. Base the kind and rate of fertilizer to apply on soil tests for best results.

If plants develop light green color during the growing season, apply a side dressing of a fertilizer containing nitrogen. Use 1 pound of 10-10-10 per 100 square feet, or per 50 feet of row, or use 2 pounds of 5-10-5. Spread the fertilizer uniformly between the rows of plants, and scratch into the soil or mulch with a rake. If there is no forecast for rain it is advisable to apply water to dissolve the nitrogen and carry it down to the roots.

If you prefer an organic fertilizer as a side dressing, use 10 pounds of 1-1-1 dried manure per 50 feet of row. The nitrogen from this material is not as rapidly available as from chemical fertilizer, and therefore it is a less effective side dressing treatment.

Compost

Compost is a relatively fine, homogeneous organic material secured

from the decomposition of various types of plant refuse such as leaves, lawn clippings, weeds, old vegetable plants, and garbage. The composting process reduces the volume of the plant material to a third or less of its original amount, and usually destroys the viability of any weed seeds which may be present.

Composted humus material is particularly useful for making soil mixtures for the production of seedling plants for the vegetable garden. Compost is also useful as a mulch when more plant refuse is available than can be used as a mulch in its fresh or undecomposed condition.

If you make compost regularly, it will be helpful to construct 2 long bins of planks or concrete blocks. Make the bins about 4 feet wide, 4 feet high, as long as desired, and open at one end. Plant refuse may be accumulated in one, while the composting process is taking place in the other.

Composting is a disintegration process caused by bacteria and fungus organisms. This results in a considerable reduction in bulk, which may be helpful if you have large amounts of organic refuse and a relatively small garden. There is usually no objectionable odor during the process when layers of soil are added to the pile. If the compost is thoroughly decayed, there is little possibility of disease or insect problems from using it.

Start the compost pile with a 6-inch layer of plant material. Coarse plants such as corn stalks should be cut into pieces about 8 inches long. Sprinkle with 1 pint of 5-10-5 or 5-10-10 fertilizer per square yard of pile. If you prefer to use an organic fertilizer, scatter 5 pounds of 1-1-1 dried manure on each layer of organic material. If an alkaline compost is desired, add 1 pint of ground limestone per square yard of surface area. Add a layer of soil about 1 inch thick.

Repeat as many layers of plant material, fertilizer, and soil as needed to use all available plant refuse. The top should be lower in the center to cause water to move in rather than drain off.

Top, using soil-compost mixture under and around plants. Bottom, you don't need big backyard to have a compost bin. Here, city apartment resident improvised by using cinder blocks and part of roof structure.

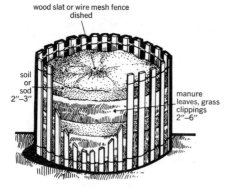

Layering of a compost pile.

Water the pile as often as necessary to maintain a relatively high moisture content, promoting decomposition. The pile should be forked over after about 3 months. The plant materials should decompose into good compost in about 4 to 5 months in warm weather, but may take longer under cool or dry conditions. It is not necessary to use bacteria pills or other substances to increase the rate or effectiveness of the decomposition process.

A plastic trash bag may be used to make a small quantity of compost from relatively fine material such as leaves, lawn clippings, or chopped garden refuse. Make layers of plant trash, fertilizer, and soil as in a compost pile. Add 2 quarts of water to dry material, and 1 quart if it is quite moist or succulent. Tie the bag and turn every few weeks to cause the moisture to move back and forth through the organic material.

Reducing Pesticide Needs

The homeowner who wishes to follow organic methods exclusively can have a more rewarding garden by planting only those crops which are not susceptible to damage by pests, or by selecting resistant varieties. The latter are usually described in seed catalogs, and information on them is available from your county Extension office. There is no conclusive evidence to indicate that crops grown organically are more resistant to pest damage than similar vigorously growing crops which have received commercial fertilizer.

Although some vegetables may be severely damaged by diseases or insects, many types can be grown successfully without the use of pesticides. Some of the best crops for the home garden are asparagus, beets, carrots, celery, Swiss chard, collards, cress, endive, kohlrabi, leeks, lettuce, mustard, onions, parsley, parsnips, peas, peppers, radishes, salsify, spinach, sweet potatoes, turnips and watermelon.

They may of course be damaged by diseases or insects in some seasons but, by planting only these vegetables, you can usually have a productive garden with little or no spraying. When soil insects are troublesome, and the gardener is unwilling to use synthetic pesticides, it may be wise to avoid growing certain root crops.

All home gardeners should be concerned about the use of pesticides. Some of these materials are more hazardous than others and their use should be limited to those crops which would otherwise be severely damaged by diseases or insects. A good rule to follow is to use the least toxic pesticide that will do the job.

Several non-hazardous organic or biodegradable pesticides are available for the home gardener. Our remarks here concern mechanical and biological controls.

Aphids occasionally are a serious pest on some garden plants. Lady beetles, available from many organic gardening supply houses, can be released to feed on aphids, but there is a chance of only temporary control because they may soon leave the area.

Cutworms often destroy certain vegetables in the garden. Newly set cabbages, tomatoes, and other plants may be protected with a cardboard collar which encircles the plant a half inch out from the stem, extending an inch into the ground and 2 inches above.

Slugs may cause damage, especially in a mulched garden, which provides excellent living conditions for this pest. A ring of wood ashes or sharp sand around plants helps control slugs. Also, shallow aluminum pans sunk into the ground with the rim level with the soil and filled with beer will attract and kill slugs. They drown in the liquid.

Spotted garden slug.

Bacillus thuringiensis is a bacterial disease effective against the larvae of a number of moths and butterflies. It is useful for controlling cabbage loopers on broccoli, brussels sprouts, cabbage, cauliflower, collards and kale.

Sweet corn may be damaged by borers in the ears, but this is usually confined to the tip, which can be cut away when the corn is prepared for cooking. Also, there will be less trouble if you select some of the new hybrid varieties with tight husks. All damage by corn borers can be prevented by treating the silk with mineral oil before it turns brown. This is a harmless organic compound.

Beans may be severely damaged by Mexican bean beetles unless the pest is controlled with insecticides.

Broccoli, brussels sprouts, cabbage, and cauliflower may become inedible because of cabbage loopers. Use of *Bacillus thuringiensis* is an effective, non-pesticide method of controlling this insect.

Pumpkin and squash vines may be killed by squash bugs and vine borers. They may be controlled with a pesticide.

Cucumber varieties such as Gemini and Victory, which are resistant to several diseases, may be grown quite successfully without spraying unless cucumber beetles become a problem. A pesticide will control the beetles.

Eggplant is very difficult to raise in most gardens unless the plants are sprayed with a pesticide to control flea beetles and Colorado potato beetles.

Many potato varieties are severely damaged by mosaic and late blight. Kennebec is resistant to these diseases, but may need a pesticide spray if potato beetles become a problem.

Tomato varieties are available which are resistant to fusarium and verticillium wilt and nematodes. Such varieties should be selected for the home garden. The plants may need to be sprayed to control blight.

End of One Season Is Start of Another
by Charles W. Marr

The end of the growing season is not a time to forget about gardening activities until next year, but rather when you should carry out several good practices that keep a successful gardener busy throughout the year.

First, as soon as any crop is harvested or killed by the first frost, the remaining crop residue should be removed or tilled into the soil. Not only does this improve the garden's appearance, but it reduces disease or insect problems that may build up in the residue. Sanitation is the first defense against many common pest headaches.

If a serious disease or insect problem was present through the season, remove residue of the affected crop from the garden and destroy it. Other crop residue can then be tilled directly into the soil, but many gardeners collect these materials in a compost heap.

Compost is a mixture of soil and decayed organic material or humus that is used as a source of plant nutrients and as a soil conditioner. As the organic materials are decomposed by various soil fungi and bacteria, the plant nutrients are "released" and the rich, dark humus provides porosity and tilth to garden soils.

In most compost "heaps", alternating layers of organic materials and soil (with some garden fertilizer added) are piled together, allowed to stand for several months with periodic mixing, and used as needed as mulch or for tilling into the soil. The soil and fertilizer serve as a source of innoculant of the various bacteria and fungi and provide nutrients for their initial growth. Several "activators" available commercially provide the same function as the soil and fertilizer.

Around the home, materials that can be used to make compost include grass clippings, leaves, garden residues, weeds, and table scraps (be careful these do not attract pets or animals to dig in the compost pile). In many areas farm manures, sawdust, shells or pods, or other organic industrial wastes may be available free or at low cost.

If properly prepared, the compost pile will become hot during the decomposition process and kill harmful plant diseases, insects, or weed seeds. However, using garden refuse as a source of organic matter is advisable only for the experienced compost maker. For the novice, it may be wise to destroy plants with serious disease or insect problems rather than run the risk of spreading these through improperly prepared compost.

County Extension offices, garden manuals or books, and commercial dealers are possible sources for detailed information on making compost. Composting is an excellent gardening practice and further reading on this subject will improve your composting skills.

The importance of cleaning up at the end of the season is often overlooked as an important step in successful garden management.

Planting cover crops after the garden cleanup offers several advantages. In areas where strong winds may erode soil from the bare garden in winter, cover crops help hold the soil in place. Secondly, plowing under the green cover crop in spring provides organic matter. Properly grown cover crops may contain 1 to 2 percent nitrogen, ½ to ¾ percent phosphorus, and 3 to 5 percent pot

---

Charles W. Marr is Extension Horticulturist, Vegetable Crops, Kansas State University, Manhattan.

ash—making them equivalent to several low-analysis fertilizer materials.

The stage of maturity of the cover crops and the initial soil fertility level will influence the actual amounts of nutrients added to garden soils. Cover crops offer potential for important nutrient sources as well as improving the tilth and porosity of garden soils. However, many gardeners who use the garden space for fall gardening late in the year may not be able to use cover crops successfully.

Cover crops should be sown in those special instances where blowing soil is a problem, and where garden cropping patterns do not interfere with the cover crop planting. Probably the most common cover crops are annual ryegrass (1 to 2 pounds per 1,000 square feet) or rye (3 to 4 pounds per 1,000 square feet), although oats, barley, or various legume crops may be used. In early spring the succulent green growth can be tilled under, and it will decompose quickly.

In areas where strong spring winds can cause damage to emerging seedling plants, gardeners can till strips in the cover crops for planting, giving wind protection to the newly seeded rows. The row centers can then be tilled later in the year.

Another way of protecting gardens from strong winter winds and reducing water use during summer months is development of a shelter-belt or windscreen on the west and north side of the garden. Trees used in windbreak plantings will compete with garden crops for water and nutrients, so there should be a space between the garden and the windbreak.

Many gardeners use either temporary windbreaks or garden crops planted as windbreaks. A temporary fence or barrier such as bales of straw or old hay can be installed for protection from winter, spring, or summer winds. In addition, rows of tall, hardy crops such as sweet corn, okra, or tomatoes in cages will provide wind protection for tender crops that are planted nearby. In the Mid-

Once brilliance of autumn has faded, leaves can be composted for use as mulch in next season's garden. Photos show compost containers. Right, rectangular wood-frame container, small and suitable for city gardeners, and (below) for those with more space, circular wire bin.

Arthur L. Verdi

west, asparagus frequently is planted along the garden's west or north side. The fern makes an excellent summer windbreak and will remain through fall and winter for wind protection.

Another practical advantage of a winter windbreak is to serve as a snow fence. Accumulated snow is an additional source of water in areas where moisture is deficient. A heavy accumulation of snow in many areas provides adequate moisture for spring seeded vegetables. The snow also will protect parsnips stored in the ground over winter.

Tools, Equipment

A key yearend task is the maintenance and repair of garden tools and equipment. Small tools should be cleaned, sharpened, and put into storage in good condition. A light film of oil on trowels, shovels, hoes, and similar tools will provide rust protection. Winterize the engines on power tools as suggested by the manufacturer. Sprayers or dusters should be thoroughly cleaned and dried so water does not stand in them through the winter months.

Winter is an excellent time for checking power tools and sprayers for repairs needed to put them in good working condition. There is ample time to order parts, make needed repairs, and check to be sure the equipment operates properly before spring.

If possible, store garden tools and equipment in an indoor storage location such as a garage, basement, storage shed, or barn. Where only outside storage is available, keep the engines of power tools or the working parts of other tools covered to protect them.

Store seeds in a cool, dry location if extra seed was left from this year's planting. A glass jar or similar container makes an excellent dry storage container for seed packages.

Hotbed, coldframe, or plant growing structures should be cleaned thoroughly and equipment checked for needed repairs. These are some of the items that will be used very early in spring and should not be overlooked in the equipment repair list

Many gardeners face a feverish flurry of activity the day before the first forecast frost. Frost-tender crops such as tomatoes, peppers, eggplant leafy greens, and beans should be harvested immediately. They then can be stored in a refrigerator and used as needed.

Many gardeners harvest green and partially colored tomatoes, wrap each in a piece of newspaper, and store in a carton in a cool basement (50° to 60° F). Tomatoes can be stored in this manner for several months and will ripen for use in several days when placed at room temperatures. Tomat vines can be hung in a warm shelte to finish ripening for immediate use

Crops in the cabbage family ar

Tomato vines can be pulled up just be fore frost and hung in basement c garage for fruit to ripen.

usually not injured by the first slight frost, so they can be allowed to remain until temperatures in the mid-twenties are forecast.

Protection can be provided against light frosts by using burlap bags, old blankets, or canvas as a cover. The vegetables may ripen in the garden during warm fall days while frosts are increasingly likely at night. Produce not processed by the first heavy frost should be prepared for storage.

One of the great joys of gardening is the year-round use of vegetables produced in the garden. A number of vegetables can be stored without a preserving process.

In many areas of the South, root vegetables such as carrots, beets, parsnips, turnips, or salsify stay in place in the garden and are dug up as needed. A protective mulch cover of straw or compost may be used. In northern areas where the soil is apt to freeze, these vegetables can be stored in place until early winter with a mulch layer, then dug and put into a cellar, basement, or pit storage area. The popularity of pit or cellar storage of produce is increasing.

Your county Extension office may have suggestions on several types of easy-to-construct vegetable storage structures which can be used for potatoes, root crops, cabbage, and onions.

Pumpkin and squash can be stored in a fairly warm (55° to 60° F), dry location in the home on shelves or racks for long periods of time. They should be harvested when the rind is hard, with a short portion of the stem left on each.

Sweetpotatoes require a warm humid storage location. They are subject to injury if exposed to temperature below 50° F, so special care is needed.

Review Garden Plan

Another important step at the end of the season is to review your garden plan, and see how to improve it next year. If you didn't keep records of crops and varieties through the season, make some notes now. By next spring, you may forget the names of varieties you want to choose again unless you have a written record.

Gardeners should keep a list of varieties that did well, besides recording where the seed (or plant) was purchased and how much was bought. Evaluate the production of each crop, and make notes if more or less space may be required next year to meet your family's needs.

This is also an excellent time to match notes with other gardeners. Comparing results from certain crops or varieties may eliminate your choice of a poor variety next year and introduce you to a new variety or crop that may be an improvement.

Check your soil test reports and fertilizer and lime recommendations. Bring them up to date or get soil samples for testing.

Many a cold winter's evening can be warmed by glancing through seed catalogs and suppliers' lists in preparation for next season. Fall and winter is a good time to drop a postcard

It's a good idea to review your garden at end of season. Records of which varieties did well, and how much space was allotted to each vegetable, can lead to more successful crop next season.

Sallie Peyton

William E. Carnahan

to garden seed and supply distributors for early mailings of their catalogs.

For those who prefer to buy seeds from local dealers, this is an excellent time to notify them of your next spring's needs for varieties, supplies, or equipment you were not able to find last spring. This allows the dealer time to locate the items before the spring rush of garden sales begins.

Problems in finding garden supplies can often be avoided completely with a visit to the dealer during winter months. While you are there, sharing successes and failures with the dealer can provide important information for him to pass along to other customers.

To sum up, the end of the season is a time to clean up the garden area, compost valuable organic materials, plant cover crops and provide winter protection for the garden soil, to clean and repair garden tools, store garden produce for later use, update soil test reports, and review last year's successes and failures. Many old and wise green thumbers claim a successful gardener is one who keeps at it 12 months a year.

The end of the growing season is not an end to garden-related activities. The late fall and winter months are important for achieving a productive and profitable garden.

Left, winter nights are good time to meet with family and decide what vegetables to plant come warmer weather. Right, seed catalog arrives in mailbox.

Help! Help! Where You Can Find It
by Barbara H. Emerson

The vegetables and fruits most commonly grown by home gardeners are widely adaptable. However, new gardeners are often puzzled by what later becomes second nature.

This country is so large that one encounters in it a wide range of growing conditions. Temperature, soil, rainfall, growing season length, and other climatic factors vary. A gardener moving from one section to another may have to adjust earlier knowledge to the new situation.

As gardeners become more experienced, they are often aware of things that were overlooked in the beginning years, or develop a keener interest in special aspects of really fine gardening. Fresh problems and different possibilities appear. New varieties are introduced and may behave differently. Answers are constantly sought.

By far the largest number of gardeners turn to other gardeners with their questions and requests for help. Relatives, neighbors, or other friends are nearest, easiest to talk with, and familiar with the conditions being discussed.

This is usually good. Such people have already learned much through their own experiences. But they, too, may need help with certain problems.

Unlike any other country in the world, every resident of the United States has available at little or no direct cost a constant source of dependable gardening and food preservation assistance—the Cooperative Extension Service. This is a publicly supported, informal, out-of-school educational organization of the U.S. Department of Agriculture (USDA) and the land grant university (State college) system.

There are home economists and agricultural Extension agents or farm advisors in most counties, and State Extension specialists on various subjects. All have close communication with their State university's professional workers for detailed information.

The Cooperative Extension Service provides personal consultation by phone or at the county office, and information materials such as bulletins, circulars, fact sheets, and newsletters prepared by county agents and university or USDA personnel. Most of the publications are free, although there is a nominal charge for a few of the more elaborate ones.

USDA publications are available from the Superintendent of Documents, U.S. Government Printing Office, Washington, D.C. 20402. The Superintendent of Documents provides free price lists for various subjects. For example, List No. SB001 on "Home Gardening of Fruits and Vegetables" covers culture and storage of fruits and vegetables. Some USDA publications are available from members of Congress.

You also can obtain USDA publications from the Consumer Information Center, Pueblo, Colo. 81009. Available publications are listed in a quarterly catalog called *Consumer Information*, which is free on request.

A soil testing service may be available through the county Extension office or commercial testing laboratories. For a few dollars a gardener can get precise advice about fertilizing and improving soil in which vegetables and fruits are grown.

Some county Extension offices offer clinics, workshops, or other programs in home gardening and food

---

Barbara H. Emerson is Technical Information Coordinator for Amchem Products, Inc., Ambler, Penna.

preservation. They may also provide information by radio, TV, or tapes.

Vegetable and fruit demonstration or trial grounds are maintained by the horticulture, vegetable crops, or pomology departments of many State universities.

All this information is as close as your telephone. "County Extension Service," "Cooperative Extension Service," or "Agricultural Extension Service" is usually listed in the phone directory under county government offices or the State land grant university.

Each State has at least one agricultural experiment station. These stations are responsible for much of the research done on home gardening problems. Your county agent can guide you to what is helpful at the experiment stations and similar institutions such as research centers and plant introduction stations. One particularly helpful related organization is the New York State Fruit Testing Cooperative at Geneva, N.Y. 14456.

There are master gardener programs in Washington, New Hampshire, Vermont, and Illinois. Other states have similar programs or are setting them up.

These courses are conducted by Extension Service personnel to further educate expert gardeners who, when finished with the certification course, can volunteer their advice to novice gardeners.

Some State-wide programs, such as in Pennsylvania, California, and Massachusetts, furnish paid experts, as do a sizable number of cities with large community gardening programs. Typical agencies most likely to be helpful to community gardeners are the Parks and Recreation Departments, Housing Development which may provide assistance and land, and Public Works which can commission open spaces.

Teachers in local Vo-Ag schools are another source of garden information. Some private and State-supported colleges have daytime courses

Photos by E. Blair Adams

Left, horticulturist Bill Scheer visits with Master Gardeners in Washington tomato demonstration plots. Right, Washington State University "self-information" center. Information on publications as well as where to go with a problem plant can be obtained at centers.

open to the general public. An increasing number of adult evening schools include courses in vegetable gardening and fruit growing.

Correspondence courses that are helpful are offered by some colleges and universities and listed in the *Directory of American Horticulture* which is published by the American Horticultural Society, Mount Vernon, Va. 22121. The price is $5.

This directory also lists addresses and often phone numbers for most professional, semi-professional, and specialized trade associations in the field of horticulture; garden club associations; horticultural libraries generally open to the public; major garden centers; and national, State, and regional horticultural organizations.

A county agent is generally familiar with gardening and other horticultural groups in his area, and can help you find the one that comes closest to meeting your needs.

Botanic gardens, horticultural organizations, and garden centers with full-time staff usually answer garden questions, especially from members, or they refer inquirers to an appropriate source. Although more concerned with ornamental horticulture, they can supply good help in growing vegetables and fruits. They offer lectures and workshops on the subject, and maintain libraries with useful books, magazines, and catalogs. Several of them sponsor and aid community vegetable gardens. So do a few garden clubs.

Local public libraries are increasing the number of titles they offer in the categories of growing and preserving food crops. If their holdings are limited, they can be supplemented through interlibrary loans from State libraries. Ask your librarian about this. In certain rural areas bookmobiles bring desired books directly to borrowers, or the books may be mailed.

Lists of recommended books for

William E. Carnahan

further reading follow various chapters throughout this Yearbook.

There are also good commercial sources of information. Most comprehensive are the nursery or garden supply centers large enough to employ one or more trained horticulturists. They have the advantage of knowing the local conditions and plant performance.

A good test of their information's authoritativeness is to pose a problem whose solution you already know. If the reply parallels or improves your understanding, you may assume you're in good hands. Beware of glib or dodging answers. An honest, "I don't know" or "I'm not sure" is better than a quick reply without basis.

The same test can be applied to books. Unfortunately, there are a number that look appealing but were prepared so quickly there is little between their covers upon which you can rely. A knowledgeable employee

One way to cultivate a greener thumb is by joining a garden club. In photo, members of Beltsville, Md., Garden club get advice from expert.

91

William E. Carnahan

can help you tell the slick from the sound.

The catalogs and gardening guides of reputable mail order suppliers often contain a wealth of detailed information from their years of experience with the crops in question. This is true of wholesale growers or plant breeders as well as retail seedsmen, fruit nurserymen, and pesticide and fertilizer manufacturers. The better ones often have an experienced staff person to answer inquiries, knowing that business growth depends on satisfied customers.

Some manufacturers of equipment and supplies for preserving food distribute highly useful booklets.

Even though there are still areas of inadequate knowledge to challenge research workers, gardeners have at their disposal enough help to assure a tremendous amount of satisfaction in producing food and beauty. The key to success is taking advantage of these excellent resources.

Gardening information can come from many sources, including USDA and State Extension publications and seed company catalogs.

Gardener's Glossary
Compiled by Robert A. Wearne

Acclimate—Plants conditioned or becoming conditioned to a new climate or different growing environment. (See Hardening Off).

Acid (Sour) Soil—Soils with a pH below 7; most fruits and vegetables grow best when the pH is between 5.2 to 7.1.

Aeration—Free movement of air through the root zone of plants; prevented in compacted or waterlogged soils.

Aerobic—Pertaining to organisms which grow only in the presence of oxygen, as bacteria in a properly prepared compost.

Alkaline (Sweet) Soil—Soil with a pH above 7; some fruits and vegetables will grow in mildly alkaline (7.4–8.0) soils, such as asparagus, beans, leeks, okra, grapefruit, lemons.

Alluvial Soils—Recently deposited water-laid materials which have been changed very little by weather elements. Found on flood plains and valleys.

Anaerobic—Growing in the absence of oxygen, or not requiring oxygen. Known only in bacteria.

Annuals—Plants living one year or less. During this time the plant grows, flowers, produces seeds, and dies. Examples: beans, peas, sweet corn, squash.

Axils (leaf)—The angle or upper side where the leaf is attached to the stem.

Bare Root—Deciduous plants such as apple trees sold with their roots bare, not in a ball of soil.

Biodegradable—Materials readily decomposed in the soil by micro-organisms such as bacteria and fungi.

Blanching—Excluding light to reduce the green color or chlorophyll in plants or plant parts, as with celery, Witloof chicory, or cauliflower.

Bolting—Production of flowers and seeds by such plants as spinach, lettuce, and radishes, generally occurring when days are long and temperatures warm.

Broadcast—Scattering seed or fertilizers uniformly over the soil surface rather than placing in rows.

BTU—British Thermal Unit, a heat unit.

BTU/hr—Quantity of heat needed per hour to maintain a given temperature.

Cambium Layer—The layer of cells that lie between the bark and the wood of a tree.

Robert A. Wearne is a Horticulturist with the Extension Service.

Chelate—Molecular form in which some nutrients, such as iron, are easily absorbed by plants.

Chill Requirement—Number of hours that deciduous fruits require below 45° F (7.2° C) before normal growth will resume in spring. Without adequate chilling, blossoming and foliage development is delayed. Example: Elberta peaches require 900 hours below 45° F and Flordabelle only 150 hours.

Chlorophyll—Green coloring matter within the cells of plants.

Chlorosis—Lack of green color in leaves; may be caused by nutritional deficiencies, environmental conditions, or disease.

Clone—A group of plants derived from an individual plant by vegetative propagation such as grafting, cutting, or divisions rather than from seed.

Clove—One of a group of small bulbs produced by garlic and shallot plants.

Coldframe—An enclosed, unheated but covered frame useful for growing and protecting young plants in early spring. The top is covered with glass or plastic and located so it is heated by sunlight.

Compost—Decayed vegetable matter such as leaves, grass clippings, or barnyard manure. It usually is mixed with soil and fertilizer. Valuable as a mulch in a garden or for improving soil texture, and in potting soils.

Cool Crops—Vegetables that do not thrive in summer heat, such as cabbage, English peas, lettuce, or spinach.

Corm—Enlarged fleshy base of a stem, bulb-like but solid, in which food accumulates. Propagated by division of the cloves. Examples: Dasheen (Taro), garlic, and shallots.

Cotyledon(s)—Seed leaf or leaves containing stored food for initial seedling growth.

Cover Crop—Generally an annual grass or legume such as clover planted to protect the garden from wind and water erosion. Known as "green manure" because it may be plowed or turned under to provide organic matter essential to the soil.

Crop Rotation—Growing annual plants in a different location in a systematic sequence. This helps control insects and diseases, improves the soil texture and fertility, and decreases erosion.

Crown (Plant)—Growing point above the root where the tops or shoots develop as with lettuce, spinach, carrots, and celery, rhubarb.

Crucifer—The mustard family. Radishes, cabbage, cauliflower, broccoli, and turnips are members.

Cucurbit—The gourd family to which cucumbers, muskmelon, watermelon, pumpkin, and squash belong.

Cultivar—This means "cultivated variety" and may be used in place of the word "variety" to indicate a specific horticultural selection.

Cultivate—To loosen the top inch or two of soil, by hand with a hoe or by using a mechanical cultivator. Primarily to control weeds.

Cure—To prepare for storing by drying the skins. Dry onions and sweet potatoes are typical examples.

Cutting—Plant stem including a node that is cut or snapped off and used to start a new plant.

Damping Off—A disease causing seedlings to die soon after germination, either before or after emerging from the soil.

Deciduous—Trees or shrubs which lose their leaves annually.

Determinate Tomato—Stem growth stops when the terminal bud becomes a flower bud. Tomato plants of this type are also known as self-topping or self-pruning.

Division—Propagation of plants by cutting them into sections as is done with plant crowns, rhizomes, stem tubers, and tuberous roots. Each section must have at least one head or stem. Example: Rhubarb.

Dormant Spray—Pesticide applied to a plant before growth starts (late winter or early spring) to control insects and diseases.

Drill Row—Small planting furrow made with a hoe, trowel, stick or mechanical drill in which seeds are planted.

Drip Irrigation—Watering plants so that only soil in the plant's immediate vicinity is moistened. Water is supplied from a thin plastic tube at a low flow rate. The technique sometimes is called trickle irrigation.

Early—Vegetables that mature sooner than others of the same species.

Emulsifiable Concentrate—Pesticide chemical mixture which contains an emulsifier to which water may be added to form an emulsion.

Emulsifier—Chemical which aids in suspending one liquid in another.

Emulsion—Mixture in which one liquid is suspended as tiny drops in another liquid, such as oil in water.

Espalier—A plant (for example, an apple tree) trained to grow on a trellis or flat against a surface such as a wall or building.

Evaporative Cooling—Air evaporates water and in the process the air loses heat to the water. Water plus heat equals vapor.

Everbearing—Plants such as strawberries which bloom intermittently and thus produce fruit during the entire growing season.

Fertilization—(1) Union of pollen with the ovule to produce seeds. This is essential in production of edible flower parts such as tomatoes, squash, corn, strawberries, and many other garden plants. (2) Application to the soil of needed plant nutrients, such as nitrogen, phosphate and potash.

Fill Dirt—Soil used to change the grade or elevation of an area.

Flat—Shallow wooden or plastic box, in which vegetable seeds may be sown or cuttings rooted.

Foliar—Refers to leaves.

Foot-candle—Standard measure of light. The light of one candle falling on a surface one foot away from the candle.

Friable (Soil)—Generally refers to a soil that crumbles when handled. A loam soil with physical properties that provide good aeration and drainage, easily tilled. Friable condition is improved or maintained by annual applications of organic matter.

Fumigation—Control of insects, disease-causing organisms, weeds, or nematodes by gases applied in an enclosed area such as a greenhouse or under a plastic cover laid on the garden soil.

Fungicide—A pesticide chemical used to control plant diseases caused by fungi such as molds and mildew. (See Pesticide).

Furrow—Small V-shaped ditch made for planting seed or irrigating. (See Drill Row).

Germination—Sprouting of a seed, and beginning of plant growth.

Grafting—Joining or insertion of one plant part called the scion upon another plant part called the rootstock so that the cambium layers of each piece make contact to produce new growth.

Green Manure—Crops such as legumes or grasses that are grown to be plowed or spaded into the soil to increase humus content and improve soil structure. (See Cover Crop).

Greens—Vegetables such as spinach, kale, collards, turnip greens.

Growing Medium—Soil or soil substitute prepared by combining such materials as peat, vermiculite, sand, or weathered sawdust. Used for growing potted plants or germinating seed.

Growing Season—Period between last killing frost in spring and first killing frost in fall.

Growth Regulators—Synthetic or natural organic compounds such as indoleacetic acid, gibberellin, or napthalene acetic acid that promote, inhibit, or modify plant growth processes. Commonly used in rooting cuttings.

Hardening Off—Adapting plants to outdoor conditions by withholding water, lowering the temperature, or gradually eliminating the protection of a cold frame, hot bed, or greenhouse. This conditions plants for survival when transplanted outdoors.

Hardy Plants—Plants adapted to winter temperatures or other climatic conditions of an area. Half hardy indicates some plants may be able to take local conditions with a certain amount of protection.

Heaving—Caused by alternate freezing and thawing of the soil during winter. This action can push small plants out of the soil and damage their root systems.

Heavy Soil—Soil containing large amounts of clay. Such a soil retains moisture, should not be cultivated when wet, and can be improved by adding organic material.

Hedgerow—Single row of shrubs or trees which provides a screen or wildlife food and cover, improves the landscape, or serves as a fence or a windbreak.

Heeling In—Temporary storing of bare-rooted trees and shrubs by placing the roots in a trench and covering with soil or sawdust.

Herbaceous Plant—Plants that die back to the ground each winter, such as asparagus and rhubarb.

Herbicide—Chemical used to control weeds and undesirable vegetation.

Hill—Raising the soil in a slight mound for planting, or setting plants some distance apart.

Host Plant—Plant on which an insect or a disease-causing organism lives.

Hot Caps—Waxpaper cones, paper sacks, cardboard boxes or plastic jugs with bottoms removed placed over individual plants in spring for frost and wind protection.

Hotbed—Same type of structure as a cold frame but heated, as with an electric cable.

Humidifier—Air passes through wet material and evaporates water.

Humus—Decomposed organic material that improves texture and productive qualities of garden soils.

Hydroponics—Growing plants in nutrient solutions rather than soil. Also called soilless gardening.

Hybrid F_1—Plants of a first generation hybrid of two dissimilar parents. Hybrid vigor, insect or disease resistance, and uniformity are qualities of this generation. Seed from hybrid vegetables growing in your garden should not be saved for future planting. Their vigor and productive qualities are only in the original hybrid seed.

Immune—Free from disease infection because of resistance. Not subject to attack by a specified pest. Immunity is absolute.

Indeterminant Tomato—Terminal bud is always vegetative, thus the stem grows indefinitely. Indeterminant plants can be trained on a trellis, a stake, or in wire cages. (See Determinate Tomato).

Indigenous—Native to a particular region. Opposite of exotic.

Inflorescence—Entire floral structure of a plant.

Inoculation—Treatment of seed with bacteria that stimulate development of bacteria nodules on plant roots. Used on legumes such as peas and beans.

Inorganic—Mineral content of the soil. In reference to fertilizers, those produced chemically. Not arising from natural growth.

Insecticide—Chemicals or agents used to control insects either on contact or as a stomach poison.

Internode—Region on a plant stem between the nodes.

Interplanting—Getting maximum production from a garden by planting early maturing vegetables between rows of slow maturing vegetables. An example is radishes or onions between rows of sweet corn.

Irrigation—Applying water to the soil by sprinklers, trickle or flooding.

K—Symbol for potash.

Lathhouse—Structure built of wood lath for protecting plants from too much sunlight or frost.

Layering—Way of propagating plants vegetatively. A stem is bent down and buried in a rooting medium to induce root development along the buried portion.

Leaching—Loss of soluble fertilizers, or removal of excess soluble salts, by percolating action of water downward through the soil.

Leader—Central and dominant stem or trunk of a tree or shrub from which the side branches develop.

Leaf Mold—Partially decayed leaves useful for improving soil structure and fertility.

Leggy—Weak-stemmed and spindly plants with sparse foliage caused by too much heat, shade, crowding, and over-fertilization.

Legume—Plant that takes nitrogen from air with the nitrifying bacteria that

95

live on its roots. Examples are garden peas and beans.

Lifting—Digging a plant for replanting or winter storage.

Light Soil—Soil that is easy to cultivate, retains little moisture, and has sandy or coarse texture.

Lime—Compound containing calcium and/or magnesium, applied to soils to reduce acidity.

Loam—Soil that consists of less than 52% sand, 28% to 50% silt, and 7% to 27% clay, resulting in a soil texture ideal for gardening.

Manure—Animal waste used as soil conditioner and fertilizer.

Micro-climate—Climate of a small area or locality as compared to a county or State. For example, the climate adjacent to the north side of a home, or influence of a lake on a portion of a county.

Micro-organism—Any microscopic animal or plant that may cause a plant disease or have the beneficial effect of decomposing plant and animal residue that becomes humus.

Mildew—Plant disease caused by several fungi, recognized by the white cottony coating on plants.

Minor Elements—See Trace Elements.

Miscible Oils—Oils that mix with water. Used to control scale insects.

Mist—Applying vaporized water to cuttings in the propagating stage.

Mites—Extremely small sucking insects that infest various plants.

Monoecious—Plants that have male and female sex organs in different flowers on the same plant, such as cucumbers and squash.

Mosaic—Virus disease that damages or kills plants, often giving the foliage a mottled appearance. Some mosaics can be spread by sucking insects, and some by handling or tools.

Mulch—Materials such as straw, leaves, lawn clippings, sawdust, black plastic sheets, or newspapers laid on the soil surface to conserve moisture, maintain an even soil temperature, and control weeds.

Nematode—Microscopic, worm-like, transparent organism that can attack plant roots or stems to cause stunted or unhealthy growth.

Nicotine—Tobacco extract used as insecticide for controlling sucking insects such as aphids.

Nitrogen—One of plant nutrients essential for growth and green color in plants. Available in both organic and inorganic forms. Designated by the letter N.

Nitrogen Fixation—Transformation of nitrogen from the air into nitrogen compounds by nitrifying bacteria on the roots of legumes.

NPK—Symbols for three of primary nutrients needed by plants. N is for nitrogen, P for phosphate, K for potash or potassium. Percentage of these nutrients in a fertilizer package is always listed in that order.

Node—Region of a plant stem that normally produces leaves and buds.

Nutrient Solution—Liquid containing some or all essential plant nutrients required for growth. Type of solution used in hydroponic culture.

Oil Sprays—Compounds of mineral or vegetables oils used to control scale and other insects on trees and shrubs.

Organic Matter—Portion of soil resulting from decomposition of animal or plant material. Helps to maintain good structure and micro-organisms in soil. Tends to give soil a darker color.

P—Symbol for phosphate.

Pan—Shallow flower pot used for germinating seeds and for forcing and growing bulbs.

Parasite—Plant or insect that attaches itself to another organism and obtains food from the host. Dodder is an example of a parasite plant.

Parthenogenic—Fruit produced without fertilization of the ovule(s). Usually seedless. (See Fertilization 1).

Patented—Plant varieties protected by a government patent, granting exclusive rights to the patent holder.

Peat or Peat Moss—Partially decomposed plant life taken from bogs and used as rooting medium, soil conditioner, or mulch.

Peat Pot—Made of compressed peat and often used for starting and growing plants that can be later planted in the garden without removing the pot.

Perennials—Any plant which normally lives more than two years. Examples are artichoke, asparagus, raspberry, and rhubarb.

Perlite—Volcanic or silica material expanded by heat treatment. Used as a soil amendment and in media for rooting cuttings. (See Rooting Media).

Pesticide—General term for any chemical used to control pests.

Pesticide Residue—Material that remains on a plant after pesticide application.

pH—Chemical symbol used to give relative acidity or alkalinity of the soil. The scale ranges from 0 to 14, with 7 the neutral reading. Readings of less than 7 indicate acid soil, readings above neutral indicate alkaline soil.

Phosphate—One of the three major plant nutrients, designated by the letter P.

Photoperiod—Length of the light period in a day.

Photoperiodism—Effect of differences in length of light period upon plant growth and development.

Pinching—Removing the terminal bud or growth to stimulate branching.

Plant Food—See Plant Nutrient.

Plant Nutrient—Substance or ingredient furnishing nourishment and promoting growth in plants. Examples: nitrogen, phosphorous, potassium, iron and sulfur supplied by the soil, organic matter and fertilizers.

Plant Residue—Plant parts such as stems, leaves, and roots remaining in or on the soil after a crop is harvested.

Plant Variety Protected—Plant varieties protected by the Government plant variety law granting exclusive rights to the holder.

Plunge—To cover or sink a plant container to the rim in sawdust, soil, peat moss, or similar materials.

Pollen—Reproductive material, usually dust-like, produced by male part of a flower.

Pollination, Open—Transfer of pollen from flower of one plant to flower of the same or different plant by natural means.

Pollination, Self—Transfer of pollen from male part of one flower to female part of the same flower, or to another flower on the same plant.

Post-emergence—Applying a herbicide after the plants emerge above soil level.

Potash—One of three major plant nutrients essential for plant growth. Same as potassium. Designated by letter K.

Pot-bound—Plants whose roots completely fill a container and surround the soil ball in which they are growing, restricting normal top growth of the plant.

Pot Herb—Plants grown or used for greens.

Potting Mixture—Combination of soil and other ingredients such as peat, sand, perlite, or vermiculite designed for starting seed or growing plants in containers.

Pre-emergence Application—Applying a herbicide to the soil to kill weed seeds before they germinate, or after a crop is planted but before it germinates and seedlings emerge above the soil's surface.

Propagation—Increasing the number of plants by planting seed or by vegetative means from cuttings, division, grafting, or layering.

Pruning—Removing branches or twigs to control the size or shape of a plant, to control fruiting, to remove dead or broken branches, or to strengthen or improve appearance of a plant.

Puddle—Immersing bare roots of trees and shrubs in a mixture of soil and water during transplanting to prevent the roots from drying out. It may also mean changing the soil structure to a land mass if soil is worked or cultivated when too wet.

Resistance—Ability of a plant to restrict disease or insect damage or withstand severe climatic conditions.

Respiration—Chemical process by which the plant absorbs oxygen from the air and releases water and carbon dioxide into the air.

Rest Period—Normal period of inactivity in growth of a plant.

Rhizome—Horizontal underground stem distinguished from a root by the presence of nodes and internodes and buds and scale-like leaves.

Ridging—Pulling the soil into a low ridge at the base of plants. Potatoes are commonly ridged or hilled to keep the tubers covered and prevent greening caused by exposure to sunlight.

Ringing—Removing a narrow strip of bark from around a branch or tree trunk to encourage fruiting. Only bark is removed, and the ring does not extend into the cambium layer.

Ripe—Stage of maturity at which fruits and certain vegetables should be harvested and preserved.

Rogue—Off-type or diseased plant. Or removal of such plants from the garden.

Root-bound—When plants have grown in a container too long. The roots become a mass of fibers and no longer support desired top growth. Same as pot-bound.

Root Crop—Vegetables grown for their edible roots, such as beets, carrots, radishes and turnips.

Rooting Media—Materials such as peat, sand, or vermiculite in which cuttings are placed during the development of roots.

Root Knot—Growth on plant roots caused by nematodes.

Rootstock—Root system upon which named varieties of fruit have been grafted. For example, apple varieties are grafted onto dwarfing or size-controlling rootstocks.

Runner—Slender, elongated and prostrate branch that has buds and can form roots at the nodes or at the tip. An example is a strawberry runner.

Rust—Plant disease caused by a fungus and characterized by a round red or yellow lesion. May be found on beans, spinach, sweet corn or raspberries.

Scale—Small sucking insects protected by a shield-like or cottony covering and found on various fruit trees.

Scion—Piece of a plant, either bud or graft, that is inserted in an established tree or rootstock.

Seed Bed—Garden soil after it has been prepared for planting seeds or transplants by plowing and disking, rototilling, spading or raking.

Seed Leaves—(see Cotyledon).

Seedling—Young plant developing from a germinating seed. It usually has the first true leaves developed.

Sets—Small onion bulbs used for early planting.

Shade Device—Plastic netting, latex paint, whitewash, cheesecloth or other material used to reduce light and heat entering a plant-growing structure.

Short Season Vegetables—Vegetables ready for harvest after one to two months following planting.

Sidedressing—Applying fertilizer on soil surface close enough to a plant so cultivating or watering would carry the fertilizer to the plant's roots.

Silt—Soil particles that are between sand and clay in size.

Slips—Same as "cutting". A way of vegetatively propagating plants.

Small Fruits—Fruits produced on bushes, vines, or low growing plants as compared to fruits produced on trees.

Softwood Cutting—Cutting taken from a woody or herbaceous plant before it has matured.

Soil—Upper layer of the earth's surface, composed of organic matter, minerals, and micro-organisms, and capable of supporting plant life.

Soil Aeration—Mechanical loosening of the soil to facilitate water and air circulation.

Soil Amendment—Any material added to soil to improve physical and productive qualities.

Soil-borne Fungi—Any fungus living in the soil and capable of causing plant disease.

Soilless Culture—Same as hydroponics. Growing plants in nutrient solutions and not in soil.

Soil Sterilization—Treating soil by fumigation, chemicals, heat or steam to destroy disease-causing organisms.

Soil Survey—Systematic examination of soils in the field and in laboratories; publishing of descriptions and classifications; mapping of kinds of soils; and interpretation of soils according to their adaptability to various crops such as fruits, vegetables, and trees.

Soil Testing—Laboratory chemical analysis of soil to determine the supply of available nutrients such as phosphorus, potassium, or calcium, the percent of organic matter, and the soil acidity or alkalinity. Tests for minor elements can also be made.

Soil Texture—Proportional amounts of sand, silt, clay, and organic matter in a given soil.

Sour Soil—Same as an acid soil or one that has a pH below 7. See pH.

Specimen Plant—A plant with some outstanding quality such as flowers, leaves, fruit, or branching habit and located as a focal point in the landscape.

Sphagnum—Mosses which grow in bogs and when decomposed become peat moss.

Spreader—Materials added to pesticide sprays to aid in distribution and coverage of the plant. Can also mean the mechanism used for spreading seed and fertilizers on the soil.

Staking—Tieing plants such as tomatoes to a stake to provide support.

Starter Solution—Fertilizer solution applied to plants at time of transplanting.

Sticker—Material added to pesticide spray to improve adherence to plant surface.

Stolon—Slender, prostrate subterranean stem. It may produce a tuber such as a potato.

Stone Fruit—Plants whose fruits contain a pit or stone such as cherries, peaches, plums, or apricots.

Stool or Stooling—Sprouts that arise from the base of a plant, sometimes called suckers. They may be used as a method of propagation.

Stress (Water)—Plant(s) unable to absorb enough water to replace that lost by transpiration. Results may be wilting, halting of growth, or death of the plant or plant parts.

Stunt—Diseases caused by certain viruses that dwarf a plant and make it unproductive.

Sub-irrigation—Application of water from below the soil surface, usually by a perforated hose or pipe.

Subsoil—Layers of soil below the top soil. It often is less fertile and contains less organic matter.

Sucker—Stem that rises from a root or rootstock and should be removed from grafted trees.

Sunscald—Caused by the sun warming trunks and large branches during winter, resulting in cracking and splitting of the bark. It can be prevented by shading or whitewashing tree trunks and larger branches. Sunscald may also occur on fruit exposed to direct sunlight.

Susceptible—Inability of plants to restrict activities of a specified pest, or to withstand an adverse environmental condition.

Syringe—Applying a mist-like spray to seedlings or transplants to reduce or replace moisture lost by transpiration.

Systemic—Pesticide material absorbed by plants, making them toxic to feeding insects. Also, pertaining to a disease in which an infection spreads throughout the plant.

Tamping—Lightly firming soil over seeds or around newly-set transplants.

Tankage—Fertilizer prepared from slaughterhouse refuse that has been sterilized and pulverized.

Tendril—Slender twining organ found along stems of some plants such as grapes, which helps the vine to both climb and cling to a support.

Thermostat—Electrical control device. It can turn on or off a heater, fan, or heating cable at a set temperature.

Thinning—Removing small or young plants from a row to provide remaining plants with more space to grow and develop.

Till—To prepare and use land for crop or plant growth by plowing, fertilizing and cultivating.

Tilth—Physical condition of a soil. "Good tilth" indicates soil has right proportions of sand, silt, clay, and organic matter so it is easily worked or cultivated.

Tolerant—Ability of plants to endure a specified pest or an adverse environmental condition, growing and producing despite the disorder.

Topdressing—Applying materials such as fertilizer or compost to the soil surface while plants are growing.

Topworking—Grafting one or more varieties to the branches of a tree.

Trace Elements—Minerals such as boron, manganese, iron or zinc normally required only in small amounts by plants. Also known as "minor elements."

Transplanting—Digging up a plant and moving it from one location to another.

Transpiration—Loss of water through openings called stomata on the leaves of plants.

Tree Dressing—Paint or paste used to cover and protect wounds of a tree caused by limb breakage or pruning.

Trellis—an open lattice structure used for shade or to support plants.

Trenching—Deep digging of garden soil and mixing in compost, manure, or some other soil conditioner.

Trowel—Garden tool with a wide blade and short handle used for digging and transplanting.

True Leaf—An ordinary leaf, which functions in the production of food by a plant.

Tuber (Stem)—Thickened or swollen underground branch or stolon with numerous buds or eyes. Thickening occurs because of the accumulation of reserved food. Example: A potato or Jerusalem artichoke.

Tuberous Roots—Thickened roots, differing from stem tubers in that they lack nodes and internodes, and buds are present only at the crown or stem end. Example: Sweet potato.

Variety—Closely related plants forming subdivision of a species and having similar characteristics. (See Cultivar).

Vector—Any agent such as an insect or animal that transmits, carries, or spreads disease from one plant to another.

Vegetative Growth—Growth of stems and foliage on plants as opposed to flower and fruit development.

Vegetative Propagation—Increasing the number of plants by such methods as cuttings, grafting, or layering.

Ventilation—Changing the air in a building by fans or natural methods.

Vermiculite—A mica product expanded by heat, forming a lightweight soil additive. Often used in synthetic soil mixes or as a rooting medium for cuttings.

Viable—Alive, such as seed capable of germinating.

Virus—Pathogenic organism too small to be seen with a compound microscope but capable of causing plant diseases. Often spread from infected to healthy plants by sucking insects such as aphids or thrips, or through pruning or handling of plants.

Water Sprout—Rapid growing succulent shoot which may appear on a trunk or limb of a fruit tree. It frequently appears after a tree has been heavily pruned.

Wettable Powders—Pesticide blended with a filler and wetting agent to permit mixing with water.

Wetting Agent—Material included in pesticide solutions that reduces surface tension and helps to completely cover the surface or foliage area of the plant being sprayed. (See Spreader).

Wilting—Drooping of leaves and stems due to lack of water. Can result from root damage, disease, injury, or hot drying winds.

Windrow—Hay, grain, leaves, etc. swept or raked into rows to dry.

Metric Multipliers

Approximate Conversion Factors

Symbol	When you know	Multiply by	To find	Symbol
		Length		
in	inches	* 2.5	centimeters	cm
ft	feet	30	centimeters	cm
yd	yards	0.9	meters	m
		Area		
in^2	square inches	6.5	square centimeters	cm^2
ft^2	square feet	0.09	square meters	m^2
yd^2	square yards	0.8	square meters	m^2
		Mass (weight)		
oz	ounces	28	grams	g
lb	pounds	0.45	kilograms	kg
		Volume		
tsp	teaspoons	5	milliliters	ml
tbsp	tablespoons	15	milliliters	ml
fl oz	fluid ounces	30	milliliters	ml
c	cups	0.24	liters	l
pt	pints	0.47	liters	l
qt	quarts	0.95	liters	l
gal	gallons	3.8	liters	l
ft^3	cubic feet	0.03	cubic meters	m^3
yd^3	cubic yards	0.76	cubic meters	m^3
		Temperature (exact)		
°F	Fahrenheit temperature	5/9 (after subtracting 32)	Celsius temperature	°C

Symbol	When you know	Multiply by	To find	Symbol
		Length		
mm	millimeters	0.04	inches	in
cm	centimeters	0.4	inches	in
m	meters	3.3	feet	ft
m	meters	1.1	yards	yd
		Area		
cm^2	square centimeters	0.16	square inches	in^2
m^2	square meters	1.2	square yards	yd^2
		Mass (weight)		
g	grams	0.035	ounces	oz
kg	kilograms	2.2	pounds	lb
		Volume		
ml	milliliters	0.03	fluid ounces	fl oz
l	liters	2.1	pints	pt
l	liters	1.06	quarts	qt
l	liters	0.26	gallons	gal
m^3	cubic meters	35	cubic feet	ft^3
m^3	cubic meters	1.3	cubic yards	yd^3
		Temperature (exact)		
°C	Celsius temperature	9/5 (then add 32)	Fahrenheit temperature	°F

* 1 in = 2.54 cm

PART 2

Home Garden Vegetables

Fred S. Witte

Planning Your Vegetable Garden—
Plots, Pyramids, and Planters

by George and Katy Abraham

If you are fortunate enough to have plenty of space for a garden, you can have a traditional type with enough space between rows to run a garden tractor.

Before you plant your garden, do some planning. Most people make their garden too big and by late summer it may be a weed patch. For the conventional garden with ample space to use a garden tractor, a plot of 50 feet by 50 feet is enough for a family of 4. With 5 or more members in households who plan on doing freezing and canning, a space of 50 feet by 100 feet is not unreasonable.

However, if you plan to use muscle power, a hand cultivator and a hoe, plus some plastic mulch to keep weeds down, you can put rows only half as far apart and make your garden half the size. With this much space you not only can grow lettuce, tomatoes, radishes, beets, carrots, onions, snap beans and chard, but you also can raise bush squash (both summer and winter) and some corn as well.

Corn needs to be planted so that you have at least two rows side by side (varieties must mature at the same time) for cross pollination. It is actually preferable to plant four rows together. The rows do not need to be long ones but can be arranged in blocks to aid cross pollination. Remember it is the seed which you eat—if there is no pollination you get no corn.

If space is limited, it is best to cut

George and Katy Abraham of Naples, N.Y., do a column, The Green Thumb, for 126 newspapers. They also broadcast regularly over TV and radio, and have written seven books on horticultural topics.

out corn, squash and pumpkins, although a few bush squash do very well in small spaces and even in containers.

Mini Gardening Is Easy

As the United States moves into its third century, we find the American gardener is no longer limited to conventional straight rows and regulated distances in which to raise vegetables. Indeed, the method of culture is limited only by the imagination. This makes it possible for the urban gardener—like his country cousin—to have the satisfaction of raising juicy tomatoes, snappy green beans and pungent radishes.

What's wrong with a window box or a balcony planter with a few flowers in front for show, and beans, carrots, onions, lettuce or radishes in back for food? Many gardeners are doing just that. An incredible amount of vegetables can be grown in small spaces with a little extra plant food and a good supply of ingenuity. Tomatoes, eggplants and vine crops add color to foundation plantings.

Cucumbers and melons can be trained up railings, trellises and fences. One cucumber vine will do very well in one cubic foot of soil if it is fed once every two weeks with a liquid plant food, or if a slow release plant food is added at the time of planting.

Even though the gardener has no patch of ground he should not be deterred. Container-grown vegetables are just as tasty. Some containers that make good mini-gardens are waste paper cans, half barrels, square boxes, cement blocks (set so openings face the sky), and pails.

We have tried galvanized water tanks, cut in half, an eave trough,

bushel baskets and beverage boxes. You can make or buy tower gardens (called vertical gardens) made of 2-inch by 4-inch mesh wire fencing rolled in a circle and lined with sisal craft paper, into which has been poured one of the soilless mixes. Openings are cut in the paper, and seeds or plants inserted at appropriate intervals. You can also buy or construct pyramid gardens using metal, wood or plastic. If you have a sloping piece of property, you can build a terrace garden using the same principle.

Vegetables grow faster in the warmth reflected from walks, drives and concrete pads. If plants are growing in containers, drainage holes provide places for evaporation and you will have to water the containers more often when they are setting on concrete or blacktop.

Even though drainage holes aid in evaporation, containers must be well drained. Heavy rains can cause water to stand around roots. No vegetables do well in waterlogged soil. It is better to have a well drained container that must be watered oftener than one that holds water and shuts out air to the roots.

Fresh vegetable lovers who have patches of ground available with hard packed cement-like soil can grow their produce in raised beds framed with 1-inch by 6-inch boards. They then can make their own soil mix to fill the beds. The raised beds can be any size, but the most convenient are those easily reached from all sides. A 4-foot square is one we find handy.

Used car tires make dandy small circular beds. In fact, when you have a slope the tires can be pegged in place with stakes, filled with soil and used to grow any number of small crops.

If the sun touches any of your garden spots for only a short time during the day, there's a solution. The trick is to resort to aluminum foil or chrome reflectors or mirrors. White houses reflect light, as do white gravel mulches. An equivalent of six hours of sun a day (whether reflected or direct) is adequate to grow most crops if the light intensity remains high. Vegetable beds should not be shaded by tall trees or high buildings. We've had peppers, tomatoes and lettuce produce with only three hours of sun and good reflected light the rest of the day. You can also put your containers on wheels and move them with the sun.

Left, bags, baskets, and buckets can become minigardens. Right, limited space for a garden poses no real problem. The authors made this "vertical" garden by lining a wire mesh tower with sisal craft paper and filling it with soil. Growing are squash, tomatoes, peppers, onions, and lettuce, besides some flowers.

Vegetable Planting

Vegetables	Plants or seed per 100 feet	Spacing (Inches)		Number days ready for use
		Rows	Plants	
Asparagus	66 plants or 1 oz.	36–48	18	(2 years)
Beans, snap bush	½ lb.	24–36	3–4	45–60
Beans, snap pole	½ lb.	36–48	4–6	60–70
Beans, Lima bush	½ lb.	30–36	3–4	65–80
Beans, Lima pole	¼ lb.	36–48	12–18	75–85
Beets	1 oz.	15–24	2	50–60
Broccoli	* 40–50 pl. or ¼ oz.	24–36	14–24	60–80
Brussels sprouts	* 50–60 pl. or ¼ oz.	24–36	14–24	90–100
Cabbage	* 50–60 pl. or ¼ oz.	24–36	14–24	60–90
Cabbage, Chinese	* 60–70 pl. or ¼ oz.	18–30	8–12	65–70
Carrots	½ oz.	15–24	2	70–80
Cauliflower	* 50–60 pl. or ¼ oz.	24–36	14–24	70–90
Celeriac	200 pl.	18–24	4–8	120
Celery	200 pl.	30–36		125
Chard, Swiss	2 oz.	18–30	6	45–55
Collards and kale	¼ oz.	18–36	8–16	50–80
Corn, sweet	3–4 oz.	24–36	12–18	70–90
Cucumbers	½ oz.	48–72	24–48	50–70
Eggplant	⅛ oz.	24–36	18–24	80–90
Garlic (cloves)	1 lb.	15–24	2–4	140–150
Kohlrabi	½ oz.	15–24	4–6	55–75
Lettuce, head	¼ oz.	18–24	6–10	70–75
Lettuce, leaf	¼ oz.	15–18	2–3	40–50
Muskmelon (cantaloupe)	* 50 pl. or ½ oz.	60–96	24–36	85–100
Mustard	¼ oz.	15–24	6–12	30–40
Okra	2 oz.	36–42	12–24	55–65
Onions	400–600 plants or sets	15–24	3–4	80–120
Onions (seed)	1 oz.	15–24	3–4	90–120
Parsley	¼ oz.	15–24	6–8	70–90
Parsnips	½ oz.	18–30	3–4	120–170
Peas, English	1 lb.	18–36	1	55–90
Peas, southern	½ lb.	24–36	4–6	60–70
Peppers	⅛ oz.	24–36	18–24	60–90
Potatoes, Irish	6–10 lb. of seed tubers	30–36	10–15	75–100
Potatoes, sweet	75–100 pl.	36–48	12–16	100–130
Pumpkins	½ oz.	60–96	36–48	75–100
Radishes	1 oz.	14–24	1	25–40
Salsify	½ oz.	15–18	3–4	150
Soybeans	1 lb.	24–30	2	120
Spinach	1 oz.	14–24	3–4	40–60
Squash, summer	1 oz.	36–60	18–36	50–60
Squash, winter	½ oz.	60–96	24–48	85–100
Tomatoes	50 pl. or ⅛ oz.	24–48	18–36	70–90
Turnip greens	½ oz.	14–24	2–3	30
Turnip, roots	½ oz.	14–24	2–3	30–60
Watermelon	1 oz.	72–96	36–72	80–100

* Transplants

Keep Soil Moist

Keep in mind that because growing space is restricted, container-grown vegetables will need more feeding and more water than those grown in open ground. Once fruit starts to form on tomatoes, peppers, or vine crops they will need even more water, as will the vegetables growing underground. Soil should be kept moist for a good yield. Also keep in mind that vegetables such as onions and radishes will get unbearably hot tasting if they are allowed to grow dry.

True garden enthusiasts make successive sowings of salad crops such as radishes, onions and lettuce about ten days apart (if they have the space) so that when one containerful has been eaten another is already mature enough to enjoy.

Many lettuce varieties will grow all season, providing tender outer leaves constantly if the small center leaves are left to grow. Buttercrunch is a favorite.

A perennial question is how thick to plant so that a maximum crop can be harvested in the space available, wihout crowding the plants into inefficiency. The large chart gives an approximate measure for space needed by various vegetables to help the gardener determine how many plants a certain size area will accommodate. Plant breeders have developed many mini-vegetables in dwarf forms for small-space gardeners. Look for them in seed catalogs and at garden centers.

Chart for Small Space Gardeners

For those who must grow their vegetables in small spaces this chart gives the approximate number of plants a square foot of container space will accommodate. Containers should be at least eight inches deep for medium sized and small vegetables but a foot to 18 inches deep for vegetables such as tomatoes, eggplants or corn. In the case of corn a 4-foot- square space is desirable to provide good cross pollination.

Measurements do not have to be exact and circular containers work as well as square or rectangular ones.

Vegetable	Approximate number of plants per square foot
Beans	3–4
Beets	25 [1]
Broccoli	3
Brussels sprouts	2
Cabbage	2
Carrots	100 [2]
Cauliflower	2
Chard, Swiss	9
Corn (dwarf)	4
Cucumber (standard)	1 [3]
(dwarf)	2 [3]
Dandelion	6
Eggplant	1
Endive	4
Garlic	36
Kale	4
Kohlrabi	4
Leeks	64
Lettuce (head)	4
(leaf & semi-head)	6
Muskmelon	1 [3]
Mustard greens	9
Onions (cooking)	16
(hamburger)	9
(green bunching)	100 [4]
Parsley	16
Parsnips	25
Peanuts	4 [3]
Peas	25 [3]
Peppers	4
Potatoes	1
Sweet potatoes	1
Radishes	144 [5]
Rutabaga	5
Spinach	4
Summer squash (bush)	1
Winter squash (bush)	1
Tomato (regular)	1 [3]
(dwarf)	2
Husk tomato (Physalis)	2
Watermelon (dwarf)	1 [3]

[1] Thin at 1-inch diameter for "greens" and let remainder grow.
[2] Thin every other one when "fingerlings" and let others grow.
[3] Train on trellis.
[4] Can thin to eat and let others grow into cooking onions.
[5] Thin small ones to eat and let others grow.

It's not easy to figure exactly how profitable your gardening enterprise will be but one thing is certain, it's good for your waistline and your general well-being. Few things around the home give as much satisfaction— as well as a bit of a boost to the budget—as your garden. Studies show that for the time you spend in the garden, you get a net return of $3 to $5 per hour. One State university cites figures to show that the average home garden in its State will be worth $165. The average gardener will spend around $35 on his garden, which means he will be getting around 75 cents "profit" for every $1 he spends. And he eats better to boot.

Location: The closer to the kitchen you can locate your garden plot, the better. It's mighty handy to be able to take a few steps outside and snip off a handful of herbs or lettuce.

It makes little difference which way the rows run, although running them lengthwise of the garden makes cultivation easier. If rows run east and west, plant your large crops on the north side of the garden so that they will not shade the small crops. Keep in mind that crops should get at least six hours of sunlight daily. Try not to locate the garden near trees, buildings, ridges or anything that will block out the sun.

While certain flowers have been bred to grow in shade, vegetables are sun worshipers and do not do well near trees, both because of the shade and the competition from tree roots. A hardwood tree 1½ feet in diameter at shoulder height gives off as much as 125 tons of water in a single season, robbing plants not only of sun but also of water for nearby soil.

If you must garden in semi-shade, try increasing the light intensity by installing aluminum sheets or other reflectors to accent the sun's rays, as small space gardeners often do.

Steer clear of black walnut trees. Gardens should not be planted within 30 feet of trees in the black walnut family since they produce a toxic substance called juglone through the roots. Many vegetables such as tomatoes, corn, peppers and others will become stunted, wilted or even die, when their roots come in contact with walnut roots.

Soil Texture: Sometimes a soil is heavy clay, but don't let this discourage you from planting a garden. There are ways of making it more friable (workable). Here is a simple test for soil fitness: Grab a handful of moist earth and squeeze it tightly for ten seconds. If the soil breaks in several places when dropped from a 3-foot height, it's workable. Soil that forms a mudball that will not break into pieces when the test is applied is apt to be too difficult to work in its present state.

Another way to tell if your soil has good "tilth" or working quality is to feel it with your fingers. If the soil has a nice "loamy" texture, it's great for plants. Loam (called "loom" by old timers) is simply a well balanced mixture of large, medium and small particles of sand, silt and clay.

Organic Matter

If you find you have a problem soil, the best conditioner you can get is organic matter, such as compost, peatmoss, sawdust, leaves, rotted dead weeds and plants removed in cleaning up the yard (avoid any diseased materials). Lawn clippings, wood chips, kitchen scraps, barnyard manures, and green manures—which include crops such as winter rye, buckwheat and legumes—are turned under to rot in the soil. Humus opens the clay and encourages earthworms to be more active helpers. The earthworms in a single acre of ground may pass more than 10 tons of dry earth through their bodies annually. They mix organic matter with the subsoil. They also build up topsoil and their burrows aerate the earth.

Don't add sand to a claylike soil to loosen it. The result may be a concrete-like mixture harder than the original clay. Limestone has a loosening effect on a heavy clay soil, coagulating the fine particles into larger ones, allowing air and water to pass freely. Ground limestone or dolomitic limestone can be used. Ground limestone is less expensive and is easy to apply. Dolomitic lime has 20 to 30 percent magnesium, plus 30 to 50 percent calcium, and is available in both hydrated and ground stone types. Since magnesium is another element needed for plant growth, many prefer dolomitic limestone.

Any form of lime can be used, but remember the more concentrated forms such as the hydrated or burned lime forms should be used in lesser amounts. The big problem most gardeners run into is using the 3 forms in equivalent amounts. Roughly speaking, 100 pounds of ground limestone is equal in action to about 74 pounds of hydrated lime or 56 pounds of burned lime.

Fertilizer: A good well rotted compost pile is a valuable adjunct for any gardener. We have already mentioned how it breaks up a heavy soil. It also adds nutrients to some degree but it cannot be counted on to feed vegetables all the necessary nutrients. Any balanced fertilizer with the big three (nitrogen, phosphorus, and potash) can be added to the garden in fall or spring before you plow. Many gardeners use a liquid plant food, usually applied at planting time and again two or three times during the growing season. If a regular dry fertilizer is applied for a mid-summer snack, take care that it doesn't touch the plants. It can be applied as a side dressing a few inches away so that it can be washed into the soil near the roots, but if it touches the plant in the dry state it will burn plant tissues.

Many slow release fertilizers on the market can be applied at planting time, and because of the rosin cover over each particle they will be released at intervals during the growing season.

Pesticides: Don't douse your vegetables with sprays or dusts so thickly that it takes many washings to get them clean enough to eat. Most small gardens are easily de-bugged by hand-picking each day. Vigilance is the watchword. A few minutes each day checking your vegetables will usually be all that's needed to keep out bad bugs.

Remember that only 10 percent of all bugs are bad. Our natural predators, including beneficial insects, birds, frogs and toads will eliminate most pests if we don't kill them off with sprays. However, should an infestation get the best of you, check it with the latest control methods recommended by your county Extension agent, State university, or the U.S. Department of Agriculture.

Crop Rotation: Farmers for centuries have found it good business to shift their crops around each year. Home gardeners cannot rotate their crops that easily because of limited space. But with careful planning, you can maintain a certain amount of rotation. Some diseases—such as root knot, clubroot, fusarium wilt, and cabbage yellows—will build up in the soil if one crop is planted in the same spot year after year. Another advantage of shifting crops around into different spaces is that vegetables such as peas or beans are legumes and can take nitrogen out of the air and put it into the soil.

Corn is a gross feeder and takes a large quantity of nutrients from the soil. However, you might have to compensate for this deficiency by extra feeding, since corn plantings may need to occupy the same area year after year since they might shade shorter vegetables if moved to a different location.

Radishes, cabbage, tomatoes or let-

William E. Carnahan

tuce can easily be rotated with beans and peas, alternating their locations each year.

Many disease problems can be prevented by practicing good sanitation. Plant parasites often overwinter in refuse from last year's crops. Pulling vines, stalks and overripe fruits is an effective way to control plant diseases and bacteria. If these parts of plants show any disease they should not be put on the compost pile but sealed in plastic bags and sent to the disposal.

Use Good Seed: Another way to prevent disease problems is to use treated seed or certified seed and raise your own plants. If you buy plants be sure they are from a trustworthy grower. Avoid any plants with root swellings or lumps.

Gardeners are fortunate to have plant breeders who are constantly developing disease-resistant varieties. Study the seed catalog. It lists resistant varieties and although they are sometimes a bit more expensive, they are worth it.

Succession Planting: To get the most from your garden it's smart to practice succession planting. This means getting two crops from each garden row—one which can be harvested in early summer, the other in fall. Often the fall garden is more productive, and fall-grown vegetables are usually of higher canning quality than those which mature during the hot dry periods of midsummer. Take beans, for example. They mature early and are finished by early summer. The vines can be dug and that same space planted with broccoli, cauliflower, Brussels sprouts, string beans or turnips.

Summer drought and early killing frost are two factors you have to keep in mind in planting the late garden. Midsummer heat can be offset by watering and mulching. To beat Jack Frost, concentrate on hardy vegetables like spinach, chard, turnips, beets, and any of the cabbage family, plus fall crops of lettuce.

Most of these vegetables can be planted as late as early July in most areas and still produce a fair crop. Long season crops such as tomatoes, peppers, or eggplants will continue to bear until frost if well cared for. Lima beans and okra will produce a partial crop when planted as late as the first of July. Garden peas such as Wando are a dependable fall crop.

Onion sets may be planted any time during the summer for green bunching onions. Seed can be sown in spring or early summer for a fall crop. Top sets from winter onions can be planted for fall use. Those not used may be left in the row and will usually over-winter for spring use.

In many areas beets and carrots can be left in the garden all winter if a light mulch is added. Parsnips are a welcome spring delicacy but take a long growing season so should be

Planting calendar helps in working out succession planting schedule for garden.

Vegetable Yields

Vegetables	Average crop expected per 100 feet	Approximate planting per person Fresh	Storage, canning or freezing
Asparagus	30 lb.	10–15 plants	10–15 plants
Beans, snap bush	120 lb.	15–16 feet	15–20 feet
Beans, snap pole	150 lb.	5–6 feet	8–10 feet
Beans, Lima bush	25 lb. shelled	10–15 feet	15–20 feet
Beans, Lima pole	50 lb. shelled	5–6 feet	8–10 feet
Beets	150 lb.	5–10 feet	10–20 feet
Broccoli	100 lb.	3–5 plants	5–6 plants
Brussels sprouts	75 lb.	2–5 plants	5–8 plants
Cabbage	150 lb.	3–4 plants	5–10 plants
Cabbage, Chinese	80 heads	3–10 feet	—
Carrots	100 lb.	5–10 feet	10–15 feet
Cauliflower	100 lb.	3–5 plants	8–12 plants
Celeriac	60 lb.	5 feet	5 feet
Celery	180 stalks	10 stalks	—
Chard, Swiss	75 lb.	3–5 plants	8–12 plants
Collards and kale	100 lb.	5–10 feet	5–10 feet
Corn, sweet	10 dozen	10–15 feet	30–50 feet
Cucumbers	120 lb.	1–2 hills	3–5 hills
Eggplant	100 lb.	2–3 plants	2–3 plants
Garlic	40 lb.	—	1–5 feet
Kohlrabi	75 lb.	3–5 feet	5–10 feet
Lettuce, head	100 heads	10 feet	—
Lettuce, leaf	50 lb.	10 feet	—
Muskmelon (cantaloupe)	100 fruits	3–5 hills	—
Mustard	100 lb.	5–10 feet	10–15 feet
Okra	100 lb.	4–6 feet	6–10 feet
Onions (plants or sets)	100 lb.	3–5 feet	30–50 feet
Onions (seed)	100 lb.	3–5 feet	30–50 feet
Parsley	30 lb.	1–3 feet	1–3 feet
Parsnips	100 lb.	10 feet	10 feet
Peas, English	20 lb.	15–20 feet	40–60 feet
Peas, southern	40 lb.	10–15 feet	20–50 feet
Peppers	60 lb.	3–5 plants	3–5 plants
Potatoes, Irish	100 lb.	50–100 feet	—
Potatoes, sweet	100 lb.	5–10 plants	10–20 plants
Pumpkins	100 lb.	1–2 hills	1–2 hills
Radishes	100 bunches	3–5 feet	—
Salsify	100 lb.	5 feet	5 feet
Soybeans	20 lb.	50 feet	50 feet
Spinach	40–50 lb.	5–10 feet	10–15 feet
Squash, summer	150 lb.	2–3 hills	2–3 hills
Squash, winter	100 lb.	1–3 hills	1–3 hills
Tomatoes	100 lb.	3–5 plants	5–10 plants
Turnip greens	50–100 lb.	5–10 feet	—
Turnip, roots	50–100 lb.	5–10 feet	5–10 feet
Watermelon	40 fruits	2–4 hills	—

sown in May in most areas, then left in the garden over winter.

Mixing Crops

In our garden we prefer to mix crops rather than planting vegetables in blocks. The only exception is corn, which must be planted in blocks in order to get proper pollination. We scatter our four rows of beans between rows of lettuce, radishes, beets and onions, instead of planting all the rows side by side. Recent experiments show this cuts down on insect activity. That is the reason many people intersperse flowers in their vegetable gardens. We feel it is rather attractive to have a few plants of marigolds, nasturtiums, calendulas, or zinnias scattered among the vegetables.

Tools: Everyone needs a trowel or two, a spade fork and a hoe. You don't need a garage full of tools to produce a good garden. Small plots can be spaded by hand. Larger plots can be plowed and disked or dragged by your nearby farmer neighbor or large-scale gardener in the neighborhood.

A garden hoe is one of the best weedkillers you can get. Make sure it is sharp enough to clip off weed seedlings. Don't be in a hurry to buy power tractors or tillers until you're sure you want to do enough gardening to justify power equipment.

If you consistently plant a very large garden, a riding tractor with attachments could serve you well. When your plot is average or small a gasoline-driven, hand-operated rototiller type machine may be useful for fitting up a garden and keeping it maintained during the growing season.

If you like good exercise, a small plot is easily maintained through hoeing, hand weeding, and using a push cultivator.

A black plastic mulch can save you lots of cultivating and weeding. Don't use clear plastic because light enters it and enables weeds to grow. Place the plastic sheet flat on the ground, and fasten the edges down with soil or stones. Then make slits in the plastic. If you sow seed you can make a long slit and sow the seed directly into the row under it. You also can sow seed before laying the plastic. After plants are up a few inches, lay the plastic down lightly and cut slits or holes where the plants are so they will grow up through them.

If soil is moist when mulch is laid, plants will need little if any extra water because the moisture is trapped underneath. Sufficient water will seep in around the holes but even in very dry weather black plastic (and other mulches) hold water around roots. The plastic hastens ripening by increasing the soil temperature, and you don't have to worry about weeds, slugs or having to cultivate the plants.

Green Thumbers can turn to many sources to answer their queries about vegetable gardening: (1) Their county Extension office, (2) Current books on vegetable gardening, (3) Bulletins from the U.S. Department of Agriculture, (4) Farm and garden programs over radio and television, (5) State colleges of agriculture, and (6) Long-time gardeners in the area.

Growing Vegetable Transplants: Lights, Containers, Media, Seed

by Franklin D. Schales

Most experts on vegetable plants agree that the ideal vegetable transplant should be stocky, have good color, be disease-free, and be at the proper stage of development for best growth when set in the garden.

How can you grow such a transplant? Ideally, the best place to grow vegetable plants is in a greenhouse equipped with automatic controls for heating and ventilating. There are several kinds of hobby-size greenhouses available from commercial sources. If you prefer to build your own, plans are available from several State Agricultural Experiment Stations.

Hot beds and cold frames also are suitable for plant growing, but require more attention than greenhouses since they are not constructed to allow for automatic ventilation. Also it is more difficult to work in these structures than in a greenhouse.

It is possible—though more difficult—to grow vegetable transplants indoors if you do not have a greenhouse, hot bed, or cold frame. The best place to grow plants indoors, if you are depending on sunshine as the only source of light, would be in a large window facing south or southwest. This should be in a room where it is possible to have the night temperature no higher than 60° F. The window should not be shaded by trees or otherwise since the plants will require all the light that reaches them. High night temperatures will result in tall, soft, and spindling plants.

Cool white fluorescent lamps provide good supplementary light for plants. These are available in pre-wired lamps and ballasts in various sizes and types. Best illumination is obtained if the lamps are spaced 2 inches apart, center to center. Since fluorescent tubes are relatively cool they may be placed close to plants without danger of burning them. An adjustable support for the lamps makes it possible to adjust lamps to differing plant heights.

If you construct a chamber for growing plants, all inside surfaces should be painted white or made of reflective materials to increase the light available.

Length of lighting each day should be controlled manually or by using a time clock. Usually 12 to 16 hours light each day is sufficient for growing plants.

Vegetables for Transplanting

Many kinds of vegetables can be satisfactorily transplanted. Usually these are classified as being either warm season or cool season vegetables, depending on tolerance to cold weather.

All the cucurbits—which include cucumber, squash, watermelon, and muskmelon—are warm season vegetables. For satisfactory results with them, plant the seed in containers that will be set in the garden without disturbing the plant's root system.

Other warm season vegetables such as pepper, eggplant, and tomato may be transplanted bare-root. However, it is also best if these are grown in a type of container that will allow transplanting with the root ball essentially intact.

Cool season vegetables include cabbage, cauliflower, brussels sprouts, broccoli, lettuce, and onion. These may be transplanted in containers or bare-root.

Franklin D. Schales is an Associate Professor of Horticulture at the University of Maryland, Salisbury.

111

Containers

Those wishing to grow vegetable transplants have a wide choice of containers to choose from. Compressed peat pellets, peat pots, plastic pots and fiber blocks are some of the types available for growing single plants in. Multi-plant containers include various size cell packs and open containers of various sizes in which more than one plant can be grown.

With most of the containers available, choose the larger sizes in the single-plant growing containers and don't crowd plants too close in the multi-plant containers. As a general rule, allow 6 to 9 square inches per plant for most vegetable transplants.

Compressed peat pellets will not allow this much space if placed against each other in the flat. However, they can be spaced further apart, and are satisfactory if the plants are set in the garden before they become too large. The space between the peat pellets should be filled with peat moss or a soilless growing medium.

If peat pots are used, when the plant is set in the garden be sure that the entire pot is buried. If the top of the pot is exposed to air and sunlight, it will act as a wick, removing moisture from around the plant roots. Sometimes it's a good idea to break the bottom out of peat pots when setting plants in the garden.

Remove plants from plastic pots or trays before putting them in the gar-

Terence O Driscoll

Top, containers for starting plants. Left, tomato sprout emerging from a starter container.

Terence O'Driscoll

den. If the plant has a very extensive root development, with almost a solid mass of roots, you should slightly break the root ball apart, or make several shallow cuts with a knife along the edge of the root ball. This will stimulate new root development into the soil after the plants are set.

Other plant-growing containers are available such as milk cartons and clay pots, but the types previously discussed are either more readily available or have proven more satisfactory for most growers.

Regardless of the container used, it is important that it drain excess water freely. Waterlogged vegetable plants will not grow properly.

Growing Media

The materials—such as soil, sand, peat moss—that plants grow in are known as growing media.

Basic requirements of satisfactory plant growing media are that they:
- Have good water drainage
- Have adequate waterholding capacity
- Be free of harmful substances such as herbicide residue
- Be reasonable in cost and readily available
- Be free of weed seeds, insects, and diseases

Not too long ago practically all plants were grown in soil that had been improved by adding sand and peat moss. These were added to improve aeration, drainage, and waterholding capacity, since most top soil is not suitable for growing plants satisfactorily without being modified somewhat. The main reason for this is because plants are normally grown in small containers, making it necessary for the media to be in nearly perfect condition to support satisfactory plant growth and development.

Many problems are associated with making a satisfactory soil mixture. One is simply finding good topsoil to use. Even if good topsoil is avail-

Vegetable seedlings growing in peat pots.

113

able, it must be heated or chemically treated to kill weed seeds, insects, and disease-causing organisms. Also, if the top soil comes from a cultivated field there is the possibility of herbicide residue which might be harmful to certain plants. For example a herbicide that is registered for and safe to use on corn might leave a residue harmful to tomato plants.

For the average home gardener wishing to start vegetable plants, the best growing medium is one of the soilless mixes available commercially; or you may purchase ingredients to make a soilless mix. Ingredients for one mix are listed below. Quantities shown will make up 2 bushels of growing medium.

Sphagnum peat moss	1 bu
Horticultural vermiculite	1 bu
Ground dolomitic limestone	1 lb
Superphosphate	4 oz
Calcium nitrate	2 oz
Calcium sulfate	7 oz
Fritted trace elements (FTE 503)	1 oz
Chelated iron (Sequestrene 330 Fe)	2 g

If you live where bark from Southern yellow pine is available, the small particle sizes (half inch or smaller in diameter) may be substituted for peat moss.

Unless you have ready access to the ingredients shown, it is simpler to purchase a ready mixed medium for plant growing. In calculating your needs for a ready mixed medium, bear in mind that 1 cubic foot of medium will fill approximately 275 peat pots 2¼ inches square, 60 four-inch round pots, and 20 packs measuring approximately 5 by 8 inches by 2¾ inches deep.

Media for seed germination should be somewhat fine textured, drain well, and be free of weed seeds and diseases. Most commercially available mixes are suitable for germinating seed. However, it is important not to contaminate the plant-growing medium with dirty tools, containers, and so on when using it. The best way to avoid problems with weeds and diseases is to keep all materials used from becoming contaminated.

In most plant-growing media the nutrients added at the time of mixing may not be adequate to grow the plant to proper transplanting size. If the plant foliage color becomes yellowish green, more nutrients are probably needed. This situation may be corrected by dissolving 1 ounce (2 level tablespoons) of a water soluble fertilizer such as 20-20-20 in a gallon of water and applying this as a regular watering at 7 to 10 day intervals. Wash off with plain water any fertilizer solution that remains on the plant foliage.

Sowing Seed

Before purchasing seed, determine the best varieties for the area, as well as the quantity you need. Usually the small packet size will provide ample seed for home gardener needs.

Most vegetable seed, except warm season cucurbits, should be sown in rows pressed into the growing medium with a board which makes a flat bottom trench about a half inch wide. Enough seed should be evenly distributed in the trench to obtain about 8 to 10 plants per inch of row. Depth of covering will vary, depending on seed size. Most vegetable seed will germinate properly if planted a quarter inch deep, provided proper temperature and moisture levels are maintained.

Before planting seed, water the medium thoroughly and allow it to drain overnight. Check the medium daily after planting and water lightly if it appears to be drying out. Take care to avoid overwatering, since seeds germinate poorly or not at all in water logged media. Water very gently all seedling containers and small plants.

As soon as seedlings emerge they should be grown at a somewhat lower temperature than that required for germination. Most warm season vine crops, eggplant, and peppers should be germinated at a temperature of 80° to 90° F, whereas most other vegetable seed will germinate properly at 60° to 80°. Plant growing temperatures should be 60° night and 70° to 75° day for warm season crops. Cool season crops and tomato plants may be grown with night temperature as low as 45° to 50°.

All vegetables discussed in this section except the cucurbits may be, and usually are, sown too thick to make a satisfactory transplant unless "spotted" out into another plant-growing container. Do this when the seedlings are about 1 inch tall and while still in the cotyledon stage—that is, before the first true leaves have developed appreciably.

To spot out, first fill with moist media whatever container is being used to grow plants in, or soak compressed peat pellets in warm water if these are to be used. The media should be moist but not waterlogged. With a round pointed object press a hole into the center of the pot as deep as the root system on the seedlings. Then carefully remove the seedlings, lifting them out with a flat wood label and gently separating them. Most of the medium they were growing in may fall off the roots. However, this should cause no problem so long as the seedling is immediately placed into the prepunched hole and the medium in the growing container pressed around the roots. Take care to avoid injuring the seedling's stem and roots. Immediately after spotting out, water the seedlings carefully.

With some transplants it is advisable to keep the plants in a shaded

"Spotting" out seedlings into individual containers.

area for a day before exposing them to full sun. However, no later than one day after transplanting, place the plants where they will receive maximum sunlight.

Larger seeded crops such as cucumber and muskmelon may be direct seeded into the container they will be grown in. Expanded peat pellets or peat moss pots filled with a growing medium are suitable for this. Press 2 or 3 seeds ¼ inch deep in the medium and cover with the same medium the plants are to grow in. Water, lightly with *warm* water and keep in a location where the daytime temperature is at least 75° F and night temperature does not go below 65°. After the seedlings emerge they may be thinned to 1 or two per pot by pinching off or very carefully pulling out excess plants. Seeds of these frost sensitive crops should not be planted more than 3 weeks before the average last frost-free date for your area.

Some general precautions to observe in planting seed include:
- Buy good seed of recommended varieties
- Plant at the proper rate and depth
- Cover seed with the same media in which seed are planted
- Do not use plain sand for germinating seed
- After an initial soaking and drain, water sparingly until seedlings become established
- Use room temperature water for all watering
- Observe safe dates for setting in the garden to determine seeding dates

Water Management

Watering properly requires practice. When the plants are small it is easy to over water. Do not have plant containers on a tray with water standing in it. If you use a tray, top water the plants until some water begins to run out the bottom of the container and stop. Do not water again until the plants show a need for water. Water of suitable quality for household use is satisfactory for watering plants.

When a soilless growing medium is used, do not be misled into thinking the plants need water if the surface of the medium looks dry. This will normally occur before the moisture level in the root zone is low enough to result in moisture stress for the plant. A better method to determine when water is needed is to squeeze a sample from the top half inch of medium between the thumb and finger. If water squeezes out easily there is adequate water. If the medium feels slightly moist but water is very difficult to squeeze out, additional watering is needed.

If the plants are where one can observe them several times daily, a good method is to water only when slight wilting occurs. When watering, apply enough to completely soak the medium to the bottom of the container.

Do not water plants with very cold water. Water temperature should be as near room temperature as possible. Applying very cold water to plants on a bright day may result in wilting, stunting of growth, and injury to leaves.

The time needed to grow a plant to suitable size for transplanting will vary with the type of vegetable and the season.

Cucurbits should be direct seeded, 2 seeds per pot in the container they will be grown in to transplanting size. Usually 3 weeks after seeding these crops will be ready to set in the garden.

The other vegetables discussed will require 5 to 7 weeks, with perhaps as long as 8 weeks from seeding to transplanting for pepper and eggplant.

The stage of development is more important than size of transplant. Tomato, pepper, and eggplant transplants should be set in the garden at late flower bud to early bloom stage. If fruits are present on the transplants, these should be removed. Otherwise development of the fruits will continue, resulting in a marked reduction in plant vigor and fruit production. Cucurbits should be transplanted when the first true leaf is 1½ to 2½ inches across.

Hardening Transplants

Some plant growers "harden off" vegetable transplants before setting them outdoors. You do this by one or more of the following: reducing growing temperature, withholding water, and increasing light intensity. Frequently this is done by placing the plants outside during favorable weather in the last 2 to 3 weeks before setting in the garden. Take the plants indoors if frost is expected. Water within a few hours after the plants start to wilt.

In no case should cucurbits be hardened. In most instances it is doubtful that hardening is needed for any properly grown vegetable transplant that is to be set in the home garden which has reasonably good protection from wind and blowing sand. It is much better to produce stocky, healthy, vigorous plants to set in the garden than to have tall, weak plants less able to withstand the rigors of growing in the garden.

Common Pests

One of the most common diseases the plant grower faces is damping off. This disease attacks germinating seeds and small seedlings, and may result in loss of an entire seed flat. The best way to combat damping off is to prevent the organisms causing it from being introduced to the plant-growing area. This may be done by using all new materials each year such as growing containers, media, etc. Also, take care in watering to prevent excessive wetness. Do not use any non-sterilized soil or items contaminated with soil. If you use a hose to water the plants, keep the end of it off the greenhouse floor.

Other bad news for transplants includes foliage diseases and soilborne wilt diseases that may have been introduced if you use contaminated soil. These are not likely to be a problem if you give proper attention to sanitation and selection of disease-resistant vegetable varieties.

The most common insects likely to

Fred S. Witte

Plants that have been outside "hardening" and are ready for transplanting in garden.

be encountered are aphids, white fly, and leaf miner. Often these are present on ornamental house plants, so one means of control is to disinfest house plants before starting the vegetable plants. Also, plants brought in from other greenhouses may be infested with insects.

Control begins with prevention to the greatest extent possible. If other plants in the area are infested with insects, either remove the plants or kill the insects with an approved material. Make sure that any plants brought in from another source are free of insects.

If an infestation occurs, determine what the insect pest is and use the proper insecticide at the correct rate for control. Observe all safety precautions with insecticides. Plants growing in the house should be taken outside for treating, weather permitting.

Insect control is also important for controlling certain diseases, especially virus diseases, which may affect tomatoes, peppers, and some cucurbits. Aphids often transmit these diseases from host plants such as weeds and some house plants to susceptible vegetable crops.

The Complex Art of Planting
by Charles W. Reynolds

Before you begin to plant the garden, plan carefully which vegetables to grow, how much of each, when to plant, and where to plant them in the garden area. Choose crops your family likes, those that can be expected to do well in your area, and those for which you have adequate space. Make a sketch on paper showing the location, the amount (row length), and time to plant each crop. Group the crops according to time of planting, growth habit, and time to maturity.

Here are some points to consider in planning the garden:

—Plant perennial crops (those which live for many years) along one side of the garden where they will interfere least with preparing the rest of the garden.

—Group early quick-maturing crops together, and plant tall-growing ones to one side where they will not shade shorter crops.

—Allow adequate space between rows for the type of cultivation you will use. Rows can be closer together for hand cultivation than for use of small tractors.

—Make repeated or succession plantings of crops like snap beans and sweet corn to provide a supply over a large part of the season.

—Keep the garden producing for the whole growing season. Follow early short season crops with others planted for midseason or fall use.

—If adequate space is available, make special plantings of selected crops for canning, freezing or other storage.

Good soil preparation is essential for gardening success. Preparation may include adding organic matter, liming to correct soil acidity, fertilization, plowing or spading, and smoothing the soil by disking or raking.

Adding organic matter improves the tilth of most soils. It makes them easier to manage, as well as improving drainage of clay soils, water retention of sandy soils, and aeration of the soil. If available, add 20 to 30 bushels of barnyard manure per 1,000 square feet (poultry manure at half this rate). Or add well-rotted compost prepared at home in a compost pile from leaves, grass clippings and waste plant material from almost any source. Compost started during the summer months will be ready for use the next year.

Green Manure

Green manure or cover crops of rye or ryegrass alone or mixed with a legume will protect the soil from erosion and add organic matter when turned under.

If garden soils are acid, lime may be needed. Soil acidity is expressed as pH. A pH of 7.0 is neutral. A pH lower than 7.0 indicates acidity, a higher one alkalinity. Most vegetable crops grow best with a pH of 6.0 to 6.5, slightly acid. If the pH is 5.0, for example, lime is needed to make the soil less acid.

Soils should be tested occasionally to determine the need for lime and for fertilizer. Such tests are available to gardeners through local Agricultural Extension Agents at little or no cost. Amounts of lime and fertilizer to use are suggested depending on results of the soil test.

If needed, broadcast lime evenly over the garden area at the recommended rate and mix well with the topsoil as you prepare the soil for planting. If the soil test shows a need for magnesium, add dolomitic lime

---

Charles W. Reynolds is Professor of Vegetable Crops, University of Maryland, College Park.

which contains magnesium as well as calcium.

Heavy soils—those having considerable clay—may be improved physically if you plow or spade the garden in the fall and leave it rough over the winter. Alternate freezing and thawing improves the tilth, making the soil easier to manage. However, if there is danger of erosion, such as on sloping land, this may not be a good practice.

Prepare garden soils by plowing or spading to a depth of at least 6 to 8 inches. This should mix into the topsoil any organic matter added, crop residues or cover crops that are present, and the lime and fertilizer applied. Do not work soils when they are wet, especially those with considerable clay. This causes damage to the tilth or physical structure of the soil which may last for a long time. Plow, spade, or cultivate only when the soil is dry enough to crumble easily.

Fertilizing

On most soils the yield and quality of vegetables will be improved by adding commercial fertilizer even if you make generous applications of manure or other organic matter. Commercial fertilizers for gardens usually contain nitrogen, phosphorus and potash. Kinds such as 5-10-5, 5-10-10, 10-10-10, and 10-6-4 are widely used. Ammonium nitrate, ammonium sulfate, or urea may be used when only nitrogen is needed such as for sidedressed applications.

In general, leafy vegetables need large amounts of nitrogen. Pod or fruit crops respond well to phosphorus, and the root crops require more potassium. Many vegetable crops respond to as much as 40 to 50 pounds per 1,000 square feet of N-P-K fertilizer such as the grades listed above. About half of this should be broadcast and mixed into the topsoil during soil preparation before planting.

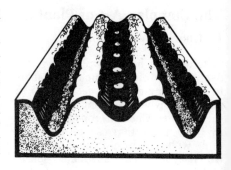

The remainder can then be applied in bands 2 or 3 inches to the side of the row at planting or as one or more sidedressings after the crop emerges.

Starter solutions are high-analysis water-soluble fertilizers mixed with water for use at transplanting. A small amount of this dilute solution around the roots of newly set plants provides readily available nutrients while the transplant is becoming established.

Use of starter solutions reduces the loss of plants following transplanting, promotes rapid early growth, and improves yields.

There are many excellent varieties or cultivars of most vegetable crops. Dozens of new ones become available each year, some of which are definite improvements over older kinds in yielding ability, quality, and disease resistance. Try some of the new ones especially those recommended by your local Agricultural Extension Service. But continue to depend upon those that have proven suitable for your conditions in the past.

Get the best seed you can locate. Cheap seed is hardly ever a bargain. Some kinds left over from previous years may have poor germination and low vigor. If you are not sure of old seed, discard it and obtain new seed.

Banding fertilizer. Place band 2 or inches to each side of seed and about or 2 inches deeper.

from a dependable source. Never save your own seed from hybrids; resulting plants will not be true to type.

Planting Seeds

Straight rows make the vegetable garden more attractive and easier to keep free of weeds. To have straight rows, tie a heavy string tightly between stakes at each end of the row. To band fertilizer beside the row, open furrows 2 or 3 inches on each side of the string, apply the right amount of fertilizer in the furrows, and cover with soil. Then make a shallow trench or furrow along the string with the hoe, hoe handle or other suitable tool. Make furrow at the right depth for the seed being planted. Sow seed uniformly and not too thickly by shaking out of seed packet or dropping with the fingers.

Most gardeners tend to plant too thickly. If such plantings are not thinned, plants grow tall and spindly and yield poorly.

Cover seed by raking soil into the furrow with the corner of the hoe or with a rake. Avoid pulling clods or lumps of soil into the furrow. Firm the soil over the seed furrow with the hoe or rake. Do not pack. Mark each row with a small garden stake showing what was planted.

Plant most vegetable seeds moderately shallow. Small seeded crops such as lettuce or carrots should be planted about a quarter inch deep. Larger seeds like beans or sweet corn should be about an inch deep, slightly deeper in dry weather. A good rule to follow is: plant seeds at a depth two to four times the width of the seed. Beets, Swiss chard and New Zealand spinach are exceptions; plant these shallow. The "seeds" as purchased are actually dry fruits containing several tiny true seed.

Distances between rows and between plants in the row vary widely among vegetable crops. Planting distances are related to size of plants at maturity and to the type of cultivation to be used. Rows may be closer together for hand cultivation, wider if small tractors are to be used.

Use of seed tapes gives accurate spacing of seed within the row with no need for thinning. These are water-soluble tapes with seed of a given crop enclosed within the tape at appropriate spacings. To use these, open a furrow at the right depth, unroll the tape in the bottom of the furrow,

William E. Carnahan

Spacing bean seeds evenly in furrow kept straight with tie line. Stake on either end of line can serve as marker to indicate what crop was planted and when.

and cover with soil. The tape disintegrates in the soil, leaving the seed accurately spaced. Seed cost is increased, but the hard work of thinning is avoided. For use in tapes, seed must have a very high germinability. Seed of several small seeded vegetables are available in seed tapes.

Planting in Hills

Seed of the vine crops—cucumbers, cantaloupes, pumpkins, squash and watermelons—are often planted in "hills". Several seeds are planted at each planting site with the sites a few feet apart. After the seedlings are up, they are thinned to two or three in each "hill". Mixing a shovelful of stable manure or compost in each hill improves growth and yield of these crops substantially. The soil at each hill may be mounded or raised somewhat or it may be left nearly level with the rest of the area.

Some vegetable crops such as broccoli, cabbage, eggplant, pepper, and tomatoes are started in hot beds or greenhouses and transplanted to the garden in order to provide earlier maturity. If plants are grown in individual pots or other containers there is little or no shock or injury from transplanting. If plants are not in individual containers, transplanting will cause less injury if plants are removed from the plant bed with a ball of soil around the roots.

Dig plants from plant bed and plant in the garden immediately. Stretch string tightly along row in well prepared soil. Open holes at proper distances with a trowel, bulb planter, or mattock. Set the transplants a little deeper than they were in plant bed. Fill soil around roots, firming slightly.

Add ½ to 1 cup of water or starter solution around each plant to moisten

A child's garden is not always of verses. Here, two young ladies lend hand at transplanting time in family garden.

Planting Chart for Vegetables

Crop	Depth to plant (inches)	Planting distances Between rows (inches)	In the row (inches)
		Cool Season Crops	
Asparagus (crowns)	6–8	36–60	12–18
Beets	¼–½	15–24	2–3
Broccoli	¼–½	24–36	12–18
Brussels sprouts	¼–½	24–36	18–24
Cabbage	¼–½	24–36	12–18
Cabbage, Chinese	¼–½	18–30	8–12
Carrots	¼–½	15–30	2–3
Cauliflower	¼–½	24–36	18–24
Celery	⅛	18–36	4–6
Chard, Swiss	¼–½	18–36	6–8
Chives	½	15–24	6–8
Collards	¼–½	24–36	18–24
Cress, upland	¼–½	15–30	2–3
Endive	¼–½	18–36	12
Garlic (cloves)	1½	18–24	3
Kale	¼–½	18–36	8–12
Kohlrabi	¼–½	18–36	4–6
Leeks	½	12–30	2–3
Lettuce, heading	¼	18–30	12
Lettuce, leaf	¼	12–18	4–6
Mustard	¼–½	18–24	3–4
Onions, plants		15–24	3–4
Onions, seed	½	15–24	3–4
Onions, sets	1–2	15–24	3–4
Parsley	¼	15–24	6–8
Parsnips	½	18–30	3–4
Peas	1–2	8–24	1
Potatoes	4	30–36	12
Radishes	½	12–24	1
Rhubarb, crowns		36–48	36–48
Rutabagas	¼–½	18–30	3–4
Spinach	½	12–24	2–4
Turnips	¼–½	18–30	2–3
		Warm Season Crops	
Beans, lima	1–1½	24–36	3–4
Beans, snap	1–1½	24–36	1–2
Cantaloupes	1	48–72	24–30
Cucumbers	1	48–60	12–18
Eggplant	¼	30–42	18–24
Okra	1	36–48	12–18
Peas, southern	1	24–36	4
Peppers	¼	30–42	18–24
Pumpkins	1	60–96	36–48
Spinach, New Zealand	½–1	30–42	15–18
Squash, summer	1–1½	48–60	18–24
Squash, winter	1–1½	60–96	36–48
Sweet corn	1–2	30–36	10–12
Sweet potatoes		30–36	12–15
Tomatoes	¼	36–60	18–24
Watermelons	1–1½	60–96	36–60

and settle soil around roots. After the water soaks in, rake dry soil around plants to level and cover wet area. Some gardeners may prefer to form a slight mound of soil in a ring around new transplants and add water after transplanting.

A paper cylinder wrapped around the stem extending from just below the soil to about two inches above will discourage cutworms.

Cucumbers, squash, and cantaloupes will mature earlier if grown in peat pots or other individual containers and transplanted with an intact ball of soil around the roots. They are easily injured and may not survive transplanting as bare rooted plants, however.

Set plants in late afternoon or on a cloudy day. Plants set during the heat of the day will wilt badly. Shading for a few days during hot weather helps the plant become established. To prevent cold injury to transplants in early spring, protect plants with hotcaps, plastic row covers, baskets, or other available material. Covers should be removed or opened for ventilation on bright, sunny days.

Windbreaks may improve survival

Move transplant to garden with soil ball and roots intact. After planting, cover soil ball with light layer of garden soil, forming mound around plant to hold water. Water thoroughly so moisture penetrates roots and mounded soil.

of transplants in cold weather of early spring or in summer heat. Unplowed strips of a small grain like rye give protection from cold winds and blowing sand in the spring. Short branches of trees or privet hedge stuck into the soil near rows of midsummer transplants provide shade and protection from hot drying winds.

Time of Planting

Vegetables may be divided into two large groups—warm season and cool season crops.

Warm season crops require warm weather for germination, growth, and development. They are injured or killed by freezing temperatures. They should not be planted outdoors in the spring without protection until the danger of frost is past. Those planted in summer for fall maturity should be planted early enough to mature before killing frosts in the fall.

William E Carnahan

Cool season crops grow best in relatively cool weather and may perform poorly in summer heat. They tolerate temperatures below freezing if properly hardened. They may be injured or killed, of course, by hard freezes. Cool season crops may be safely planted outdoors two to four weeks before the expected date of the last frost in spring. They continue to grow well past the earliest frost in fall, but should be started early enough to mature before hard freezes are expected.

Good gardeners plan, plant, and manage the garden to provide fresh vegetables over the whole growing season. When early crops are harvested, prepare the soil again and plant others to mature in summer or fall. Make several plantings of bush beans or sweet corn to provide a fresh supply over much of the summer. Plant broccoli, radish, turnips, kohlrabi, spinach, and other cool season crops not only in early spring but also in late summer for fall use.

Mulching

Mulches help to control weeds, regulate soil temperature, conserve moisture, and reduce soil and disease injury to fruiting vegetables such as tomatoes. Organic mulches include straw, grass clippings, wood chips and shavings, spoiled hay, etc. Black plastic, aluminum foil, and heavy Kraft paper are newer materials.

Do not add organic mulches until the soil has warmed up well and has been cultivated to control weeds, and the vegetable seeds have germinated and made several inches of growth. Earlier application keeps soil cooler and delays crop maturity.

Spread straw, hay, and leaves 3 to 4 inches deep around plants and between rows. Spread sawdust and wood chips no more than 2 inches deep.

For fall crops, organic mulches may be applied soon after planting because the soil is warm.

Organic mulching materials require nitrogen for decomposition and will compete with the crop for the available soil nitrogen. To insure sufficient nitrogen for crop growth, add a moderate amount of a complete fertilizer or a source of nitrogen only, such as ammonium nitrate, when the mulch is applied.

Black plastic mulch is unrolled over the prepared and fertilized rows in early spring. Edges of the material are fastened down, usually by covering with soil, and seed or transplants are planted through holes cut in the material. Plastic mulch tends to increase the soil temperature and results in earlier maturity of spring planted crops. It has been very beneficial when used with tomatoes, eggplants, cucumbers, cantaloupes, and summer squash.

Aluminum foil, or black plastic with a thin coating of aluminum foil, repels some kinds of insects such as aphids. By repelling aphids, it reduces damage from diseases they spread.

Mulches need not be exclusively organic. Here, black plastic is used with tomatoes. After plastic sheeting is in place, holes are made for inserting transplants. Plastic provides warmth, retains moisture, and thwarts weeds.

Vegetables in Containers Require Enough Sun, Space, Drainage

by Kathryn L. Arthurs

Growing vegetables in containers can be fun as well as challenging, and for those of us with little or no ground space available it provides a good alternative. All you need to grow container vegetables is enough sun and adequate space for a good-sized container.

Most types of vegetables lend themselves to container gardening. All you have to do is find the varieties that have been hybridized for container growth, or those that can be adapted to confined quarters. Some crops like corn that produce large root systems will need a very big container. Other plants with indeterminate growth habits—such as pole beans, cucumbers, and tomatoes—require a support system. Still other vegetables will grow in average-size pots or in hanging baskets.

Container-grown vegetables make few demands when it comes to location. Absolute requirements are 5 hours or more of full sun, enough space to set the container, and adequate air circulation. A nearby water supply equipped with a hose and a soaker/sprayer nozzle attachment is a real convenience, but not essential.

Once these demands are met, you can place containers anywhere—on a patio or deck, terrace, balcony, window box, garage roof, walkway. If you have no available ground space, consider growing vegetables in hanging baskets.

Drainage can be a problem in container gardening. With smaller containers, wherever possible use drip saucers to catch excess water. A large container without a saucer that sits directly on a solid surface (a cement or brick patio, for instance) may benefit from being elevated slightly. If the container stays in contact with a solid surface, water can accumulate, causing root rot as well as possibly staining the patio surface. You can use short lengths of wood to raise the pots one or two inches off the patio.

Types of Containers

Large containers are the best for growing vegetables. As long as the plants have ample root space, you can introduce most vegetables that normally grow in the ground.

For growing vegetables, a minimum-size container is a 6-inch diameter pot with a soil depth of 8 inches. This size can sustain lettuce, herbs, peppers, radishes, and other shallow-rooted vegetables. Root crops—such as beets, carrots, radishes, and turnips —need depth and enough surface space to fill out to their mature size. Thinning these crops will be essential.

Each vegetable determines the best size and style container it needs for an adequate harvest. Very large containers are required for regular-size tomatoes, for squash, pole beans, cucumbers, and corn. Half barrels, wooden tubs, or large pressed paper containers work well.

Adequate drainage is another requirement for growing vegetables in containers. Most commercial containers come with drainage holes, but you may find these insufficient. Since most vegetables in containers need daily watering, and fast draining is crucial, consider increasing the size or number of the drainage holes or slots. Wooden containers can have new drainage holes drilled. Existing holes in clay and ceramic pots can

---

Kathryn L. Arthurs is a garden writer/editor for a Western publishing company. She lives in Palo Alto, Calif.

be enlarged by carefully chipping away the edges, or additional ones may be drilled with a masonry bit.

If a container lacks drainage holes, you can provide a drainage layer of rocks, pebbles, or pot shards to hold any excess water until it can be used or evaporated. Since vegetables need daily watering (depending on the individual plant and your climate), the drainage layer should fill from a quarter to a third of the total container volume. Keep in mind that this drainage layer won't guarantee success; containers with ample drainage holes are best for growing vegetables.

There are many types of containers you can purchase or make yourself that can be used for growing vegetables: red clay pots, wooden containers, pressed paper pots, plastic pots, and raised beds. Each type has advantages and drawbacks. Study your individual needs carefully, then select the containers that best meet them.

The container gardener's stand-by, red clay pots, have much to recommend them. They are readily available in a wide range of sizes and shapes; they are porous, allowing excess moisture to evaporate through their sides; they "weather" well; and their weight keeps them from being top-heavy. They are attractive and blend into most garden or patio designs.

On the minus side, clay pots are breakable and expensive. Filled with damp soil, large pots will be heavy and difficult to move. Their porosity lets the potting mix dry out quickly, requiring more frequent watering.

Tubs, half barrels (originally used to age wine or whiskey), rectangular or square boxes, and hanging slatted baskets all come in wood. Redwood is probably the most commonly used type of wood, with cedar a close second. Both woods resist damage by termites and "weather" well. Wood, like clay, is porous.

Wooden containers are available in a wide variety of sizes and shapes and are relatively inexpensive. Some wooden containers will deteriorate; those that are reinforced with metal bands are sturdier than containers held together with nails or glue. Like clay pots, half barrels and large tubs will weigh a lot when planted. Check plants in wooden containers daily for water needs.

Pressed paper pots, a recent innovation in plant containers, come in many sizes, are inexpensive and lightweight. Their weight can be a disadvantage if wind is a problem or if the

Stair-step benches hold containers filled with vegetables. Sunny brick patio is ideal location.

vegetables grown in them are top-heavy.

Plastic is a common material used in smaller containers and hangers. Plastic pots are lightweight, inexpensive and non-porous. Most plastic containers come in green or white, colors that can be visually jarring in a garden. They are breakable.

Since plastic pots are non-porous, moisture is retained in the potting mix. This will be a problem only if drainage holes are inadequate or you tend to overwater. Plastic hanging baskets make good choices because of their weight and water retention. Hanging plants dry out more quickly than other container plants and need a firm support to hang from.

A raised bed lacks one of the basic qualifications for container gardening: it is stationary. It does, however, restrict the growing area and provide for good drainage.

A Good Potting Mix

When you garden in containers, you want a potting mix that is fast draining, yet provides enough water retention to keep the soil evenly moist in the root zone area. A mix that drains too fast won't provide enough moisture, and one that holds too much moisture may cause the roots to rot.

Most home gardeners who grow vegetables in containers find a "soilless" commercial potting mix works well. These mixes are easy to use, lightweight, fast-draining, and free from soil-borne diseases and weed seeds. Since they come in varying-sized bags, you can buy as much as you need at the time. The unused portion can be stored in its bag until you want to use it again.

If you choose to make your own mix, a good potting soil for containers consists of equal parts sharp sand (be sure to buy washed sand), good garden soil, and organic material (peat moss, leaf mold, fir bark, or sawdust). To be sure your homemade mix is free from disease and weeds, heat it in a low temperature oven for about 1½ to 2 hours. This should kill any bacteria, pests, or weed seeds present.

Other good soilless potting mixes specially formulated for container gardening are the University of California mix and the Cornell mix. Information on each mix can be obtained by writing the University of California, Division of Agricultural Sciences, Berkeley, Calif. 94720, or to Cornell University, Department of Floriculture, Ithaca, N. Y. 14853.

Some commercial mixes are extremely lightweight. These are excellent to use in hanging baskets, in very large containers that you want to move around, or where sheer weight could be a problem, such as on a balcony or in a window box. An ultra-lightweight mix also has some disadvantages. If wind is a problem in your area, top-heavy containers may topple over. Top-heavy plants, such as corn, tomatoes, and eggplants, may not get enough soil support for their root systems.

If you find your commercial potting mix isn't absorbing water (the water runs through the container rapidly and many particles float on the surface without absorbing any moisture), try using a few drops of liquid detergent in the water. The detergent acts as a wetting agent. Or you can use a commercial wetting solution. Once these stubborn mixes begin to soak up water, your problem should be solved.

A soilless commercial mix contains few if any nutrients. Vegetables grown in these mixes will need regular fertilizing with a complete fertilizer formula.

Container vegetables have needs that differ from vegetables grown in the ground. Fertilizing, watering, general maintenance, and harvesting demand close daily attention and are crucial to the plant's well-being. Vegetables in containers are at the gardener's mercy.

Planting Techniques

Most vegetables grow as well from seed as from transplanted seedlings. However, if your containers will be conspicuous (on a balcony, patio, or in a window box), planting seedlings will give you an instant display. Some vegetables, such as tomatoes, peppers, eggplants, and squash, may be difficult to grow from seed. Using seedlings will speed up their growing process.

If you plant seeds in larger containers, you can still have attractive pots while the seeds are sprouting. Plant annual or herb seedlings as a border.

Limiting the number of plants to each container is very important. Estimate the number of plants a container can sustain. Measure root crops by the space they'll occupy when fully matured. Bush squash and vine crops such as melons and cucumbers should each have a good-sized space. Corn needs cross-pollination, so plant several stalks to each container.

Beans and tomatoes with indeterminate growth habits will need supports. Beans can climb up poles or a trellis. Tomatoes can be staked or enclosed in a wire cage.

Other vegetables can be grown singly or in groups, depending on the container size and eventual size of the plant at maturity.

To plant vegetables by seed, fill the container to within 1 inch of the rim with damp potting mix, then sow seeds according to their package directions. Be sure to plant more than you want, since it's unlikely you'll get 100 per cent germination. When the seeds have sprouted and each seedling has mature leaves, thin the plants to the desired number.

To thin seedlings in a container, cut off the seedling's stem at the soil level with scissors, a knife, razor blade, or pruning shears. Pulling unwanted seedlings out may disturb or destroy surrounding root systems.

To plant vegetable seedlings, prepare the container as before. Remove the seedling carefully from its pot. (Seedlings grown in peat pots can be planted directly, pot and all. Break off the upper rim so the soil level is uniform.)

If the roots are tangled or potbound, loosen them with your fingers. Dig a small hole in the potting mix and plant the seedling. Try not to bury the plant stem or change the soil level. Tomato seedlings are an exception; you can bury tomato plants up to half the stem length as long as there are at least 2 sets of mature leaves above the soil.

To help transplanted seedlings establish themselves, use a transplant starter solution. Follow label directions.

Vining plants and vegetables, such as tomatoes, may need to be staked or trellised. Any support structure must be sturdy. Stakes, poles, and trellises should be set in place when the seedlings are little to avoid disturbing their root systems (wire cages used for tomatoes should be set up at this time too). Some vining plants, like pole beans, will attach themselves to the support. Others, like tomatoes or cucumbers, need to be tied. Use twine or plastic tape for tying; be sure not to tie stems too tightly or cut the stem. The most stable support systems are those attached to the container itself.

Watering

Watering is probably the most critical task a container gardener performs. More plants grown in containers fail from improper watering than from any other single cause. Plants given too much water may develop root rot. Vegetables that receive too little water may wilt and die.

Improper watering can also cause blossoms to drop.

Ideally, potting mix in a container should be evenly moist throughout—not waterlogged. Plants need ample moisture to prevent "water stress."

Many gardeners water containers in the morning, adding water until it comes out the drainage holes. This method is recommended only if your potting mix is fast-draining and the container has adequate drainage holes.

With watering in the morning, foliage should be dry by evening, helping prevent diseases. If you live in a hot, dry climate, check your containers again in the early afternoon. Vegetables in containers will dry out faster than those in the ground.

The best way to water container plants is by hand—either with a hose that has a sprayer attachment or a watering can. More inventive gardeners may want to try automatic watering systems, but these can be costly.

A few words of warning: Hoses without a sprayer/mister nozzle can disperse water with enough force to create holes in the potting mix. This can damage root systems. Also, if your hose sits in the sun, let enough water run through it until the water is cool or lukewarm. Hot water isn't good for plant roots.

Mulching and Fertilizing

Mulching, especially in larger containers, can help keep moisture in the soil longer. You can use any of the organic mulches, such as wood chips, compost, or sawdust, very effectively. Plastic mulches will work, but they aren't too attractive.

Vegetables grown in containers are trapped. Once they use the nutrients available in the potting mix, the root systems have nowhere else to go. Frequent and regular fertilizing is the answer.

The container gardener will find many kinds of complete fertilizers specially formulated for use on vegetables. Common N-P-K breakdowns are 18-20-16, 18-12-10, or 10-10-10. Fish emulsion is also commonly used.

These fertilizers can be applied in a liquid solution in conjunction with watering, scratched or dug in dry form into the soil surface, or, in the case of timed-release fertilizers, sprinkled on the soil surface. Whichever type you choose, follow the label directions carefully.

Since containers with vegetables should be watered daily, nutrients can leach out of the soil rapidly. Consider applying fertilizer at half strength twice as often; this should assure your vegetables consistent fertilizer.

Timed-release fertilizers are also a good solution. Their capsules are constructed to release a tiny amount of fertilizer each time the vegetable is watered, and you only apply this type once a season.

Container gardeners are fortunate since each vegetable is isolated by its pot, and no one crop is concentrated, lessening the chance of a pest infestation. Unfortunately, pests can still present a problem. Most insects, such as whiteflies and aphids, can be discouraged with blasts of water. Tomato hornworms can be hand picked. Snails and slugs can be baited with a chemical.

If pest damage becomes intolerable or your crop is being damaged, use a spray formulated to kill the damaging pest. Be sure any chemical sprays you choose are recommended for use on vegetables.

Choosing Vegetables

Vegetables that grow best in containers share certain characteristics. They will grow in confined spaces, usually have determinate growth habits, need a minimum of added support, and produce a large enough crop yield to make your efforts worthwhile.

Some vegetables, such as asparagus and corn, have such large root systems that trying to grow them in containers—if you can locate pots large enough—is very difficult. A low crop yield per plant, again asparagus and corn are good examples, is another deterrent to container culture.

Listed here are the vegetables, and specialized varieties of more difficult vegetables, that are recommended as best adapted to life in containers. Most have compact growth habits and relatively high crop yields. Some have been specifically hybridized for container growth.

Other varieties of recommended vegetables can also be adapted to containers, but because of their size or growth habits will require more work and attention. With the popularity of growing vegetables in containers on the upswing, seed hybridizers should continue to find new, more adaptable vegetable varieties.

Artichoke. "Green Globe" is a consistent producer. Use very large containers.

Beans. Use bush forms in containers for best results. Pole varieties need supports (poles in teepee shape or trellises will work); plants may also be topheavy. Snap beans to try are "Green Crop", "Tender Crop", "Bush Romano", "Bush Blue Lake", "Royalty" (purple pod). Lima bean varieties include "Henderson Bush" and "Jackson Wonder Bush".

Beets. These root crops will need at least 10 to 12 inches of soil depth and about a 3- to 4-inch space between each plant. Two good varieties are "Little Egypt" and "Early Red Ball".

Brussels sprouts. This cool weather crop needs a large container, but produces a heavy yield per plant. Two compact varieties are "Jade Cross" and "Long Island Improved".

Cabbage. Regular varieties aren't recommended for containers. You can try dwarf varieties such as "Dwarf Morden" and "Earliana". Chinese cabbage is a good container crop; plant "Michihli" or "Burpee Hybrid".

Carrots. Be sure to use containers with enough depth (at least 12 inches up to 20) for root formation and a very light mix for good growth. Any variety will grow in containers. Some of the shorter varieties are fine: try "Danvers Half Long", "Little Finger", "Short & Sweet", and "Tiny Sweet".

Chard. Any variety will grow in a large container, at least 2 feet deep. Use a potting mix with enough support for a large root system.

Collards. If you harvest the outer leaves consistently, you can have a continuous supply of greens. "Vates" is a compact variety.

Corn. Because of its size, low crop yield per plant, and need for cross pollination, corn isn't a good container vegetable. If you still want to try it, plant dwarf or midget varieties. Plant at least three stalks per container. Some varieties to consider are "Golden Midget", "Golden Cross Bantam", "Midget Hybrid", and "Fireside Popcorn".

Cucumber. Cucumbers need a large container, and some can adapt to a trellis support. The varieties that form small vines or are bushlike are best: "Little Minnie", "Tiny Dill", "Spartan Dawn", and "Cherokee 7". "Patio Pik" and "Pot Luck" were developed for containers and can be used in hanging baskets.

Eggplant. Eggplant needs a large container to grow well. Since warm soil is required for good growth, plant seedlings. Any variety will grow in containers; smaller varieties are "Morden Midget" and "Slim Jim."

Herbs. All of the herbs can be grown in containers.

Endive. Plant any variety in early spring; reseed containers again in August for a fall crop.

Kale. Plant any variety in a large container. Harvest outer leaves to extend the crop.

131

Kohlrabi. This vegetable's unusual appearance makes it a conversation piece in a container. Any variety is fast-growing.

Lettuce. Because lettuce is a cool weather crop, being able to move it to a shaded or protected spot is a plus. Try growing leaf lettuce in containers; harvest the outer leaves for a continuous harvest. Any variety can be container-grown.

Melons. Because of the plant size and low yield, growing melons in containers is impractical. You can try some of the midget varieties in large containers. A midget cantaloupe is "Minnesota Midget". "Yellow Lollipop", "Red Lollipop", and "Little Midget" are small-size watermelons.

Mustard Greens. Extend the harvest by picking outer leaves. Mustard greens are a good container crop.

Okra. Plant this Southern favorite in a large container. Plants have a high crop yield. Try "Dwarf Green Long Pod", "Clemson Spineless", and "Red River".

Onions. While most onions can be grown in containers, the larger types make unattractive displays. Chives and green bunching onions (scallions) are good pot plants.

Peas. Peas grown in containers demand a lot of attention, need large containers, and produce a small yield for your time and effort. If you want the challenge, you can try "Little Marvel", "Green Arrow", "Dwarf Gray Sugar", and "Mighty Midget"

Rhubarb. In larger containers, any variety will do well and make an attractive display. Move pots to a garage or sheltered area during a freeze.

Radishes. All varieties make excellent container plants. You can use them as borders in large containers.

Spinach. Another cool season crop, spinach can be grown in boxes or large containers. New Zealand spinach (not a true spinach) grows well in pots and recovers rapidly from cutting.

Squash. Not a good container crop because of its size, but you can attempt it if you use very large pots and plant bush varieties. One new hybrid, "Scallopini", forms a compact bush plant.

Tomatoes. Many varieties have been hybridized especially for containers. Use medium to large containers since most tomatoes need some support. Use stakes or a wire cage as a support; be sure the wire squares are large enough to allow for harvesting.

Some tomato varieties to try are "Tiny Tim", "Small Fry", "Patio Hybrid", "Sugar Lump", "Tumbling Tom" (recommended for hanging baskets), "Stakeless", "Burpee's Pixie", "Salad Top", "Sweet 100", and "Toy Boy".

Harvesting

Since container gardening is a small-scale operation, most crops will be harvested for a specific meal. This allows you to pick them just before meal preparation begins so they will be at their freshest.

Pick leafy crops carefully, such as chard, lettuce, or collards. Remove only the outer leaves to keep the plant producing. Root crops, such as radishes or carrots, should be pulled out without disturbing their neighbors. Crops that have fruit ripening continuously, like tomatoes and beans, should be picked so as not to ruin or destroy future fruit.

Try not to pick more of a crop than you can use. If you harvest too much, keep your vegetables in the crisper section of a refrigerator. It is unlikely that a container crop would produce enough to make canning or freezing worthwhile.

Play It Cool With Cole Crops (Cabbage, Etc.); They Attain Best Quality If Matured in Fall

by Philip A. Minges

The closely related vegetables commonly referred to as cole crops include cabbage, cauliflower, brussels sprouts, broccoli, and kohlrabi. Being frost-tolerant, they are valuable for extending the harvest season for gardens after frost or cool weather has eliminated the popular warm season vegetables.

In fact, this group of vegetables develops best quality and remains edible longer in the garden when matured during the moderately cool weather and shorter days of fall. When maturing in hot weather, the harvest period is relatively short, quality often is less desirable, and yields are likely to be lower.

The crops are adapted to all sections of the United States provided proper planting dates are selected. In areas with short growing seasons, spring plantings for summer and early fall harvest work well. In intermediate areas, early spring plantings for summer harvest are possible while summer plantings for fall harvest are ideal. In areas with mild winters, late summer or early fall plantings for fall and winter harvest are common.

For broccoli, cabbage and cauliflower, planting two or more varieties of differing maturities—for example, a fast and a slow maturing one—can easily extend the harvest season from a single planting from a week or so to a month or more.

Local information on varieties and preferred planting dates should be sought by gardeners. Poor selection of varieties and/or planting dates may lead to poor results due to premature seeding caused by undue exposure to cool temperatures early in the growing season and other problems.

As a group these vegetables rank fairly high in nutritional value and are quite adaptable for use fresh or cooked and for preserving. Cauliflower is suitable for freezing, canning or pickling. Broccoli and brussels sprouts are excellent for freezing. Cabbage when kept cool and moist will store for several weeks after harvest and, of course, it can be preserved as sauerkraut.

Cole crops can be grown on a wide range of soils. Fertile, deep, well-drained, sandy and silt loams are the most desirable. Have a soil test made; your county Extension office can tell you how to have this done. Good drainage is particularly important where the garden will be continued into or through the winter months.

A soil pH level of around 6.5 or slightly above is desirable for efficient use of fertilizer and soil nutrients, and for reducing development of a soil-borne disease called clubroot. If the soil is acid as indicated by a pH reading below 6.0, apply lime before preparing the garden. In humid areas with acid soils, apply it at a rate of 10 pounds per 100 square feet every 3 to 5 years. This should maintain a good soil pH once the pH has been brought up to the desirable level.

In some parts of the Eastern United States and generally in the Western portions, soils tend to be alkaline and therefore don't require lime. In many desert areas a problem may be excessively high pH and/or a high salt content. Avoid highly saline soils if possible, or correct them by leaching

---

Philip A. Minges is Professor of Vegetable Crops, Cornell University, Ithaca, N.Y.

—perhaps in combination with the addition of sulfur. There are a few exceptions, but minor element problems seldom are serious when the pH level is in the range of 6.0 to 7.5.

Fertile soils may supply sufficient amounts of nitrogen, phosphorus and potassium for cole crop needs, making fertilizer applications unnecessary. In the Western States nitrogen is often the only limiting nutrient, though phosphorus applications may benefit late fall and winter plantings.

In the humid Eastern States a complete fertilizer usually is advisable. On new garden sites or where fertility is known to be low, 3 to 4 pounds per 100 square feet of a 5-10-10 fertilizer is advised. On more fertile soils 1 to 1½ pounds may be adequate. A sidedressing with nitrogen at a rate of about a half pound of ammonium nitrate per 20 feet of row, applied 2 to 4 weeks after planting, may be beneficial on sandy soils, when heavy rains occurred shortly after planting, or during relatively cool weather.

Planting and Culture

Cole crops can be established in the garden either by setting transplants or by seeding. Using transplants is the rule for cabbage, broccoli, cauliflower and brussels sprouts in many areas while kohlrabi normally is seeded. Using transplants saves 2 to 3 weeks growing time in the garden. But for fall crops of cabbage, broccoli and cauliflower, direct seeding with later thinning is very satisfactory. To obtain similar maturity, seeding should be done about 2 weeks earlier than setting transplants.

A desirable transplant is 4 to 5 weeks old with 3 to 5 true leaves, and is free of diseases. Older transplants sometimes will head a bit earlier but otherwise there is little advantage; in fact, head sizes often average considerably smaller when old, large plants are set out. You can buy transplants at local garden supply outlets, or grow them at home. In direct seedings, the plants being thinned can serve as transplants if carefully dug. When reset immediately, they will reach harvest stage some 10 to 14 days later than the undisturbed plants —thus spreading the harvest period from a single seeding.

Transplants to be used for early spring planting should be grown at temperatures no lower than 60° F to lessen the hazard of premature flower stalk formation.

Either flat or bed culture is suitable. Beds have an advantage when soil drainage is likely to be slow during the late fall and winter months, and where furrow irrigation is practiced. In the Western States, beds about 6 to 8 inches high and 30 to 42 inches center to center are common. On the wider beds, 2 rows 12 to 14 inches apart can be used for most of these crops except brussels sprouts, for which 1 row is usual. Fertilizer can be broadcast before forming the beds or banded later along the sides. Forming the beds a few days ahead of planting to allow the soil to settle a bit is a good practice when time permits.

In transplanting, dig holes deep enough to accommodate the root system (or soil block), place the plant in the hole, and then firm some soil around the roots or soil block. Often it helps to apply a cup of starter solution containing some soluble fertilizer, and in areas where root maggot is troublesome a chemical for control of this pest.

After the water has drained down, pile more soil around the plant so the stem is covered a bit higher than it was in the plant-growing container.

In direct seeding, open shallow furrows and drop seeds about 1 inch apart. Partially cover the seed with soil and firm it over the seed, then finish covering with loose soil to a total of ½ to 1 inch in depth. If the soil is dry and rains are not expected, water immediately. In hot, dry weather, you may need to sprinkle or otherwise apply water a few times to insure adequate germination. When the plants show 2 to 3 true leaves, thin them to the desired spacings.

A row spacing of 30 to 36 inches is suitable. Slightly closer rows will do when space is limited, especially for kohlrabi. Common spacings in the row are 15 to 24 inches for cabbage, cauliflower, and broccoli; 30 to 36 inches for brussels sprouts; and 4 to 6 inches for kohlrabi.

You can control weeds easiest when they are small, using a hoe or small hand cultivator. Approved herbicides can be applied before planting and worked into the soil when rather large plantings are contemplated. Consult the label for rates and other instructions. Weedy gardens are never as productive as those kept weed-free.

All cole crops develop most rapidly and with best quality when adequate soil moisture is provided throughout the season. In humid areas, irrigation during dry periods can be helpful. In areas with normally dry summers, irrigation is essential until fall and winter rains begin. Light irrigation should start at transplanting or seeding and the amount gradually increased as the plants grow until about 1 to 1½ inches of water is applied weekly.

Pest Control

As a rule, fungicides and insecticides are essential for controlling certain diseases and insect pests on cole crops. For some diseases other precautions are desirable, such as use of disease-free seed or transplants and resistant varieties.

Insect pests tend to be localized but flea beetles, aphids and cabbage worms are common over the country. Flea beetles are most damaging on

young plants, so control for these comes early. Cabbage maggots which feed on the roots, causing plants to wilt prematurely during hot weather, also must be controlled at planting time. Aphids and worms usually cause trouble later in the season. Slugs or snails are a bothersome pest in many areas.

Most fungicides are non-toxic to humans and animals, and there are a number of fairly safe insecticides that give reasonable control of insect pests.

Cole crops generally are biennials or have similar tendencies. What this means is the plants can be induced to initiate seedstalks and flowers by exposure to cool temperatures over a period of several days. If that occurs early in the growing season, the crops may fail to produce a usable product.

The threshold temperatures appear to be in the range of 50° to 55° F, and the plants increase in sensitivity to the cold as they increase in size and age. Thus in areas with mild winters, plants set out too late in the fall or too early in the spring may go to

To produce high quality cabbage the gardener must control weeds, insects, and diseases.

seed prematurely. In the case of broccoli and cauliflower, crops where the developing flower stalk is the edible product, the result is a very small curd or head. The use of slow-bolting varieties coupled with proper planting dates will avoid these problems, unless the early part of the growing season is cooler than normal.

Varieties

Hybrid varieties (F_1) are available for cabbage, broccoli, cauliflower and brussels sprouts. Usually they are more uniform in plant size, maturity and size of head than standard varieties and often the total yield is greater. The uniformity of maturity may not be an advantage for the home gardener, except when the crop is grown primarily for processing.

Varieties differ in many characteristics including color, days to maturity, uniformity, yields, ease of bolting, product size, length of time the product remains usable, and resistance to diseases and disorders. Generally there are not as great differences in flavor or eating quality as among varieties of other vegetables. And most varieties in this grouping are rather widely adapted, the exceptions being the winter types of cauliflower and late varieties of broccoli which are limited to areas of mild winters.

Broccoli—The green buds and flower stems are the edible portion. The center shoot is large, ranging from 5 to 10 inches or more across depending on the variety, growing conditions and other factors. It should be harvested before the buds begin to separate or start to show yellow color. The heads remain in edible condition fairly long in cool weather, but pass prime maturity very quickly when it is hot. After the center head is cut, smaller side shoots develop which can extend the harvest season up to a month or more.

Broccoli is of best quality if consumed soon after harvest, though it

Center head of broccoli with side shoots. Shoots can prolong harvest season month or more after head is cut.

will keep a few days under high humidity and low temperatures (near 32° F).

Spartan Early, Coastal, Italian Green Sprouting, Early One and De-Cicco are among the fast growing standard varieties that are suitable for the Central and Eastern portions of the United States. Waltham 29 is a popular late one. Medium strains of Green Sprouting, Topper and Pacifica are popular on the West Coast. Early hybrids (F_1) include Green Comet, Gem, Bravo and Premium Crop. From 50 to 85 days are required from transplanting to harvest, or 65 to 110 days when seeded in the garden. Late summer and early spring plantings require more time than plantings in periods of longer days and higher temperatures.

Cabbage—Types of varieties vary in color from green with smooth or

savoyed leaves to reddish purple, and in shape from flat to pointed. The intermediate round types are the most common.

There are numerous acceptable varieties. Some Fusarium-resistant standard varieties are Resistant Golden Acre, Resistant Wakefield, Greenback, and Resistant Danish (in order of maturity from early to late). Other popular standard varieties include

Early Marvel, Golden Acre, Copenhagen Market, Early Round Dutch (slower bolting), Red Acre, Danish Ballhead and Chieftain Savoy.

Some Fusarium-resistant hybrids are Wizard, Market Victor, Gourmet, Market Prize, King Cole, and Excel. Other hybrids include Emerald Cross, Stonehead, Ruby Ball, Red Head, Savoy King, and Savoy Ace.

The heads are usable as soon as they become fairly firm. Early varieties grown under favorable conditions will reach the harvest stage 55 to 70 days after transplanting. Later varieties may require 110 to 120 days or more. In warm weather, heads may split open fairly soon after reaching the harvest stage. In cool weather they may remain good for several weeks. In cold regions, cabbage should be harvested before hard freezes occur. It can be stored at high humidity and low temperature (32—40° F) for several months.

Cabbage planted early for summer harvest often will develop small heads on the stem after the center head is removed. These are quite edible and can be used to extend the harvest period.

Cauliflower—The Snowball group of varieties is the most commonly used in home gardens. Snow King, Snow Crown and Snowflower are fast-growing hybrids. Self Blanch and Snowball Y are examples of good standard late varieties. The purple headed type, somewhat of a novelty, turns green when cooked and resembles broccoli. In southern California and probably in other warm winter areas, the winter type will perform satisfactorily. November-December, February, and Mayflower are common varieties.

The ideal time to set out cauliflower plants is late July or August,

Top, early hybrid Premium Crop broccoli. Left, Savoy cabbage.

as late September and October often provide the most desirable weather for developing good quality. Covering the developing heads helps in producing pure white curds, but the slightly yellowish curds obtained without covering are generally of equal quality. Exposed curds (heads) may be injured by frosts. So as the frosty season approaches, protect the heads by tying together the inner leaves or breaking an inner leaf or two over the head.

The heads are ready to use as soon as they reach suitable size. They should be cut before the parts begin to separate or become "ricey". As with broccoli, cauliflower tastes best soon after harvest.

Brussels Sprouts—This crop is best grown for fall harvest by setting out plants in June or early July. Catskill, Jade Cross (F_1), and Long Island Improved are common varieties. In warm weather the sprouts tend to be loose and of poor quality, but they firm up and become milder in flavor as the cool weather arrives. Sprouts 1 to 1½ inches in diameter are desirable. In harvesting, remove the leaf beneath the sprout and cut or break off the sprout. Harvesting can continue as long as the sprouts develop.

Debudding, cutting out the growing point in late August or early September after the plants are 15 to 20 inches tall, tends to induce the sprouts on the plant to be ready at about the same time. This practice may be helpful in areas where winter sets in early.

Kohlrabi—Kohlrabi, a "stem turnip", can be eaten fresh or boiled or added to soups and stews. It can be planted in the spring, but usually is best for use in the fall (and winter in the South) after frost-tender vegetables are gone. Seeded in the garden, kohlrabi is ready to use in 55 to 65 days. Harvest kohlrabi when it is 2 to 4 inches in size and the flesh is still tender.

Suitable varieties include Early White Vienna and Early Purple Vienna.

Chinese Cabbage—This vegetable, though a different species than the cole crops, has similar culture. It can be used raw or cooked.

Plants tend to go to seed rapidly when planted in the spring, so late June and early July seedlings for fall production are most suitable.

Michihli forms a tall slender head. Wong-bok, Hybrid G, and Burpee Hybrid have shorter, blockier heads. Crispy Choy is a non-heading or looseleaf type.

Green and purple types of kohlrabi dug after partially overwintering in a garden in Oregon.

The Popular, Cultivated Tomato And Kinfolk Peppers, Eggplant

by Allan K. Stoner and Benigno Villalon

Tomatoes, peppers and eggplant are all members of the same family. Since they require virtually the same climatic and cultural conditions to grow in the home garden, they will be discussed together. These are considered warm season crops. Thus they are suited to spring, summer, and autumn culture over most of the North and upper South and they will grow in the winter in the extreme South.

Tomatoes are probably the most popular garden vegetable grown in the United States. This can be attributed to their unique flavor, attractiveness, richness as a source of vitamins C and A, and versatility as a food. The popularity of peppers can be attributed to the same factors, although they are usually not consumed in large enough quantities to make them an important nutritional factor in the diet.

The cultivated tomato, *Lycopersicon esculentum* Mill., originated in the Andes mountains of South America. It was introduced to other areas of the world by Indians and European travelers. The first report of the tomato in North America was in 1710 where it was grown primarily as an ornamental plant. Tomatoes began gaining wide acceptance as a food plant in the United States between 1820 and 1850.

Peppers are also native to tropical America and were grown by American Indian tribes in both North and South America over 2,000 years ago. The small red hot peppers were discovered by Columbus in the West Indies and introduced into Europe where they became popular before gaining widespread acceptance in the United States. Peppers became one of the first New World foods used commercially in Europe.

Pepper varieties grown in the United States are grouped in *Capsicum annum*, with the only exception being the red hot tabasco pepper *Capsicum frutescens* imported from the state of Tabasco, Mexico.

Eggplant, *Solanum melongena*, is believed to be native to India. It apparently moved into the Mediterranean area during the Dark Ages and was later introduced into America by the Spaniards.

Probably the most important step for the gardener in growing tomatoes or peppers is to select the proper varieties to plant. Many varieties of both crops are well adapted for home gardens.

A good garden tomato variety should possess resistance to as many of the commonly occurring diseases as possible, and resistance to growth cracks and bursting caused by alternating dry and wet weather. It should also be adapted to the local environmental and soil conditions and produce attractive fruit with good flavor and high nutritional value.

Resistance to verticillium wilt, fusarium wilt and nematodes is often indicated by including a V, F, N with the name. Nematode resistance is normally only required in Southern and some Western areas while V and F resistance is likely to be important in most areas.

Generally, gardeners should grow varieties with an indeterminate type of vine that will continue to grow and set fruit over a long period of time. Determinate varieties that set and

---

Allan K. Stoner is Research Horticulturist, U.S. Department of Agriculture, Beltsville, Md., and Benigno Villalon is Assistant Professor, Texas Agricultural Experiment Station, Weslaco.

ripen all their fruits at nearly the same time are ideal for home canning when you want a lot of tomatoes at one time.

In addition to different vine types, the gardener can choose from small "cherry" to large "beefsteak" varieties. Varieties range in ripe fruit color from yellow to orange, pink, and bright red, and vary in fruit shapes.

New tomato varieties are released by seed companies and State and Federal experiment stations each year. You can obtain information about adapted varieties from seed catalogs, local nurserymen, county agricultural agents, newspaper and magazine garden articles, and successful neighborhood gardeners.

Pepper Groupings

Pepper varieties are easily classified as sweet, mild or hot depending on the amount of the heat or pungent compound, capsaicin, present in the fruit. However, there are many different common or commercial names for the hundreds of fruit types and shapes.

The horticultural varietal grouping that follows helps in understanding pepper fruit diversity even though varieties within the groups may be completely unrelated.

Bell Group fruits are large, blocky, about 3 inches wide by 4 inches long, 3- to 4-lobed, and taper slightly. Most are dark green, turning bright red at maturity, although some turn yellow. The California Wonder sweet types are probably the most popular garden peppers in the United States. There are also some hot bells. There may be upwards of 200 open pollinated and hybrid varieties in this group.

Cayenne Group—This is the chili group characterized by slim, pointed, hot or mild, slightly curved fruit pods, 2 to 12 inches long. The largest fruited varieties in this group are the Anaheim or New Mexico chili whose pods are 6 to 12 inches long. These are used in the green stage for chili relleno or mild green sauce. The fully matured red dry pod is used in making red chopped chili pepper, ground chili powder, paprika if the variety is sweet, and oleoresin.

The Cayenne is 4 to 12 inches long, pointed, wrinkled, deep red, dry at maturity and used primarily in making hot sauce. The small hot peppers include the 2-inch pointed, slim red chili used in hot sauces, and the bullet-shaped 1½-inch to 2½-inch chubby Serrano eaten green in fresh salads or sauces. Probably the most popular of all the small hot peppers is the pungent jalapeno. It is conical, 2 to 3 inches long with a blunt point, an inch to 1½ inches wide at the shoulders, with thick walls. Jalapenos may be eaten fresh but most are canned by a process called "escabeche".

Perfection or Pimiento Group fruits are sweet, conical, slightly pointed, 2 to 3 inches wide and 3 to 4 inches long, with thick red walls. Popular pimiento varieties include Bighart, Truhart Perfection, and Pimiento L. These are primarily used for canning, stuffing olives, cheeses, etc., but can be used fresh in salads for flavor and color.

Celestial Group fruits are produced upright or erect, cone shaped, ¾ inch to 2 inches long, 3-celled, and colors may or may not change from yellowish to red or purplish to light orange-red. Different colors may appear on the same plant simultaneously, making them an attractive ornamental plant. Popular varieties include Floral Gem, Fresno chili, and Celestial.

Tabasco Group fruits are an inch to 3 inches long, slim, tapered, and very hot. Tabasco is the most popular variety in this group. Others include Japanese Cluster, Coral Gem, Chili Piquin, Small Red Chili, and very small Cayenne. They are used in sauces, for pickling, and are attractive ornamentals.

Cherry Group fruits are cherry-shaped or globose, 3-celled, borne on long slender, upright pedicels more or less above the leaves. Fruit may be orange to deep red, sweet or hot, large or small. They are attractive as ornamentals and are used for pungent seasoning. Popular varieties include Bird's Eye, Red Cherry Small, and Red Cherry Large.

Tomato Group fruits are distinctly flattened or oblate, 4-celled, and bear a striking resemblance to a tomato. These are used for pickling, canning, or fresh pepper rings. Varieties include Sunnybrook, Topepo, and Tomato.

Compared to tomatoes and peppers, there are relatively few eggplant varieties. Large fruited varieties are most commonly grown in the United States; however, many gardeners and cooks consider them inferior in quality to the small-fruited varieties. Gardeners in Northern areas with a short growing season must be especially aware of the number of days required for a variety to reach maturity. Eggplant varieties also differ in the shape and color of their fruits.

Planting, Fertilizing

Plant tomatoes, peppers and eggplants where they will receive a maximum amount of direct sunlight. A fertile, well-drained soil is required for best results. If the soil is not naturally fertile, fertilize it, preferably with a combination of manure and commercial fertilizer. All three crops are moderately tolerant to an acid soil (pH 5.5 to 6.8), but strongly acid soils should be limed according to soil test recommendations.

Fertilize tomatoes, peppers and eggplants in about the same way. However, since it is more important that peppers and eggplants start quickly and grow rapidly after transplanting, give them a little more nitrogen and potassium. If peppers or eggplants start blooming and set fruit

Teaspoon of all-purpose fertilizer (5-10-5 or 10-6-4) applied to planting hole at planting time will give your tomato plants a good start.

while the plants are too small, they will be stunted and fail to develop the plant size needed for a good yield.

On loam and heavier soils of fair to good fertility, 5 to 8 pounds of 5-10-5 fertilizer per 500 square feet should be mixed with the soil about a week before transplanting. On lighter or more sandy soils, 10 to 20 pounds of 5-10-5 per 500 square feet should be incorporated into the soil before planting. When the plants have set several fruits, apply a topdressing of the same type of fertilizer to prevent the plants from slowing down in vegetative growth. If the soil is very low in fertility, you may need to fertilize more frequently. Poor foliage color and stunted growth call for additional fertilizer.

William E. Carnahan

Tomatoes, peppers, and eggplants need water in an amount equal to that provided by a 1-inch rain each week during the growing season. If rainfall is deficient or you live in an arid area, soak the plants thoroughly once a week. If the soil is sandy, you may need to water more frequently. Heavy soakings at weekly intervals are better than many light sprinklings. Tomatoes, peppers and eggplant respond very well to trickle or drip irrigation also.

Peppers particularly need abundant water during flowering and fruit set to prevent shedding of flowers and small fruits.

Transplanting

Tomatoes, peppers, and eggplant may be seeded directly into the garden in areas with a long growing season, but transplanting into the garden generally is recommended. Prior to direct seeding, work the soil into a somewhat granular condition. After planting, keep the soil moist until the seeds germinate. If the seeds are sown thick to insure getting a good plant stand, thin the seedlings to the proper spacing by the time they have three leaves. When early tomatoes, peppers or eggplants are desired, or the growing season is likely to be too short for heavy yields, use purchased or home grown transplants. A chapter about transplants begins on Page 111.

Tomatoes can be safely planted outside on the frost-free date, but because peppers and eggplant are somewhat more exacting in their temperature requirements than tomatoes, they should not be planted in the garden until a week or more after the frost-free date. A good general rule is to transplant outside when the new leaves on oak trees are fully grown.

William E. Carnahan

Stake tomato plants before they get too large and before roots can be damaged by stake.

If there is danger of frost after the plants are put outside, protect them with paper or plastic coverings, newspapers, or boxes. Remove the covers during the day.

Set tomato plants into the garden at about the same depth as they were growing indoors. You don't need to remove the growing containers if they are made of peat or paper. If clay containers were used, knock the plants out of the pots before transplanting.

After transplanting, press the soil firmly around the plant so that a slight depression is formed to hold water. Then pour about a pint of water (to which fertilizer has been added) around each plant. Use 2 tablespoons of granular 5-10-5 fertilizer per gallon of water or a water-soluble starter fertilizer.

Plant peppers and eggplants in rows 30 to 42 inches apart and spaced 12 to 18 inches apart in the row.

Distances between tomato plants depend on the variety used and whether they are to be pruned and staked. Staked plants should be 18 inches apart in rows 3 feet apart. Unstaked plants should be 3 feet apart in rows 4 to 5 feet apart.

Staking or supporting tomatoes makes it easier to cultivate and harvest, and helps prevent fruit rots by keeping the fruits from coming in contact with soil. However, staked

Tomato plants in wire cages. Be sure openings are large enough so you can pick fruit. Cages are stored in off-season for re-use.

plants are more subject to losses from blossom-end rot than unstaked plants. Due to the woody nature of pepper and eggplant stems, you don't have to stake or support these crops.

If you plan to stake your tomatoes, insert the stakes soon after transplanting to prevent root damage. Wood stakes about 6 to 8 feet long and 1½ inches wide can be used. Push the stakes into the soil about 2 feet.

Tie soft twine or strips of rag tightly around the stake 2 to 3 inches above a leaf stem, then loop the twine loosely around the main stem not far below the base of the leaf stem, and tie with a square knot. Plant ties, made of tape reinforced with wire, may also be used to fasten plants to stakes. Six-inch mesh concrete reinforcing wire may also be used to support tomato plants by forming a circle 18 inches in diameter around the plant.

If you wish to prune staked tomato plants to 1 or 2 stems, about once a week remove by hand the small shoots that appear at the point where the leaf stem joins the main stem. Grasp the shoot with your thumb and forefinger and bend it sharply to one side until it snaps, then pull it off in the opposite direction.

Weed Control

The area around tomatoes, peppers, and eggplants should be kept free of weeds because of competition for sunlight, soil nutrients, and water. You can do this by mulching, hand pulling, or cultivating not more than 1 to 2 inches deep. Pepper and eggplant roots are particularly slow growing; thus any amount of root pruning can cause stunted growth and flowers to drop. Avoid cultivation when the soil is wet since it can lead to clumping of the soil and soil compaction.

Several insect species damage tomatoes including flea beetles,

tomato fruitworms, hornworms, aphids, leafminers, pinworms, Colorado potato beetles, whitefly, and spider mites. In small gardens some of these can be controlled by hand picking them from the plants. The others can be controlled by using approved insecticides at the proper time.

Two of the most common tomato diseases occurring in home gardens are fusarium and verticillium wilts. They are caused by fungi that live in the soil. Before the development of resistant varieties, gardeners were urged to plant in a different part of the garden each year; this is still a good idea. The best control, however, is to grow resistant varieties. Spraying or dusting is ineffective in controlling either of the wilt diseases.

Blossom-end rot is the most troublesome fruit rot for the home gardener. It is caused by a calcium deficiency and is aggravated by any kind of drought stress on the plants. Calcium, in the form of finely ground dolomitric limestone, will help prevent blossom-end rot. It must be applied before tomatoes are planted.

Other fruit rots are caused by fungi. Usually these fruit rots are not a problem when plants are staked. Most fruit rots can be controlled either by spraying with a fungicide or mulching with a suitable material such as black plastic. In areas where the leaves are frequently wet because of rain or dew, leaf spot diseases such

Tomato hornworm damages foliage and fruit on tomatoes, eggplants, and peppers.

as early blight, late blight, gray leaf spot or septoria leaf spot can be destructive. These can be controlled by applying a suitable fungicide at 7- to 10-day intervals. Wetting the foliage when watering can accentuate these diseases.

Virus diseases can cause a mottled discoloration of tomato foliage and occasionally a mottling of the fruit. Since tobacco mosaic virus is transmitted by direct contact, wash your hands and tools before touching the plants. Do not smoke while handling tomato, pepper, and eggplant plants.

Cucumber mosaic virus is transmitted by aphids that may be harbored in some perennial flowers or in nearby weeds. Cucumber mosaic can be controlled by eradicating perennial weeds and by spraying the tomato plants with an insecticide that controls aphids.

Tomatoes are subject to damage by many species of nematodes, but rootknot nematodes are the most troublesome. Affected plants become yellow and stunted and their roots can be galled, pruned, matted, or decayed. If nematodes are known to be present in damaging numbers based on the experience of previous years, they should be controlled before tomatoes are planted. Nematode control can be obtained by using an approved nematicide.

Insects attacking peppers—such as leaf miners, aphids, budworms, flea beetles, hornworms, pepper weevils, cutworms, and the pepper maggot—can be controlled with timely applications of insecticides used according to the manufacturer's directions.

Common pepper diseases include seedling damping off, bacterial leaf spot, Cercospora leaf spot, Phytophthora root rot, and mosaic virus diseases. Seed treatment and applications of fungicides or soil fumigation can help reduce losses. Several fungicides give adequate control of most leaf spot diseases.

Mosaic virus diseases such as tobacco etch virus, potato virus Y, tobacco mosaic, cucumber mosaic and tobacco ringspot virus can only be controlled by using resistant varieties. There are many pepper varieties resistant to tobacco mosaic, but very few resistant to tobacco etch and potato virus Y, and none to cucumber mosaic or tobacco ringspot virus. The release of multiple virus resistant pepper varieties can be anticipated in the future.

Phomopsis rot and verticillium wilt are two serious diseases of eggplants. The Phomopsis rot is characterized by large sunken, tan-colored or black areas on the fruits. It may also cause canker-like lesions on the lower part of the stem and leaf spots which may enlarge until the whole leaf turns brown. The disease may be carried over winter by debris in the soil from the previous crop. To control fruit rot, use clean seed, practice a 3- to 4-year crop rotation, and grow rot-resistant varieties.

Verticillium wilt of eggplant is particularly common in cooler regions and is similar in its behavior to wilt disease of tomato. It seems to persist in the soil indefinitely, and can be distributed by plants from infested seedbeds. Wilt injury ranges from stunting, with decreased productivity, to death of the plant.

Several insects attack eggplants, particularly flea beetles, aphids, lace bugs, and sometimes the Colorado potato beetle. Red spiders occasionally become troublesome on eggplants, especially during dry weather.

Harvesting

To obtain the best flavor and color, harvest tomatoes after they are fully ripe. Tomatoes can be expected to ripen 60 to 90 days after transpalnting. If picked green, they can be ripened at temperatures between 55° and 72° F. Light will increase the color of tomatoes somewhat, but light

Eggplant nearly ready for harvesting.

is not essential to ripening. When tomatoes are placed in direct sunlight to ripen, the added heat often lowers their quality.

Green sweet peppers are harvested when they reach a good usable size and still retain their dark green color. Immature peppers are soft and yield readily to mild pressure of the fingers. Red peppers, either sweet or hot, are allowed to develop full red color before picking. Hot peppers can be harvested early for green sauce or canning or allowed to ripen, then harvested.

Pepper fruit will be ready for harvesting between 70 days for early green fruits to 130 days for some of the fully mature red pods. Peppers are generally harvested by breaking them from the plant with the stems left attached to the fruits.

Eggplants may be harvested any time after they have reached sufficient size, but before the skin color becomes dull, the flesh tough, and the seeds begin to harden. Most varieties will be ready for the first harvest in 85 to 90 days after transplanting.

Harvest eggplants by cutting the tough stems with a sharp knife.

Per plant yields for the various crops will vary greatly depending on the variety, the growing season, the area of the country, and the cultural practices you follow. However, it is reasonable to expect tomatoes to yield 10 to 14 pounds per plant, peppers to yield 1 pound per plant, and you can anticipate 4 to 8 eggplants per plant.

Leafy Salad Vegetables: Lettuce, Celery, Cress, Endive, Escarole, Chicory

by Bruce Johnstone

The principal leafy salad vegetables covered in this chapter, especially lettuce, are among the most widely grown vegetables by home gardeners throughout the United States. Most of them—but not celery and chicory—are easy and fast to grow, and with the exception of celery are among the relatively few vegetables that tolerate moderate shade.

They also are adapted to small home gardens because each of them requires but little space for an average size crop. Salad crops in general also conform to the currently popular American taste for low calorie and high vitamin content foods.

Besides the leafy salad crops covered in this chapter—lettuce, celery, cress, endive, escarole and chicory—a few other leafy vegetables covered under different categories and in separate chapters also can be used advantageously as green leafy ingredients in salad making. Among these are spinach, New Zealand spinach, chard and mustard, each adding a slightly different flavor, color and texture to various salads.

Other common salad vegetables such as tomatoes, cucumbers, onions and radishes are covered in different chapters of this book and can be located through the table of contents or the index.

Lettuce

Known botanically as *Lactuca sativa* of the Composite family, lettuce probably originated somewhere

---

Bruce Johnstone of Excelsior, Minn., retired as chief horticulturist at Northrup, King & Co. He co-authored *Vegetable Gardening From the Ground Up*, a paperback book, in 1976, and *America's Most Beautiful Flowers* in 1977.

in Asia Minor and the eastern Mediterranean region. Used as a food plant for some 2,500 years, it was a favorite of Persian kings in the sixth century B.C. and later as a food plant by the Romans. In the late 15th century, it first was brought to the New World by Columbus.

Lettuce seed is rather small (25,000 seeds per ounce), germinates quickly (7 days) in cool (65°-70° F) temperature, and produces a crop comparatively fast. Loose leaf lettuce types normally produce a crop in 40 to 50 days while most heading varieties require 60 to 80 days to mature.

The loose leaf varieties are more widely grown than heading types in home gardens because they are faster to mature, easier to grow, and somewhat more shade tolerant. They also have about three times as much vitamin A and roughly six times as much ascorbic acid or vitamin C as the equivalent amount of the heading varieties. Loose leaf varieties of lettuce require less thinning and thrive under somewhat warmer and more adverse conditions than the heading types.

Because lettuce basically is a cool weather crop, seed should be sown direct in the garden in early spring in order to mature before the summer heat arrives to cause bolting and deterioration of the foliage. (Bolting is premature flowering). Five feet of row per adult in family is usually enough for each planting.

Successive plantings can be made in midsummer for autumn crops.

The seed should be scattered thinly, covered a quarter inch deep in rows as close as 8 and up to 24 inches apart, depending on space available.

Thinning is not absolutely necessary for loose leaf kinds but spacing the plants 4 to 6 inches apart is com-

147

S. Goto

monly recommended and results in larger, more easily harvested leaves. Typical loose leaf varieties available are: Black Seeded Simpson, Grand Rapids and Salad Bowl.

Heading varieties are of two main types, crisphead and butterhead. Crisphead varieties are of thinner texture, are crisp, frequently have curled and serrated edges, are harder and more durable in handling and storage. Most of the so called Iceberg types available in stores are of this class. Other typical crisphead varieties: Ithaca, Great Lakes 118 and 659.

In contrast, butterhead types are softer and more fragile in texture, have thicker leaves and a smooth, buttery substance. Butterhead types— Bibb, Buttercrunch, White Boston— have a distinct delicate flavor and usually are more perishable than the crisphead varieties.

Heading varieties have cultural requirements similar to the loosehead types of lettuce except they require a longer, cooler growing season, more careful thinning, and need full sun for best development. All lettuce types are heavy feeders and because of their limited root structure require ample and constant soil moisture. They need high nitrogen fertility in a moist soil and give best results if growth continues unchecked.

Cos lettuce (Romaine) or celery lettuce has an elongated framework, smooth outer leaves, and a blanched inner head. The leaves are more brittle than the other heading types, the midrib is heavier, and the flavor uniquely sweet and mild. Cos types usually take 65 to 70 days to mature and have the same basic planting and cultural requirements as the other heading types. Most popular varieties are Paris White Cos and Paris Island Cos.

Where there are short, hot growing seasons as in much of our Northern, Central and Midwest states, the heading varieties are most successfully grown by starting seed indoors in very early spring, then getting the transplants into the garden as soon as frost danger is past. In this way the plants can mature and form heads

Left, young gardener checks lettuce in her garden in Hawaii. Right, Bibb lettuce being harvested.

before summer heat curtails growth and development.

Harvest with a sharp knife as soon as looseleaf types are the size of your hand. Heading varieties should be full and firm. If allowed to go to seed in warm weather, leaves lose quality and become bitter.

When cultivating or hoeing lettuce, take care to keep the blade shallow and not too close to the plants to avoid injuring the root system which is sparse and close to the surface.

Homegrown lettuce is relatively free of disease although leafhoppers can be a problem, mostly in spreading virus disease. Effective chemical controls are available.

Celery

Celery *(Apium graveolens*—family Umbelliferae) is native to marshy areas from Scandinavia to Algeria and Egypt and eastward to the Caucasus and into Baluchistan and parts of India.

The two main classes of celery are the green and the golden, or self-blanching. The green type with unblanched stalks adds considerably to the appearance and flavor of both salads and casseroles and is currently more popular on American tables. This type includes Giant Pascal, Forkhook and Utah strains. For use as a canape of raw vegetables, some cooks still prefer the golden or self-blanching type with yellowish white stalks and usually a milder, blander favor. Popular golden varieties grown are Golden Plume, Cornell 19 and Michigan Golden.

Celery seed is very small (60,000 per ounce) compared with other common vegetables, very slow to germinate (15 to 21 days) and requires a long, cool growing season of 120 to 140 days to produce a crop.

Celery needs a rich, moist soil and mild, equable growing conditions without sudden cold spells or dry periods to check its growth. Muck or sandy loam soils with good fertility are ideal. These exacting conditions make celery growing by home gardeners rather difficult, especially in much of the Midwest and inland Northern areas. In coastal regions or areas near large bodies of water, the usually longer and more temperate growing conditions are more suitable for celery culture.

Because celery is such a slow growing, rather difficult crop to raise, it should not ordinarily be chosen by a beginning gardener in most areas. It is successfully produced, however, by many experienced gardeners in favorable areas who take the time and care necessary. Because of its many culinary uses from salads to casseroles to attractive canapes, it probably is well worth the effort.

Celery seed must be started very early (usually indoors) 8 to 10 weeks before spring planting time unless commercially grown transplants are available. Germination is very slow, usually 2 to 3 weeks, and can be hastened slightly by presoaking the seed

Celery display in a garden.

overnight before sowing in flats 1/16 inch deep. Seed flats must be kept moist and covered at 60° to 70° F temperature until the sprouts appear.

At this stage, they should be uncovered immediately and moved to direct sunlight and a slightly cooler situation. Seedlings must be transplanted or thinned so that developing plants are 1½ to 2 inches apart and kept in full sunlight until frost-free planting time. The young plants then can be hardened off outdoors, and set in the garden, spacing them 6 to 10 inches apart in rows 2 feet apart.

For ordinary usage, figure on a half-dozen plants per adult in family. Harvest by cutting at base of stalk with a sharp knife. The usual harvest span is from the stage when the stalk is two-thirds of full size until fully would be about a 5-10-10 ratio.

Celery requires ample and continuous soil moisture and a high fertility. If soil is not rich, fertilizer should be used. The formula depends on the individual soil type, but in most cases would be about a 5-10-10 ratio.

Celery may be attacked by leaf-eating worms and aphids (plant lice). You can control these insects with approved insecticides. Blight and mildew also may be problems; control them with an appropriate fungicide.

Endive—Escarole

Endive (Cichorium endivia—family Compositae) is native to regions of the eastern Mediterranean and was grown and used by Greeks and Egyptians before the Christian era. Closely related to chicory, endive has small seeds (27,000 per ounce) which germinate quickly (5 to 14 days) under moist conditions and in varying temperatures from 60° to 70° F.

There are two principal types of endive: Curled or Curly—with loose, narrow, medium green fringed and curly leaves; and Batavian or escarole with broader, thicker, smooth leaves that have a white midrib forming a loose head with partly blanched inner foliage.

Endive is more tolerant of summer heat and low soil moisture than most lettuce varieties, and is also slower to grow and mature (usually 85 to 95 days). The curled varieties can be cut and cropped, yet continue to produce new secondary edible leaves. These curled varieties such as Green Curled, Ruffec, and Deep Heart have a slightly bitter flavor but are very decorative and desirable in salads and for garnish. The broad-leaved Batavian or escarole varieties are somewhat milder and add a different flavor and texture to salads.

Seed is usually sown direct in the garden in the early spring ¼ inch deep in rows 2 to 3 feet apart, later thinned to 6 or 8 inches between plants. Four to five feet of row per adult in family will suffice for average table use.

For earlier harvest, seed may be started in flats indoors 6 to 8 weeks before planting time, then transplanted to the garden. Summer sowing of seed will produce autumn crops which, maturing in cooler weather, are apt to be somewhat milder in flavor and with less of the slight bitterness characteristic of summer harvested crops. Loosely tying the outer leaves upright to exclude sunlight tends to blanch the inner leaves, making them milder and reducing the bitter taste.

Harvest by cutting at base or carefully pulling entire plant when inner leaves are partly or wholly blanched. Outer leaves are apt to be bitter and usually are discarded.

Endive seldom is bothered by insects or disease problems. Sometimes, in mild damp areas, slugs or snails may appear and eat the foliage. Control them with special snail bait or slug protectant. Dry ashes around plants usually repel both slugs and snails.

Cress

Garden Cress or pepper grass (*Lepidium sativum*) belongs to the Cruciferae family and although similar in flavor to water cress and upland cress, it is far more popular and much easier to grow under ordinary gardening conditions. Water Cress (*Nasturtium officinale*) is a semi-aquatic plant requiring very cold spring water conditions to grow well. Upland Cress (*Barbarea verna*) tolerates a normal soil but is slower to grow, somewhat bitter in taste, and not commonly produced in U. S. gardens.

Garden Cress is both easy to grow and extremely fast to form edible leaves. The seeds are moderate in size (12,000 per ounce) and under moderate temperature of 65° to 70° germinate in 4 to 7 days.

Garden cress is probably the fastest seed to sprout of all garden vegetables. The young seedlings also grow rapidly and the very young immature leaves are tender, mildly pungent like water cress, and they can be cropped for table use when only a few inches high—10 days to 2 weeks old.

Garden cress is used commonly as a quick growing indoor crop, often available in preseeded kits with a medium of vermiculite, peat moss, etc., and is intended to be grown in a sunny kitchen window to produce edible leaves in 10 to 15 days. Grown this way indoors, cress can be available and used all winter long by successive plantings.

Outdoor spring and summer garden culture is also easy enough but for continued harvest one must make successive plantings every few weeks. Hot summer weather causes garden cress plants to bolt quickly and lose quality, so early cropping is necessary.

Sow the seed ¼ inch deep in rows a foot apart and harvest as soon as seedlings are 3 to 4 inches high for the best quality.

A 10- to 15-foot row usually suffices for the average family. Cut with a sharp knife as soon as leaves are formed.

Chicory

Chicory (*Cichorium intybus*—family Compositae), also known as French Endive or Witloof Chicory, is thought to be native to Europe and Asia. Although some chicory is grown for the roots which are dried, ground and used as a coffee adulerant, we will cover here the salad type and culture in which the blanched leaves are the garden crop wanted.

Chicory is related closely to endive but usually produced in a far different manner.

The seeds are small (27,000 per ounce) and they germinate in 7 to 14 days at temperatures between 68° to 85° R.

Seed ordinarily is spring sown a quarter inch deep in 15- to 18-inch rows and the seedlings thinned to eventually stand 4 to 5 inches apart. It must not be planted too early or premature flowering (bolting) will occur.

The parsnip-like roots are harvested in the fall before freezing weather, washed, and trimmed of all leaves except the single central crown bud on top. The roots are then stored under cover in a cool frost-free room.

These roots are stored and later planted for winter production of the edible shoots by setting them slantwise at a depth of 4 to 6 inches with crowns about even with the surface in a medium of sand, sawdust or a similar porous medium at temperatures of 50° to 60° F in a dark place. In 3 to 4 weeks the blanched heads or shoots appear and are ready to cut and harvest. Successive winter plantings of the stored dormant roots every 2 to 3 weeks can be made to produce edible shoots throughout the winter.

Onions Are Finicky as to Growing, Curing; And Garlic May Not Be a Joy Either

by J. S. Vandemark

Onions are grown in nearly every part of the United States. Fairly cool temperatures are important during early development, and good soil fertility and adequate moisture are essential. High temperatures help during bulbing and curing. Low humidity is desirable for curing.

Domesticated in Asia and the Middle East, onions were rapidly moved to Europe. They were grown both by the settlers and Indians after being brought to North America by Spanish explorers.

Onions are used in a variety of forms. They are eaten raw, as scallions, and in salads. Cooked onions are served broiled, boiled, baked, creamed, steamed, fried, french fried, pickled, in soups and stews, and in combinations with other vegetables and meats.

Growers learned from experience that the early development period of onions should be cool and damp to allow secondary roots to develop.

Soil fertilization should be given particular attention. Onions require about twice as much fertilizer as most vegetables. Gardeners may find it advantageous to give the row a second feeding after 40 to 60 days by placing the fertilizer in a trench 1 to 2 inches deep and 3 inches to one side of the row. Use 10-10-10 fertilizer at a rate of 1 pound per 25 or 30 feet of row for this sidedressing.

Good growth requires a loose, friable (crumbly), fertile soil. Hard compact soils tend to restrain bulb development, causing the bulb to be irregularly shaped and small.

Proper time to plant depends on the area. In the South, onions are grown in winter, from seeds, sets or seedlings that were fall-planted. In Northern and Central regions, onions are planted in spring as early as the soil can be prepared. Onions are tolerant to frosts. Seeds germinate best about 60° to 65° F; however, satisfactory results will be obtained anywhere from 50° to 75° soil temperature.

Onions must be kept free of weeds and grasses throughout the entire season as they compete poorly with other plants.

Onions From Sets

Onion sets are the surest route to success in the home vegetable garden as the emerging plant will be vigorous and strong. They may serve double duty, producing green onions or mature dry onions. Onion sets, which you can buy, consist of small dry onions up to ¾ inch in diameter grown the previous year specifically for starting plants.

Select sets early when they are firm and dormant. While it seems contradictory, round onion sets produce flat onions, while elongated or tapered sets mature into round onions. Sets are available in three colors: White, red or brown. Most gardeners prefer white sets for producing green onions or scallions; but the other two colors are acceptable.

Divide the sets into two groups, those smaller in diameter than a dime and those larger than a dime. Use large size sets for green onions; the large size may bolt (form flower stalks) and not produce a good dry bulb.

Plant the small size set for dry onion production as there is little chance of bolting. These small sets will best produce large, dry bulbs.

J. S. Vandemark is Vegetable Crop Specialist, University of Illinois, Urbana.

For dry onions, plant the sets to a depth of 1 inch in rows 12 to 14 inches apart, with the sets 2 inches apart. Soil should be worked to a medium fine condition, fertilized, and kept free of weeds throughout the season.

When you observe half or more of the tops bent over naturally, the onions may be pulled and allowed to dry. Do not break onion stems early as this interrupts their natural growth period and will result in reducing the bulb size. Onions can be eaten at any point between green onions and mature dry onions.

When planting larger size bulbs for green onions, use 12- to 24-inch wide rows. Sets can be planted close enough to touch each other and 1½ inches deep. If you hill the row slightly after the stem is up 4 inches, the green onions will have longer usable white stems. Once the tops are 6 inches high you can start using them as green onions.

Green onions become stronger in flavor as they get older. When too strong to eat raw, they are excellent for cooking.

If you notice a plant producing a flower stalk (bolting), pull and use it.

Seed is the least popular method of growing, as a longer period for development is needed. Onion seed should be planted at the same time as sets and requires similar fertilization. To assure a good stand, plant seed ¾ to 1 inch deep, at a rate of 1 to 5 seeds per inch. When the seedlings are established they should be thinned: For large dry onions 2 to 3 inches apart, for medium sized 1 to 2 inches, and for boilers and green onions, ¼ to 1 inch.

Choose varieties adapted to your purpose and section of the country. Green onion varieties from seed include Evergreen Bunching, Beltsville Bunching, and Southport Bunching. Varieties for dry onions produced in the South include Crystal Wax, Excel, Texas Grano, and Granex. In the North suitable varieties are Early Yellow Globe, Empire, Fiesta, Downing Yellow Globe, Spartan lines, and Nutmeg.

Onions From Transplants

Transplanting of young onion plants is an increasingly popular method for home gardeners because of the large size bulb produced. Plants may be obtained from your seed dealer and garden supplier. In the South and Southwest plant in fall, and in the North in early spring. To produce large dry onions, plant 4 to 5 inches apart in the row, with rows 12 to 24 inches apart. Closer spacing yields smaller size bulbs. Fertile, loose soil is essential.

Transplants are put in soil to a depth of 1 to 1½ inches and covered. Many growers use a special water-soluble fertilizer mixture immediately on planting.

This transplant fertilizer is in addition to the normal fertilizing of the garden and the necessary sidedressing discussed earlier. To prepare a water fertilizer mixture, buy a soluble fertilizer, 10-50-10, 10-52-17, or similar analysis. Dissolve one tablespoon in a gallon of water and apply at the rate of 1 cup per plant.

Crystal Wax, Bermudas, and Grano types are used in the South for fall planting. In the North, Bermudas are used when flat bulbs are desired. For large round bulbs use standard or hybrid Sweet Spanish (white or yellow). If red onions are preferred, varieties like Red Bermudas, Red Giant, Red Hamburger, and Benny's Red are used in the North, while Early Red or Creole are suitable for the South.

Special onions include the *Egyptian* onion or tree onion, a perennial planted in fall throughout the country. This onion forms small bulblets or sets where the flowers normally grow. It may be harvested, planted and used in the same manner as sets.

Hanging onions for curing.

In late winter or early spring, Egyptian onions may be pulled and used as green onions. Those left unharvested will produce new sets for future planting. If you choose, the plants also may be dug, divided, and replanted for additional propagation. Planting material, either sets or plants, are hard to get through normal channels and often gardeners obtain their starts from friends.

The *Potato* or *Multiplier* onion seldom produces seed and is propagated by bulb division. These onions are planted in the same manner as sets, usually 2 to 3 inches apart. The original bulb splits into segments; when dry, they may be used to plant as sets.

In the South, multiplier sets are planted in fall and used as green onions in spring. In the North, they are planted early in spring.

True multipliers also are hard to get at most gardening centers, and like the Egyptian onion frequently are passed from gardener to gardener.

Many homeowners do not have proper conditions to store onions for long periods; however, they usually can make out all right for three to four months. After half of the tops have broken over naturally at the neck, the onions may be pulled. When the tops have wilted, cut them off 1 to 1½ inches above the bulb.

Cure by placing in an open crate or mesh bag. This process, which prepares the onions for storage, takes from two to a few weeks depending on humidity. Clean by removing dirt and outer loose dry skins that come off when handling.

The cleaned onion may be left in a mesh bag and hung in a place such as a garage ceiling. An ideal storage spot is dry with air temperatures of 35° to 50° F. If roots reappear, the conditions are too moist, and if sprouts appear the temperature is too high; either will cause rotting and deterioration of bulbs.

The storage period is shorter for Sweet Spanish and Bermudas than for small globe-type onions.

Offshoots of Onions

Garlic, shallots, chives and leeks are members of the onion family. These are often grown at home as they frequently are hard to obtain. Fertility and cultural practices are similar to those for onions.

Many areas will find *garlic* poorly adapted to home gardens.

Garlic produces a group of cloves that are encased in a sheath, rather than a single bulb. Separate into single cloves for planting.

The larger outer cloves produce the best garlic. These are planted 1 to 2 inches deep in a well fertilized garden, in rows 12 to 24 inches apart, with the cloves 5 to 6 inches apart. Planting is done in the South and Southwest from fall through January, and in the rest of the country as early in spring as possible. Delayed planting seriously reduces yield.

Harvest when the top dries down. To prepare garlic for storage, cure the bulbs under cool, dry conditions. Garlic may be stored under a wide range of temperatures, but does best under dry conditions with a temperature range of 40° to 60° F.

Shallots, prized by French chefs and gourmet cooks, are grown for either dry bulbs or young green shoots. They are harvested and used like green onions. Frequently writers confuse shallots with green onions, but shallots are a different species of the onion family.

The shallot bulb has multiple sections like garlic. These are separated and planted the same way you plant onion sets for dry bulbs, spaced 3 to 5 inches apart, with 12- to 24-inch rows.

In the South they are most often raised and used for the green portion, being planted in fall and harvested during winter. In the North, shallots are planted as early as possible in late winter or spring and used for both green onions and dry bulbs.

Seed shallots may be obtained from specialty seed stores or by purchasing dry shallot bulbs in the gourmet section of your food store and dividing into single segments for planting. Shallots are harvested, handled and stored like onions. Gardeners frequently save their own planting material from year to year.

Chives are a perennial member of the onion family and are grown for leaves, rather than bulb or stem. A small bulbous plant, chives grow in 6- to 10-inch clumps. Attractive violet flowers appear on older plants in spring.

Chives may be propagated by either dividing the clump or starting from seed. They are generally started from seed very early in spring.

After three years, large chive clumps should be subdivided in early spring to prevent overcrowding and a decline in vigor. The fertility program for chives is similar to that for onion transplants. Remove the flowers as they will cause the plant to be-

Top, clump of great headed (Elephant) garlic. Left, chives grown in kitchen window can be harvested as needed all year long.

come semi-dormant, preventing new growth.

Harvest chives any time there are fresh, young leaves. Young leaves may be chopped and frozen for future use. In late fall in the North many gardeners dig a clump of chives, allowing the exposed clump to freeze until mid-winter, and then bring the chives indoors for a fresh winter supply.

Leeks are grown as an annual, entirely from seed rather than plant divisions. The seed is planted in the garden or started in a hotbed for 2 to 3 months before transplanting to the garden. The young plants are set out in the South in fall and in the North in early spring. Leek transplants are handled in the same manner as onion transplants as regards fertility, transplant solution, and sidedressing.

The primary production difference from onions is that leeks are blanched by banking soil along the row gradually throughout the growing season. Exercise care when the plants are young as early banking may cause decay. Common varieties are American Flag, Conqueror, Tivi and Odin.

Leeks can be used any time the stems reach the size of ¾ to 1 inch in diameter. Before freezing in the North, they are harvested and stored in root cellars or placed in a polyethylene bag in the refrigerator. In parts of the country with open winters, they are allowed to remain in the garden and eaten as desired.

Left, leeks are used for almost any purpose that onions are. Right, leek seedlings being transplanted.

Root Crops More or Less Trouble-Free, Produce Lots of Food in a Small Space

by N. S. Mansour and J. R. Baggett

Root crops are valuable and satisfying additions to the garden because they offer a prolonged harvest season, long storage life, and produce a large amount of food in a small amount of space. With few exceptions, the root crops are trouble-free and do not require the continual care of other kinds of vegetables. They are also adapted to winter storage in root cellars or simple pits.

The table gives general characteristics of some of the root crops.

Root crops are best grown in well drained, loose friable (crumbly) soil. This is especially important because these crops are among the earliest planted and the latest harvested. If the soil is heavy, it may be beneficial to form a raised bed, 4 to 6 inches high and 12 to 24 inches across the top. The width of the top depends on whether 1 or 2 rows are planted on the bed. Use of raised beds reduces soil compaction during the growing season, permits easier digging, and allows carrots and parsnips to attain greater length and be smoother in shape.

Root crops, especially carrots and parsnips, tend to have fewer misshapen roots when grown on beds and will store better in the garden without splitting or rotting. Adding sand and humus to heavy soils will improve soil fertility and permit easier digging and cleaning of roots. However, adding sand without humus can cause a concrete-like condition.

Soil pH (acidity) is usually not limiting. But radishes, turnips, and rutabagas benefit from neutral or slightly alkaline soils if club root disease is present. This disease is inhibited by an alkaline soil condition. You can reduce acidity by adding lime, at the rate of 15 to 20 pounds per hundred square feet of garden area.

A fertilizer rate of 2 pounds of 10-20-10 per hundred square feet of garden area is adequate; additional nitrogen may be required for celeriac. The fertilizer and lime should be thoroughly incorporated in the soil to a depth of 4 to 6 inches. Lime is best applied in fall. Don't use manure where carrots are to be grown, because misshapen roots can result.

Root crops may be planted as soon as the garden soil can be prepared. All have small seeds, which should be planted no deeper than ½ inch.

Crusting and drying can be minimized by mulching the seeded row with sawdust or bark dust, or using such materials as vermiculite or washed sand for covering the seeds instead of soil. Water the newly seeded rows frequently, as often as once a day, in order to prevent a crust from forming and to protect the tender seedlings from drying out until they are well established. Consult the table for appropriate thinning and spacing information.

Pests

Root crops do not often require pest control. However, radishes, turnips, and rutabagas can be seriously attacked by the cabbage maggot and control of this insect is important. The best prevention for the pest is treatment of the soil before planting with a suitable soil insecticide. This material can be spread over the garden area and gently raked into the top 1 or 2 inches of soil. Take care

---

N. S. Mansour is Extension Vegetable Crops Specialist, and J. R. Baggett is Professor of Horticulture, at Oregon State University, Corvallis.

Root Crop Characteristics

	Optimum monthly average growing temperature (Fahrenheit)		Optimum soil temperatures range for germination (Fahrenheit)	Frost tolerance	Spacing suggested in inches		Days to maturity	Time and frequency of planting	Harvest duration for each planting
	Min.	Max.			In row	Between rows			
Beets	40°	65°	50°–85°	Moderate	2–4	16–24	55–80	Early spring and early summer	2–3 months
Celeriac	45°	70°	60°–70°	Good	4–6	23–30	100–110 (56–84 for transplants)	Early spring only	3–6 weeks
Carrots	45°	70°	45°–85°	Moderate	1–3	16–24	60–85	Early spring and early summer	2–4 months
Parsnips	40°	75°	50°–70°	Good	3–6	18–30	100–130	Early spring only	3–4 months
Salsify	45°	85°	50°–90°	Good	2–4	18–30	150–155	Early spring only	1–2 months
Radishes (spring)	40°	75°	45°–90°	Good	½–1	9–18	25–30	Early spring and weekly	1 week
(winter)	40°	75°	45°–90°	Good	½–1	9–18	52–56	Early fall	3–5 weeks
Turnips	40°	75°	60°–95°	Good	2–6	12–30	45–75	Early spring and late summer	2–3 weeks
Rutabagas	40°	75°	50°–90°	Good	5–8	18–36	90–95	Early spring and midsummer	1–2 months

not to mix the insecticide too deeply into the soil since this will dilute it and reduce effectiveness.

You may need to drench turnips and rutabagas with an insecticide solution four to six weeks after planting, since the soil treatment can wear off by that time.

Control of aphids and flea beatles may sometimes be necessary.

Always consult the labels for application details of any pesticide.

Use of weed control chemicals is not feasible in most home gardens. Mulches are useful in reducing weeds after the crops have been well established. They are also useful in improving the color of carrots by preventing the green discoloration normally seen at shoulders of the root. Turnips and rutabagas should not be covered with a mulch if the typical purple color of the upper part of the root is desired.

There are usually no serious disease problems. The club root disease of radishes, turnips and rutabagas occurs only occasionally in gardens. Beets are sometimes seriously affected by canker, a non-parasitic disorder that occurs when there is boron deficiency in the soil.

Fungus and virus diseases of root crop foliage occasionally occur, but it is generally not necessary nor feasible for the gardener to attempt control.

Most root crops can be harvested over a wide range of maturity. Beets can be harvested when small or allowed to mature fully, and carrots can be used from pencil size to full grown over a period of three to six months. Spring radishes, a fast growing crop, are usable over a very short period. See table for days to maturity and harvest duration. Root crops often must be thinned if they are to mature properly and in the right length of time.

Mature root crops can be harvested and stored in a cool, moist location, but carrots, beets, parsnips and turnips are best left in the garden area until severe frost is expected. They should then be pulled, topped, and stored in a cool, moist root cellar until needed for table use. Don't wash them until they are needed. In milder winter areas, leaving the roots in the ground is a practical storage method.

Root crops are relatively efficient as good producers for the space occupied. Twenty feet of row of each crop should produce an adequate quantity for a family of 4. Radishes, however, should be planted in short rows, preferably no more than 3 to 6 feet long, at weekly intervals. This will provide a steady supply of radishes throughout the season. Turnips also have a shorter life than most root crops and are best planted several times during the season.

The root crops are moderately high in vitamin C, with carrots an excellent source of vitamin A.

Radishes are a quick and easy crop to grow in a small space.

Notes on individual crops follow.

BEETS (*Beta vulgaris*)—Somewhat more susceptible to foliage diseases than the other root crops, and less tolerant of drought or low fertility than some, beets are still a relatively easy crop to grow.

Quality factors are not critical and although there may be a preference for small roots for whole pickles, beets are edible over a wide range of maturity. Young roots are more intense in color, with less conspicuous light colored rings, more tender and finer textured.

Choice of varieties for gardens is not critical because a number of good ones of equal quality are available. However, if downy mildew is a problem, then look in seed catalogs for resistant varieties such as strains of popular Detroit Dark Red. Red beets are traditionally the most popular, but other types such as golden and white are also available. All have about the same potential flavor and quality.

CELERIAC (*Apium graveolens*, var. *rapaceum*)—A variant of common celery, celeriac produces heavy roots somewhat like a rutabaga in appearance. It is used raw in salads, pickled, or cooked in soups and stews. Good culture requires plenty of water and fertility and a long growing season. The use of transplants started indoors 10 weeks prior to outdoor planting time is superior over direct seeding.

Celeriac roots are used when they reach 2 to 3 inches in diameter, but have a long period of use and can be

Root crops in Oregon. Top right, Sakurijima Japanese radish left overwinter. Top left, three types of beets, Cylindra type, Crosby's Flat Egyptian type, and globe type, dug from garden in February. Bottom, American Purple Top rutabagas in garden in February.

stored with success. Few varieties are available.

CARROTS (*Daucus carota* L.)—Because of its nutritional value, bright orange color for the table, productivity, storage life and ease of culture, the carrot is a garden favorite.

If a problem is encountered with carrots, it is often failure to get a stand of seedlings. Carrots can not tolerate either deep planting or a dry seed bed, so the trick is to manage shallow planting with a continuously moist soil. Seeding at ¼-inch depth with a light mulch of sawdust and daily sprinkling is usually successful. If frequent, light irrigation is not possible, use a slightly deeper planting depth (to ½ inch).

Carrots are well adapted to culture in a wide row or bed instead of a single row. Seedlings should be thinned initially if there is not open space around each plant. Harvest of seedlings can begin when they are finger-size and continue through what will appear an inexhaustible supply, because the more the roots are pulled out, the larger the remaining ones will become.

Choose varieties to suit soil conditions. If the soil is deep and friable, any variety will do well and the very long market types such as Imperator or similar F_1 hybrids may be preferred. In heavy, impermeable soil, it is best to grow the shorter types such as the very adaptable Red Cored Chantenay, or even the stubby Oxheart. Nantes is a medium-long type which is tender and of good flavor, but susceptible to cracking and rotting in fall and generally not culturally rugged.

PARSNIPS (*Pastinaca sativa*)—Parsnips resemble carrots in cultural needs and the same suggestions for planting apply, except they should have more space for full development.

Parsnips are slower to mature, but have a long storage life, either in the ground or in the cellar. Drying out is

the greatest hazard in storage, so they should be surrounded by a moist medium or otherwise protected.

Roots left in the ground are remarkably resistant to decay and to freezing injury. Exposure to cold increases the sugar content and greatly enhances the flavor, but non-chilled roots are not poisonous as is sometimes supposed.

Because parsnip roots are very long, digging them in wet, heavy soil is a burden. Lightening the soil with sand and using raised beds makes digging and washing easier.

Only three varieties are normally available: Hollow Crown, Model and All America. All are satisfactory, though Model is smoother and less thinly tapered than Hollow Crown.

SALSIFY (*Tragopogon porrifolius*)—Used mostly for soup, in which it may have a faint oyster-like flavor, salsify or oyster plant is not frequently grown. Generally free from cultural problems, salsify has one disadvantage—the tendency for a branchy root which is difficult to clean and peel.

There is little or no choice of varieties in the United States. Mammoth Sandwich Island is usually listed.

Parsnip seedlings in early true stage.

SPRING RADISHES (*Raphanus sativus*)—This is the short-season, strictly annual type of radish. The term "spring radish" is somewhat misleading because these radishes can be grown throughout the season in cooler areas and in all but the hottest months in warmer climates.

Spring radishes live a very fast life, maturing and becoming pithy and unusable in a remarkably short time. Radish seeds germinate very rapidly and seedlings are fast and vigorous compared to those of most relatives in the cabbage family. They tolerate a wide range of conditions. However, hot, dry conditions encourage strong flavor and even faster maturing. Overcrowding results in small unusable roots. Overfertilization can cause excessive top growth at the expense of root enlargement. The most difficult problem for many gardeners is the cabbage root maggot. Control was discussed earlier in the chapter.

Many varieties of spring radishes are available, especially of the small red globe type. Comet, Sparkler, Cherry Belle, and Early Scarlet Globe are examples. Another distinct variety is the long White Icicle. There are also white globe types which are slightly larger and in some cases longer lasting than the red globe varieties. Try different varieties until you find what suits your particular needs.

WINTER RADISHES (*Raphanus sativus*)—These radishes are slower in growth, much larger, and longer keeping than spring types. They are almost always grown as a fall crop because the decreasing temperatures and day length discourage flowering.

When planted in spring, most varieties flower before sizable roots can develop. An exception is All Season, a long white variety of the Japanese "daikon" type. Well grown daikons are smooth, cylindrical, and up to 18 inches in length while still of excellent quality.

A giant beet-shaped variety, Sakurajima, can reach 50 pounds or more if a long fall growing period is available. Smaller varieties such as Long Black Spanish and Chinese Rose Winter are also available.

Try varieties to find the degrees of pungency desired. Depending on conditions, they vary from very hot to mild. Winter radishes retain good texture for a long period and are good for cooking and pickling as well as fresh table use.

TURNIP (*Brassica rapa*)—The turnip is second to spring radish in quick growth and short life. In hot weather the roots are often strong or bitter in flavor and become pithy almost when they reach maximum size. For this reason turnips are usually grown for harvest in spring and fall and sometimes planted several times during each of these seasons. Cabbage root maggot control is the most critical culture problem.

One variety, Purple Top White Globe, dominates home garden turnip production. There are other choices, however, including pure white, yellow, and red varieties if catalogs are searched. Special varieties for greens such as Shogoin are available, but the tops of any variety can be used.

RUTABAGA (*Brassica campestris* var. *napobrassica*)—Originating long ago from a cross between cabbage and turnip, rutabaga generally resembles turnip but has slower growth, longer storage life, less prickly leaves, firmer flesh, and a great deal more vitamin A.

Rutabagas may be left in the ground for use during winter, if climate permits, or stored for long periods. They do not become pithy if overmature, as turnips do. Cabbage root maggot control is equally critical, however.

Most rutabagas grown are of the variety American Purple Top. Others encountered, such as Laurentian, are essentially the same or of no better quality.

Greens or "Potherbs"—Chard, Collards, Kale, Mustard, Spinach, New Zealand Spinach

by Albert A. Banadyga

Greens include chard, collards, kale, mustard, spinach, and New Zealand spinach. Grown for their tender and succulent leaves and stems, greens are often referred to as "potherbs" since they are usually cooked before eating.

Quite easy to grow, greens require a relatively short growing season. Greens are good sources of some of the vitamins and minerals. They are tasty when cooked fresh from the garden. Raw greens are often added to tossed salads to give them additional color and a different and zestful flavor.

Greens are cool season crops, with the exception of New Zealand spinach. Thus home gardeners normally plant and grow them during cooler periods of the year—spring and fall.

Rapid and continuous growth is essential for both high quality and high yields. To obtain this growth, you need to provide a fairly rich soil containing adequate amounts of organic matter, a good supply of plant nutrients, and a continuous supply of soil moisture.

CHARD (*Beta vulgaris* var. *cicla*)—often called Swiss chard, is a type of beet developed for its large crisp leaves and fleshy leafstalks rather than its roots. It is of quite ancient origin, first reported in the Mediterranean region and the Canary Islands.

Chard is a popular garden vegetable, particularly in the North. It will withstand warm summer temperatures, so that a planting in the early spring can be continuously harvested throughout the summer and fall.

One serving of cooked chard has 13 calories and provides 87% of the Vitamin A and 25% of the Vitamin C required daily by the average adult.

Chard thrives best in a well-drained mellow or friable (crumbly) soil, such as a sandy or clay loam. However, it will grow well in most soils if provided with nutrients and moisture. Soil pH may range from 6.0 to 6.8. Chard is a cool season crop, will withstand light frosts, and does best if planted in the early spring about 2 to 4 weeks before the last frost.

The optimum monthly average temperature for plant growth is 60° to 65° F with a monthly minimum average of 40° and a monthly maximum average of 75°. Soil temperatures for seed germination may range from 50° to 85° with a minimum of 40°, an optimum of 85° and a maximum of 95°.

Adequate soil moisture is especially important for seed germination and early plant growth. Irrigation is particularly beneficial during dry conditions.

Popular varieties include Lucullus, Fordhook Giant, Large White Rib, and Rhubarb (red stemmed). Chard has a multiple seed, as does the beet. Thus one to six plants may emerge from each seed. There are about 1,200 seeds per ounce.

Fertilizing

Before planting, broadcast about 3 pints of a complete fertilizer, such as 10-10-10, to each 50 feet of row. Mix the fertilizer thoroughly with the upper 6 to 8 inches of soil. The seed bed should be thoroughly prepared, free of clods and trash, and slightly firmed.

Distance between rows may be 18 inches. Make a slight furrow in the row, about ½ inch deep, and plant

---

Albert A. Banadyga is Extension Horticulturist, North Carolina State University, Raleigh.

the seeds in the furrow 1 to 2 inches apart. Cover the seed with ½ inch of soil and firm lightly with the back of a rake. Seedlings should emerge in 8 to 10 days.

As the plants grow, periodically thin them out until they are about 12 inches apart. Plants pulled out in the thinning process may be used as greens.

For a very early crop, plants may be started in a greenhouse or coldframe and transplanted to the garden after danger of heavy frosts is over. Plant the seeds in cups, or other small containers, about 3 to 4 weeks before they are to be planted in the open garden. Be sure to "harden-off" the tender plants by gradually withholding water and exposing them to outside weather conditions. Hardening-off should begin about a week before the plants are set out in the open.

Control weeds by hoeing before the weeds get a good start. Or better still, mulch the young plants with such materials as straw, grass clippings, newspapers, or black plastic film.

Be on the lookout for such insects as cabbage worms, aphids, beet leaf miner, and flea beetle. Worms may be removed by hand picking, aphids may be washed off with a fine spray of water from a garden hose, or recommended insecticides may be used. Crop rotation and sanitation will help reduce damage from leaf spots and other diseases.

For rapid and continuous growth, sidedress the crop about a month after planting. Repeat at 4- to 6-week intervals. At each sidedressing uniformly distribute one pint of a complete fertilizer (or one cup of sodium nitrate) per 50 feet of row. Place the fertilizer in a band 4 to 6 inches out to the side of the plants, making certain not to get any on the plants themselves.

Leaves and stems are ready for harvest about 50 to 60 days after planting. With a sharp knife cut off a few of the outer leaves, about an inch above ground, while they are still tender and succulent. Cut carefully to

Left, Fordhook Giant Swiss chard. Right, Swiss chard, properly spaced and with outer leaves ready for harvest. These average about a foot high, and should not be left to become coarse and overgrown.

avoid injury to younger leaves and the central bud. Continue harvesting throughout the summer and fall. As you remove the outer stems and leaves, new ones will continue to form and grow from the central bud.

Harvesting should be continued, regardless of whether or not the greens will be used, or else new leaves will not be available later in the season.

A good yield, for the full season, is about one pound of chard greens per foot of row. A 30-foot row will supply an average family of 4 with an adequate supply of fresh chard greens throughout the season.

Some gardeners dig up the plants just before the first heavy freeze in fall, and store the entire plant in a protected cellar or coldframe for continued harvest into the winter. Plants are stacked upright, with roots in contact with the soil, and watered lightly to prevent excessive wilting and to encourage a very limited amount of continued growth.

COLLARDS (*Brassica oleraceae* var. *acephala*)—originated in the British Isles and Western Europe. The collard is often called a non-heading cabbage, since it does not form a true head but rather a large rosette of leaves. It belongs to the cabbage family and its culture and use are quite similar to those of cabbage. It may be grown throughout the year in the South, and as both a spring and fall crop in the North.

One serving of cooked collards has 21 calories and provides 87% of the Vitamin A, 74% of the Vitamin C, and 14% of the calcium in the minimum daily requirements of an average adult.

Collards may be grown on a very wide range of soils, but sandy, silt, or clay loams are preferred. Soil pH may range between 5.5 to 6.8, with 6.0 being ideal. The plant is a heavy feeder, often growing to a height of 3 to 4 feet.

Optimum monthly average temperatures for plant growth are 60° to 65° F with a monthly minimum average of 40° and a monthly maximum average of 75°. Soil temperature for seed germination may range from 45° to 95° with a minimum of 40°, an optimum of 85° and a maximum of 100°. Collards will withstand a greater range of temperature, both heat and cold, than most other vegetables grown in the South. If the temperature drops gradually over a period of several days, collards can withstand temperatures as low as 15°.

Collard seed will germinate in 4 to 9 days even under low soil moisture conditions. The plants grow best in well-drained soils that are provided with adequate moisture by rainfall or irrigation.

Popular Varieties

Vates, Morris Heading, Georgia, Cabbage-Collard, and Green Glaze are popular varieties. One-fourth ounce of seed is sufficient to plant 100 feet of row. There are about 8,000 seeds per ounce.

Collards are hardy and may be planted in the spring about 4 to 6 weeks before the last spring frost and again in the fall about 6 to 8 weeks before the first fall frost. Prepare the soil as suggested for chard. Broadcast 4 pints of fertilizer, such as 10-10-10, per 50 feet of row and mix it thoroughly with the soil before planting. Rows are normally 3 to 4 feet apart.

Collards may be seeded directly in the garden or transplanted. When seeding directly in the garden, plant the seed ¼-inch deep and about 1 inch apart in the row. Seedlings will emerge in about 5 days. Plants may be left at the seeded spacing, or thinned to 6, 12, or 18 inches apart depending on how they will be harvested. The plants pulled out in thinning may serve as transplants or trimmed and used as greens.

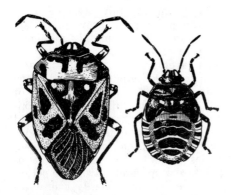

Harlequin bug. Adult, left. Nymph, right.

Transplants may be grown in protected beds in the spring and in open plant beds in the summer. Seeds are planted in individual containers, flats, or rows in ground beds about 4 to 6 weeks before time to transplant into the garden. When plants are 6 to 8 inches tall, set them in the garden row 12 to 18 inches apart, and water well.

About a month after planting in the garden, sidedress collards as suggested for chard.

Control weeds by hoeing or mulching. Insects that may cause problems include cabbage worms, aphids, harlequin bugs, and root maggots. Two diseases that may cause damage are downy mildew and black leg. General control measures are the same as suggested previously for chard.

Collards may be harvested by three general methods, or a combination of the three. The entire young plant may be cut off at ground level just as are mustard greens, the entire mature plant may be cut off at the ground, or the bottom leaves may be stripped off the plant periodically leaving the bud to grow and produce more leaves. The last method is most popular with home gardeners since it entails making only one planting, and spring-planted collards may be continuously harvested throughout the summer and into winter.

When the entire small or immature plants are to be harvested, successive plantings may be made at 2- to 3-week intervals. Immature plants may be harvested about 40 days and mature plants about 75 days after planting.

Collards tend to improve in flavor as the weather becomes cooler in the fall. Many gardeners do not harvest the fall crop until after the first frost. Leaves remain tender and edible for several weeks after they reach maturity or full size.

Yields vary from ⅓ to 1½ pounds per foot of row, depending on harvesting methods used.

KALE (*Brassica oleraceae* var. *acephala*)—is a native of Europe and recorded use dates back to 200 B.C. Like collards, it is a member of the cabbage family and is grown for its succulent leaves and stems.

A hardy vegetable, kale can be overwintered in latitudes as far north as southern Pennsylvania and in areas having similar winter conditions. It is also quite heat-resistant and may grow in the summer, but its greatest value is as a cool weather green. No other vegetable is so well adapted to fall sowing in areas having winters of moderate severity.

One serving of cooked kale has 21 calories. It provides an average adult with all his daily requirements of Vitamin A and Vitamin C as well as 13% of his daily calcium requirement.

Temperature, soil, fertility, and moisture conditions for kale are the same as previously suggested for collards.

Vates, Dwarf Siberian, and Dwarf Blue Scotch are good standard varieties. Seed may be planted in the spring 4 to 6 weeks before the last killing frost, and in the fall 6 to 8 weeks before the first killing frost. There are about 10,000 seeds per ounce.

Rows may be 18 to 24 inches apart.

Seed is planted in the row an inch apart and a half inch deep. Seedlings will emerge in 3 to 5 days. Plants may be left as thick as seeded or gradually thinned until they are 8 to 14 inches apart. The plants pulled out in thinning may be used as greens.

Pests and pest control for kale are similar to those suggested previously for collards.

Kale may be harvested in one of two ways. The entire young plants may be cut off at ground level, about 40 days after seeding. This process is used when the plants are left unthinned after seeding.

When plants are spaced 8 to 14 inches apart, the lower leaves are stripped off periodically while the bud and a rosette of leaves are left to continue growth for future harvests. This second harvest method requires about 50 to 60 days from seeding to first harvest. Leaves should be harvested before they become old, tough, and woody.

A foot of row will produce about a half pound of kale greens.

MUSTARD (*Brassica juncea* var. *crispifolia*)—also known as "mustard greens," is a short season crop grown for its tender leaves and stems. This crop had its origin in China and Asia. A different species, black mustard, is grown for its dark seed which is used in making the condiment known as table mustard.

One serving of cooked mustard greens has 16 calories and provides 91% of the Vitamin A, 74% of the Vitamin C, and 12% of the calcium in the daily requirements of an average adult.

Soil, fertility, and moisture requirements for mustard greens are similar to those previously suggested for collards. This crop will not withstand the extremes in temperature that kale

Top, collards are a nutritious crop found in many gardens of Southeastern U.S. Left, collard plant ready for harvest.

and collards will. It also bolts (goes to seed) much more rapidly, particularly in spring.

Optimum monthly average temperature for plant growth is 60° to 65° F with a monthly minimum average of 45° and a monthly maximum average of 70° to 75°. Optimum soil temperature range for seed germination is 60° to 105° with a minimum of 40°, an optimum of 85°, and a maximum of 105°.

Reliable mustard varieties include Southern Giant Curled, Tendergreen, Florida Broadleaf, and Green Wave. One-fourth ounce of seed will plant 100 feet of row. There are about 15,000 seeds per ounce.

Soil preparation and fertilization are similar to those previously suggested for collards. Seeds are planted directly in the row about 4 to 6 weeks before the last frost in the spring and about 6 to 8 weeks before the first frost in the fall. Additional successive plantings may be made at two-week intervals. Rows may be 15 to 30 inches apart. Seeds are planted in the row, an inch apart and ¼ inch deep. Seedlings will emerge in 3 to 5 days.

Plants may be left unthinned or thinned to 4 to 6 inches apart. The plants thinned out may be used for greens.

Pests and pest control are the same as described for collards.

The young tender leaves are harvested any time after they reach 6 to 8 inches in height and before they become tough and woody. Harvest begins about 35 to 40 days after seeding. Normally the entire plant is cut off slightly above the ground. In the Deep South, mustard greens may be carried over into winter and harvested by stripping the lower leaves similar to collards and kale.

One foot of row will yield about a half pound of mustard greens.

SPINACH (*Spinacia oleraceae* var. *inermis*)—was cultivated by the Persians over 2,000 years ago. The edible portion of the plant is the compact rosette of leaves before the central bud begins to elongate to form a seedstalk.

This is a hardy cool weather crop. In most of the United States it is grown as an early spring and a late fall crop. In some areas having mild summer temperatures, spinach is grown continuously from early spring to late fall. In portions of the South it may be planted in the fall and harvested during the winter and early spring.

Below, rows of spinach in a garden. Bottom, Long Standing Bloomsdale spinach.

A serving of cooked spinach has 20 calories. It provides an average adult with 100% of the Vitamin A, 56% of the Vitamin C, and 28% of the iron indicated in the minimum daily requirement.

Spinach will grow on almost any fertile soil that is well drained and has a good supply of organic matter. Avoid poorly drained soils or those that cake or crust easily. The soil pH range is 6.0 to 6.8. Spinach grows poorly on soils with a pH below 6.0.

Optimum monthly average temperature for plant growth is 60° to 65° F. Optimum monthly minimum average is 40° and optimum maximum average is 75°. Optimum soil temperature for seed germination is 70° with a minimum of 35° and a maximum of 85°.

Spinach is shallow rooted; provide adequate soil moisture for rapid and continuous growth.

Among the leading spinach varieties for fall production are Hybrid #7, Virginia Savoy, Resistoflay, Viroflay, and Chesapeake. Among the better spring varieties are Long Standing Bloomsdale and America. A half ounce of seed will plant 100 feet of row. There are approximately 2,800 seeds per ounce.

Spinach is normally planted about 4 to 6 weeks before the last frost in the spring and again 6 to 8 weeks before the first frost in fall. Two to three successive plantings may be made at 2- to 3- week intervals.

Rows are usually 14 to 30 inches apart, depending on cultivation equipment to be used. Before planting, broadcast about 3 to 4 pints of fertilizer, such as 10-10-10, per 50 feet of row and mix thoroughly with the soil. Plant seeds in the row a half inch deep and an inch apart. It takes spinach seedlings about 5 to 8 days to emerge. Thin plants to a 3- to 4-inch spacing before they become crowded in the row.

Keep weeds out by hoeing before they get a good start. Use varieties resistant to mildew and yellows. Observe plants closely for aphids, leaf miners, or cabbage worms. Use similar control measures as suggested under chard.

Spinach may be harvested from the time the plants have 5 to 6 leaves until just before seedstalks develop. This period is usually 35 to 45 days after seeding. Harvest by cutting the entire plant off, just above ground level, with a sharp knife.

One foot of row will yield a third to a half pound of spinach greens.

NEW ZEALAND SPINACH (*Tetragonia expansa*)—is a native of New Zealand, Japan, Australia, and South America. It was introduced to England in 1771 and is presently grown to a very limited extent in the United States. Not a true spinach, it does resemble spinach somewhat in appearance and is used similarly.

The plant is large, growing to a height of two or more feet in a spreading and branching habit of growth, and has thick succulent leaves. It is a warm season crop and an excellent source of fresh greens throughout the summer.

One serving of cooked New Zealand spinach has 11 calories and provides 72% of the Vitamin A, 27% of the Vitamin C, and 10% of the iron in an average adult's minimum daily requirement.

New Zealand spinach thrives in a well drained loamy soil, high in organic matter and fertility. A soil pH of 6.0 to 6.8 is desirable. Fertility and soil moisture needs are similar to those for spinach. The plant grows best with a monthly average temperature of 60° to 75° F, a monthly minimum average of 50° and a monthly maximum average of 95°. Optimum soil temperature range for seed germination is 70° to 95° with a minimum of 60° and a maximum of 100°.

Since this is a warm season crop,

New Zealand spinach.

delay planting until danger of spring frosts is over and the soil has warmed.

Space rows 3 to 4 feet apart. Before planting, broadcast 3 pints of fertilizer, such as 10-10-10, per 50 feet of row and mix thoroughly with the soil.

Seed is large and irregularly shaped with a count of about 350 per ounce. Plant seeds in the row 1 inch deep and 4 to 6 inches apart. To insure prompt germination, soak seed in warm water for 2 to 3 hours before planting.

Seedlings normally emerge in 7 to 12 days. Plants may be gradually thinned out until they are spaced 12 to 18 inches apart in the row.

Control weeds by hoeing or with a mulch. There are no insects or diseases of consequence.

About 4 to 6 weeks after planting, sidedress with either one pint of fertilizer (such as 10-10-10) or ½ cup of nitrate of soda per 50 feet of row.

Harvest may begin about 70 days after seeding. Successive harvests of the tips are made from a single planting. At each harvest about 3 inches of the tips of the branches are cut or pinched off. This results in more branching and more new succulent growth. The thick leaves as well as the tender stems are cooked.

Harvesting continues throughout the summer and until the first fall frost. Take care not to remove too large a portion of the plant at one time. During early harvests, a half to two-thirds of the branch tips may be taken at one harvest. As the plant continues to branch a greater portion of the tips may be harvested at one time.

One foot of row will yield about three-fourths pound of greens for the entire season.

Summary: Greens are easy to grow and quite nutritious. For maximum yields and quality, growth should be rapid and continuous. To insure such growth, provide for a well drained soil, add organic matter if needed to keep the soil loamy and retentive of moisure and nutrients, maintain an adequate fertilizer level, provide for adequate moisture, plant during the proper season, control pests, and harvest at peak of quality.

Beans and Peas Are Easy to Grow And Produce a Wealth of Food

by Jack P. Meiners and John M. Kraft

Beans and peas are among the most satisfying vegetables for home garden growing and eating. The seeds are planted directly in the garden, and germinate rapidly to give rise to plants that grow quickly and vigorously. This gives even the beginning gardener a feeling of accomplishment. In fact, beans and peas should be in every garden because they are easy to grow and provide a wealth of food.

When we speak merely of beans and peas we are over-simplifying the situation regarding legumes of interest to the home gardener. Actually, he can choose from several types of legumes, all of which make tasty and highly nutritious food for the family. Beans may include snap or string beans, dry beans, and lima beans. Peas include garden pea—also called English pea—and southernpea, also known as cowpea.

Beans and peas are discussed together in this section because, botanically speaking, they are related. Both are members of the legume family, which bear the characteristic butterfly-like flowers and have the capacity with the help of bacteria in the root nodules to take nitrogen out of the air. Otherwise, beans and peas differ from each other somewhat in culture. Most beans and southernpea are warm weather plants while garden pea requires cool growing conditions.

Snap Beans

Snap bean is one of several types

J. P. Meiners is Chief, Applied Plant Pathology Laboratory, Agricultural Research Center, Beltsville, Md. John M. Kraft is Research Plant Pathologist, Agricultural Research Service, Irrigated Agriculture Research and Extension Center, Prosser, Wash.

of bean known collectively as "common bean" and by the scientific name, *Phaseolus vulgaris*. The home gardener may grow several kinds depending upon how he plans to use the product.

Those grown for the immature pods are known variously as "snap" beans, "string" beans, "green" beans, "French" beans, or "garden" beans. Other types grown for the immature seeds are known as "green shell" beans, and still others grown for the mature seeds are called "dry" beans. All the types are similar in requirements for growing in the garden.

Snap beans, used when the pods are immature, need only a short growing season and thus are favored by most home gardeners.

The common bean probably originated in Central America, but has been cultivated throughout much of North and South America by the Indians since prehistoric times. Because the seeds were easily carried and stored, and because beans grew well in a variety of climates, they were widely distributed by the explorers and now are widely grown throughout the world.

The Indians probably consumed beans primarily in the dry or green shell state, and early varieties of snap beans were stringy—hence the term "string" beans. The tender, stringless and nearly fiberless varieties we know today were developed within the last 50 years.

Snap beans are a good source of vitamins A and C, thiamine, and riboflavin. They are also a good source of calcium and iron.

Snap beans are a warm season crop and easily injured by frost. The first of several successive plantings should be made at about the time of the aver-

age frost-free date for your area. Usually there is not much to be gained by planting earlier, since early planted beans require longer to mature and there is the risk that seed may rot in the cold soil, with additional delay if reseeding becomes necessary.

Snap beans grow best where the average maximum temperature does not exceed 85° F and the average minimum temperature does not go below 50°, but the most desirable range is between 70° and 80°. Very high temperatures lower the yield due to blossom drop.

The soil should be warm at planting time. Snap bean seed germinates poorly at soil temperatures below 50° F and the optimum range is between 60° and 85°.

In the lower South and Southwest, snap beans may be grown during all seasons except mid-winter, but should not be planted so that podding occurs when weather is too hot or cold.

Snap beans grow well in a wide range of soils, the best being those that are well drained and reasonably fertile. The physical nature of the soil should be friable and not interfere with emergence. Upon germination, the two large seed halves or cotyledons must emerge through the soil and they can be seriously hampered by compact or crusted soil.

If your soil is very heavy you may need to cover the seed with sand, peat, leaf mulch, or other material that will not form a crust. Should a crust form following planting, it may be necessary to break it to allow the seedlings to emerge from the soil. If you live in an area of heavy rainfall you may need to plant beans on raised beds to get proper drainage.

Snap beans are not heavy users of water but require a constant supply. One inch per week is sufficient on most types of soils. An adequate supply of moisture is important from bud formation to pod set. Excessive or too little moisture may cause blossom and pod drop. If soil is too dry at planting time, it is preferable to irrigate first, then plant.

Bush and Pole

There are two basic plant habits—bush and pole. The bush is a low, self-supporting plant that grows 1 to 2 feet in height. Pole beans are vines which must be supported by stakes or a trellis.

Use bush varieties for quick production and pole types for a longer season. Successive plantings of bush types every 10 to 14 days will provide beans for most of the growing season. Because pole beans bear over a longer period, usually one planting suffices.

Disease-free Western-grown seed should be used rather than that grown by the gardener since snap beans are

Tendercrop snap bean is mosaic-resistant and heavy yielding. It has tender, round, green pods and wide range of adaptability.

Planting Guide for Beans and Peas

Type	Pounds of seed per 50-ft row	Depth of planting, inches	Spacing between rows, inches	Spacing in rows, inches Seeds	Spacing in rows, inches Plants	Days to harvest
Snap beans, bush	¼–½	1–1½	18–30	1–2	2–4	50–60
Snap beans, pole	¼–½	1–1½	24–48	3–6	4–8	60–70
Lima beans, bush	¼–½	1–1½	18–30	2–4	4–8	65–75
Lima beans, pole	¼–½	1–1½	36	3–6	6–8	70–90
Peas, garden	¼	1–2	6 (double rows) 36–48	1–1½	1–1½	55–70
Peas, southern	¼–½	1–1½	30–54	2–4	2–4	55–80

Don Normark

subject to diseases carried on the seed. Do not soak the seed prior to planting because seeds of many bean varieties tend to crack and germinate poorly under extreme moisture conditions.

Bush-type beans are most commonly planted in rows 18 to 30 inches apart with seeds placed 1 to 2 inches apart and thinned so plants are 2 to 4 inches apart.

Pole beans may also be planted in rows 2 to 4 feet apart with vines supported by poles, a fence, or a trellis made of posts and twine. In addition, they may be planted in hills about 3 feet apart each way, with 6 to 8 seeds in each hill, and later thinned to 4 to 5 plants. A pole is placed in the center of every hill with each pole upright, or four poles are tied together in wigwam fashion. Gardeners usually train the vines to climb the supports in a clockwise direction.

Bean seed should be covered not more than 1 inch in heavy soils and 1½ inches in sandy soils.

If the vegetable garden has received a general application of manure, com-

Corn may get as high as an elephant's eye, but pole beans can top that when properly trellised.

post, or commercial fertilizer, beans should need no additional fertilizer during the growing season. Where no fertilizer has been applied, a moderate application of one such as 5-10-5 usually is all you need. Avoid heavy applications of fertilizers high in nitrogen as they may cause heavy vine growth, delayed maturity, and a small yield of pods.

On light soils or after periods of heavy rainfall, you may need to sidedress the plants with a nitrogenous fertilizer during the growing season.

Mulching of beans with organic materials or black plastic is recommended since it conserves moisture and prevents weed growth. Black plastic also hastens early season growth.

Control weeds by cultivating or hoeing. These operations should be very shallow since bean roots are close to the surface. In cultivating, throw soil against the bean stems to aid in support and development of additional roots. An adequate mulch greatly decreases weed growth.

Pest Control

Snap beans suffer from diseases and insect pests, the number and intensity of attack depending upon geographic location. In the West, virus diseases and root rots are the most serious diseases. In the East and South, bacterial and fungus leaf and pod diseases are most important.

Diseases are best controlled by preventative measures such as using disease-free, Western-grown seed which controls bacterial diseases and anthracnose; planting disease-resistant varieties when available which controls virus diseases and rust; and using seed treated with pesticide to control soil-borne diseases and insects.

Practicing cleanliness and sanitation in the garden controls many diseases and insects. Not handling or working among the bean plants when the foliage is wet from dew or rain helps control bacterial diseases. Not planting beans in the same ground year after year reduces root rots.

Seed catalogs or State agricultural experiment station publications usually indicate if varieties are resistant to particular diseases.

Insects must be controlled with insecticides applied to the plants (for aphids, Mexican bean beetle, mites, potato leafhopper), or seed (for the seed corn maggot). Contact your local county agent or State university for specific control recommendations.

Snap beans are best for eating while the pods are still young, the seeds still small, the interior flesh is firm, and the pod wall fiber content low. The proper stage for picking lasts only a few days and delay can mean a poor quality product.

Bush varieties usually yield three or four pickings whereas pole varieties yield numerous pickings. Regular and thorough picking of both types is important because it causes the plants to continue to set pods longer.

Yields can vary tremendously, but bush beans yield about 50 pounds per hundred feet of row and pole beans about 60 pounds.

Green shell beans are harvested and shelled out after the seeds are nearly fully grown, but before they have hardened and dried. Snap bean varieties may be used for green shell beans, but it is better to plant one of the horticultural varieties for this purpose. Depending on variety, from planting to green shell stage will take 55 to 75 days.

Dry beans are harvested after the pods are mature and dried or partially dried. Delay in harvesting may result in loss of seed due to shattering.

Lima Beans

Natives of the Western Hemisphere, limas probably originated in Central America. The small seeded types have been cultivated since prehistoric times in North America. The large seeded

Below, Fordhook 242 bush lima beans are vigorous, productive and heat-resistant. Right, trellised lima beans. Bottom, lima pole beans growing on strings.

types were developed in South America, specifically in Peru, hence the name "lima" from the capital city of that country.

The growing of lima beans is similar to that of snap beans. Therefore, this section will emphasize mainly the cultural requirements specific for limas, and we suggest that the prospective lima bean grower read the section on growing snap beans also.

Lima beans fall into two classes as far as the gardener is concerned—the large seeded, generally referred to as Fordhook type, and the small seeded, known as baby limas. The scientific name for both types is *Phaseolus lunatus*. In the South the lima is called butter bean.

Both large and small seeded limas come in bush or pole types similar to those of snap beans.

Lima beans are used as green shell beans when the seeds have developed to nearly full size. The pods are not consumed. Nutritionally, lima beans are high in the vitamins thiamine and riboflavin and in phosphorus and iron.

Lima beans require warmer soil and air temperatures than snap beans, and thus are planted somewhat later. A planting date of 2 weeks after the average date of the last frost probably is a good rule-of-thumb. Furthermore, lima beans need a frost-free period of 3 to 4 months with relatively warm days and nights.

Proper soil temperature is critical for seed germination and should be 65° F for quick emergence of seedlings.

Gardeners in most of the more northerly parts of the United States, including the northern New England States and parts of other States along the Canadian border, probably should not attempt to grow lima beans. Just south of this region only bush baby limas should be grown since they mature in a shorter period than large seeded bush or pole limas.

Lima beans grow best in lighter-textured, well-drained soils. They need a soil somewhat richer than required for snap beans, but not excessive in nitrogen.

Don't plant lima beans until two or more weeks after snap beans are first planted. To lengthen the growing season, plant protectors can be used over the seeds and young plants in marginal areas—or to obtain earlier growth in other regions. It is possible also for gardeners to start limas indoors and transplant to the garden. Be sure the root systems are not disturbed in transplanting.

Because of requirements for a longer growing season, successive plantings of bush lima beans should not be made except in the more southern areas of the United States.

Seed Treatment Vital

If at all possible, lima bean seed should be treated with a fungicide and insecticide before planting, especially if the soil temperature is below 65° F. Consult your seedsman, county agent, or State university as to the proper seed treatment and follow the label instructions carefully.

Lima beans may be attacked by a number of diseases and insects. Many of these pests also attack snap beans. As mentioned previously, lima beans are more subject to seed and seedling diseases, so seed treatment is essential to obtaining good stands and sturdy plants.

In the Mid-Atlantic States mildew may be a problem and can be controlled by fungicides or use of resistant varieties, if available. In the South anthracnose may be serious and should be controlled with a fungicide. Nematodes may be a problem, particularly in the South. These can be controlled through resistant varieties and soil fumigation.

Insect control through insecticides may be necessary.

Consult your county agent or State university for diagnosis and recommendation of specific control measures against pests.

Lima beans are ready for picking when the pods are well filled but still bright and fresh in appearance. The end of the pod should feel spongy when squeezed.

Bush limas can be picked for about 3 weeks and pole limas for about 4 weeks or until frost. Depending upon soil fertility, temperature, moisture, and many other factors, yields of from 20 to 40 pounds per 100 feet of row may be expected.

Garden Peas

The garden pea (*Pisum sativum* L.) is thought to have originated in Eastern Europe or Western Asia and was widely distributed in prehistoric times. It has been traced back to the stone age where dried seeds were found among relics of the Swiss Lake villages.

However, the eating of green peas was not referred to until the Norman conquest around 1066.

Green pea consumption was not common until the 18th century but with the appearance of canning and freezing, peas have become an important vegetable crop. To distinguish them from the field pea, used dried and split, or the southernpea, garden peas are often called green peas or English peas.

Peas are a cool weather, rapid maturing crop. They must be planted early for maximum yield and should be brought to maturity under cool conditions.

The garden pea thrives best when grown in the South and lower parts of California during fall, winter and early spring. Farther North, peas thrive when grown in spring or autumn. In the Northern States and at higher elevations, they may be grown from spring until autumn; however, if summer heat is too severe, the season may be limited to spring.

Sandy or rocky soils usually result in early crops, but plants on these soil types frequently suffer from water stress if there is no supplemental irrigation.

Select a seed bed site which is uniform, level and well drained. The seed bed should be worked at least to a depth of about 2½ inches. A well prepared seed bed is essential for uniform germination and seedling stands. Peas will not thrive on poorly drained or water soaked land.

An adequate range in soil pH for peas is between 5.5 and 6.7. A soil with excessive organic residue or nitrogen is not ideal for pea growth because it promotes rank vine growth at the expense of pod production.

Peas are usually planted from 1 to 1½ inches deep (heavy soils) to 2 inches deep (sandy soils) as soon as the soil can be properly worked. Peas are sown 8 or 10 to the foot.

The seed should be treated with a commercial seed protectant and planted in single or double rows. The double rows should be about 6 inches apart which allows dwarf or bush varieties to cling and hold one another up. When planted in single rows, the dwarf varieties should be sown in rows 3 feet apart and the taller varieties 4 feet apart.

If the plants are to be supported, wire netting or string trellis can be put between the rows. Tall varieties usually do better when grown on a trellis. However, left unsupported they will form a ground cover, smothering the weeds.

Staked or trellised plants will usually require some hand weeding. Always avoid deep hoeing that may injure the roots.

Peas have a number of disease and insect pests. Preventive measures are desirable—for example, protective seed treatments, never plant peas in the same location in succession, and remove old vines from the garden.

If a disease or insect problem appears that you aren't familiar with,

Picking and sampling green peas.

consult your county agent. And keep in mind the many excellent books and articles available from public libraries and garden stores, as well as the extension bulletins on vegetable gardening and pest control.

In general, harvest the pods when they appear well filled but before they begin to harden or fade in color.

Peas will yield the maximum food value when the seeds are full size. A few days before this stage, however, peas are at their prime for taste and tenderness.

The number of pickings will be at least two or three, since pea pods do not all mature at the same time but usually during a period of 7 to 10 days.

Pull the pods carefully off the vine, or plants may be uprooted. For best quality, pick peas just before meal preparation or processing because sugar conversion to starch will begin only a few hours after picking.

Edible podded peas (sugar peas, Chinese peas, snow peas) have pod walls that are tender, brittle, succulent, and free from fiber. After destringing, the young pods are cooked whole (like snap beans), or used as greens in salads. If pods develop too fast, eat the shelled peas.

Southernpeas

In the South the unqualified word "pea" usually refers to the southernpea, also known as cowpea and protopea. There are several distinct types of southern pea of which blackeye, crowder, and cream are the best known. Each of these types has its own unique appearance and flavor.

Southernpeas (*Vigna unguiculata*) are highly nutritious, tasty, and easily grown. They deserve to be much more widely grown by home gardeners.

Southernpeas are natives of Africa, brought to the West Indies by slave traders. From there they are believed

Terence O'Driscoll

to have been introduced to the United States in the 1700's. Southernpeas are consumed as fresh shelled peas, sometimes mixed with the immature pods, and as dry peas. Nutritionally, they are similar to lima beans.

The yard-long bean, or asparagus bean, a pole type of cowpea, produces pods 1 to 2 feet long. However, it is less productive than other varieties, and commonly used in Chinese cookery.

The southernpea is a warm weather crop and should be planted much later in the spring than snap beans. Adequate stands are difficult to achieve until the soil is quite warm. Although well suited to summer culture in the South, it also is adapted to northern conditions of about the same range as the lima bean.

Southernpeas are adapted to a wide range of climatic, soil and cultural conditions and are particularly drought and heat tolerant. Excess moisture causes reduction in yield.

As with beans, southernpeas may be classified as vining, semi-vining, or bush types. Also, there are short-, mid-, and long-season varieties. The bush or compact varieties that mature quickly are most suitable for home gardening, particularly outside the deep South. Southernpeas are usually planted in rows 2½ to 3½ feet apart with an inter-row spacing between seeds of 2 to 4 inches.

Heavy applications of nitrogen fertilizers should not be used for southernpeas. An application of a low nitrogen fertilizer such as 4-12-12 applied at the equivalent of 1½ to 2½ pounds per 50 feet of row usually is adequate.

Seed should be purchased from a seedsman as it is more apt to be free of disease and true to variety than if the gardener saves his own seed. If possible, the seed should be treated with a fungicide and insecticide.

Planting southernpeas.

In the South, successive plantings made about three weeks apart until mid-summer will give a continuous supply of green peas.

Various insects and diseases attack southernpeas. Some can destroy the crop while others do little damage.

The most destructive insect, especially in the South, is the cowpea curculio which feeds on the pods and seeds as they develop. Other insects also may be destructive to the plant and several species of weevils may seriously damage stored cowpea seed.

Application of insecticides is essential for successful culture of southernpeas in the South, especially during the fall. Several insects, such as the southern green stinkbug and cornworm, cause great damage during this time of year.

Specific information on insecticide application may be obtained from your county agent or State agricultural experiment station.

Diseases include wilt, root knot, and those caused by viruses. They may be best controlled by use of resistant varieties, planting disease free and treated seed, removing old vines, and not planting peas in the same location in successive years.

Southernpeas for home use as fresh-shelled peas are harvested when the deep green pod color changes—depending upon specific variety—to light yellow, silver, red or purple. The peas should be almost maximum in size but with appreciable green color still in the cotyledon.

A rule-of-thumb states that peas should be harvested 16 days after bloom, but this depends on the temperature. Depending on variety and environmental conditions, peas may be ready for picking from 55 to 80 days after planting.

For Further Reading:

Grow Your Own Green Beans, Extension Service, Ext. Cir. 886, Industrial Bldg., Oregon State University, Corvallis, Oreg. 97331. 5¢

Grow Your Own Vegetables, Cooperative Extension Service, Ext. Cir. 559, Pennsylvania State University, University Park, Pa. 16802. 10¢

Growing Vegetables in the Home Garden, H. and G. Bul. 202, for sale by Superintendent of Documents, U.S. Government Printing Office, Washington, D.C. 20402. 95¢

Home and Farm Vegetable Garden, Ext. Cir. 871, Extension Service, Industrial Bldg., Oregon State University, Corvallis, Oreg. 97331. 15¢

Home Gardens, Cooperative Extension Service, Ext. Cir. 442, Washington State University, Pullman, Wash. 99163. 20¢

Home Vegetable Gardening, Cooperative Extension Service, Ext. Cir. 457, New Mexico State University, Las Cruces, N. M. 88003. Free

Illinois Vegetable Garden Guide, Cooperative Extension Service, Ext. Cir. 1091, University of Illinois, Champaign-Urbana, Ill. 61801.

More Vegetables From Your Garden, Extension Division, Pub. 657, Virginia Polytechnic Institute and State University, Blacksburg, Va. 24061. 12¢

Plant Diseases, Cooperative Extension Service, College of Agriculture, E. M. 3540, Washington State University, Pullman, Wash. 99163. 62¢

Soil Sanitation Procedures for the Home Gardener, Cooperative Extension Service, College of Agriculture, E. M. 3844, Washington State University, Pullman, Wash. 99163. 5¢

Sweet Corn, That Home Garden Favorite For Good Nutrition and Eating Pleasure

by E. V. Wann

Sweet corn is a common item in most American home gardens. It provides a delightful addition to everyday meals, and in season the roasting ears are enjoyed for picnics and cookouts. Since the days of the Pilgrims, corn-on-the-cob has been a popular American favorite. Sweet corn— either fresh, frozen or canned—may be served as a separate dish or used in succotash (an American Indian dish), custards, puddings, fritters, souffles, and stuffed peppers, or added to soups and chowders. Sweet corn may also be used in relishes and mixed pickles.

Most gardeners will regard sweet corn as an essential item in their garden and take great pride in the good nutrition and eating pleasure it affords.

Corn (Zea mays) is a member of the grass family, which includes other cereal crops such as wheat, oats, barley, sorghum, and rice. Corn is conveniently divided into six types based on its use and kernel characteristics. These are dent corn, sweet corn, popcorn, flint corn, flour corn and pod corn. They are all of the same species but differ genetically. Dent corn, sweet corn, and popcorn are the most commonly grown for their food and feed value throughout the world.

Sweet corn is believed by most authorities to have originated in North America as a mutation from field corn. The first references to sweet corn date from 1779; an 8-rowed, red-cob type called Susquehanna, or Papoon, was introduced that year near Plymouth, Mass. In Thomas Jefferson's *Garden Book* (1810) "shriveled corn" is mentioned, which is obviously sweet corn.

By 1828 "sugar corn" was listed in New England seed catalogs. Also, evidence indicates sweet corn was being grown by the American Indians of the upper Missouri by 1833. Another early reference to sweet corn appeared in the *Travel Letters* (1821) of Timothy Dwight as being the most delicious vegetable of any known in this country.

Sweet corn as a specific crop must have come into existence at least by 1820 and reached sufficient popularity by 1828 to be in a seed catalog. The subsequent history of sweet corn is one of variety development. By 1900 there were no less than 63 varieties, and the first F_1 hybrid was introduced about 1924. Today, there are well over 200 varieties and hybrids available to sweet corn growers and gardeners.

Sweet corn differs from the other types of corn primarily by its ability to produce and retain greater quantities of sugar in the kernels. This characteristic is conditioned by a single recessive gene called *sugary-1*, symbolized su_1. Other less pronounced differences are its tender kernels at edible maturity, refined flavors, a tendency to produce suckers at the base of the plant, and wrinkled seeds when dried. Dent corn is considered the "normal" type with all the other types being genetic variation (mutations) of it. Popcorn, for example, has very hard starch in the kernels that expands explosively when heated, thus producing the fluffy white popcorn kernel.

In recent years a new kind of sweet corn has come into use that is sweeter

---

E. V. Wann is Research Geneticist and Laboratory Director at the U. S. Vegetable Laboratory, Agricultural Research Service, Charleston, S.C.

181

Well-filled ear of sweet corn.

than the standard sweet corn. Its sweetness is not conditioned by the *sugary-1* gene but by a similar genetic factor designated *shrunken-2* (sh_2). This gene conditions an even higher level in sugar in the kernels, giving them a sweeter taste and prolonging the edible state by three or four days.

The different types of corn should never be planted together at the same time. Pollen from dent corn or popcorn will contaminate sweet corn, causing the kernels to be starchy and not sweet. Likewise, the standard sweet varieties should not be interplanted at the same time with the extra sweet (*shrunken-2*) varieties, as the pollen from one will contaminate the other—destroying the quality of both. If both types are to be planted they should be separated by at least 400 yards distance, or one planted about four weeks after the other so they are not pollinating at the same time.

Climatic Needs

Sweet corn is essentially a warm-weather crop. It is easily killed by frost and may be seriously injured by prolonged temperatures several degrees above freezing. Germination and emergence of the seedlings are delayed and may be prevented by soil temperatures below 50° F. Sweet corn does best in areas having mean temperatures of 65° to 75° during the required 65- to 100-day growing season. In the Northeast and North Central States this corresponds to the months June, July and August. In central and south Florida, on the other hand, sweet corn is planted fall, winter and spring. Generally, sweet corn can be grown successfully in the proper season from Mexico to Canada and in many other parts of the world.

Sweet corn will grow satisfactorily on a wide range of soil types as long as they are friable and well drained. However, a deep, loamy, naturally rich soil is preferred. Soil should be only moderately acid (pH 5.8 to 6.8). If the pH is lower than 5.8, lime should be applied. Have soil tests made to determine the proper kind and amount of lime and fertilizer to apply. Consult your local Agricultural Extension Agent since most States have laboratories that provide a soil testing service.

Available plant nutrients are especially important early in plant growth. If the plants become nutrient-deficient and stunted, they never fully recover and the yield will be reduced. Commercial fertilizers are recommended for sweet corn on just about all soils throughout the country. Fertilizer recommendations for sweet corn vary for different sections of the country and from one soil type to another. Again, depend on soil testing for specific recommendations.

Some general fertilizer recommendations for typical soils follow:

- On light sandy soils of the Atlantic and Gulf Coastal Plains, broadcast 20 pounds of 10-10-10 fertilizer per 1,000 square feet of area before planting. Then apply a side dress of nitrogen when the corn is in the 6 to 8 leaf stage at the rate of about a half pound of actual nitrogen (N) per 100 feet of row (note that Ammonium Nitrate contains 33 percent actual N and Sodium Nitrate contains 16 percent).
- On soil of average fertility in the Northeast, apply 15 to 18 pounds of 5-10-5 per 1,000 square feet of area prior to planting and about 3 pounds of the same fertilizer banded per 100 feet of row at the time of planting.
- In the more fertile valleys of the West and Pacific Northwest and on the rich soils of the Midwest corn belt, apply in bands 3 to 5 pounds of 5-10-5 per 100 feet of row at the time of planting.

The broadcast applications are usually worked into the soil before planting. The band applications should be made when the seedbed is prepared, about 3 inches to the sides of the row of seed and 1 to 2 inches deeper than the seed is planted. The above rates are based on rows spaced 3 feet apart.

Moisture Needs

Sweet corn requires a continuous and adequate moisture supply for satisfactory growth and yield. In nonirrigated areas of the United States, sweet corn is grown with reasonable success where the rainfall from April through September is 20 inches or more and fairly well distributed. Unless the soil can retain a large supply of water, sweet corn will suffer from lack of moisture if rainless periods last more than 2 weeks during the growing season.

In the South and Southwest, after the tassels show, the plants need rain or irrigation every week. For these areas, and where soil moisture is likely to be depleted, it is advisable to provide some supplemental irrigation. Furrow irrigation is satisfactory in most soils where runoff can be controlled.

Sweet corn varieties differ in the way their growth is affected by day length. Early maturing varieties developed for the North are not recommended for the South. They are adapted to the long, cool summer days in the North and do not make satisfactory growth in the deep South. Conversely the southern varieties are not adapted to the North. When planted in the North they may not silk and tassel until they reach 8 to 12 feet in height, and it is too late for them to produce edible corn before frost. Therefore, specific varieties are recommended for different sections of the country.

Hybrids

F_1 hybrids have largely replaced the open-pollinated varieties. As with all hybrid plants, new seed must be obtained for each crop. Seed saved from the hybrid plants will not reproduce true to type and will not retain the hybrid vigor of the parent plants. Several public research agencies and private companies breed and introduce new varieties of sweet corn. As a result, a large number of excellent hybrids are available for gardeners. Some hybrids will be available for only a few years, being replaced by better ones.

Sweet corn variety trials are conducted each year by many State agricultural experiment stations, and lists of recommended varieties are published based on these trials. Contact your State agricultural extension service for a list of varieties recommended specifically for your area.

Sweet corn requires plenty of space

and is adapted only to larger gardens exposed to full sunlight. It does best planted in rows 30 to 36 inches apart with single plants spaced 12 to 16 inches apart in the row. Overcrowding the corn will reduce the ear yield drastically. Planting four or more short rows is better than one long row to insure complete pollination. If the prevailing wind is across the row, pollen will be carried away from the silks and result in poorly filled ears.

To conserve space in the garden, corn may be planted next to vine crops, such as cucumber and cantaloupe. As the vines grow, they will grow between the corn plants.

Proper seedbed preparation is important for sweet corn in the garden since herbicides are generally not used. A clean freshly worked seedbed enables the seedling to emerge rapidly and get off to a good start ahead of grass and weeds. Seed should be planted to a depth of about 1 inch in moist, heavy soils and 1 to 2 inches in light, sandy soils, depending on the moisture conditions at planting time.

It is generally a good idea to plant at approximately twice the desired stand and thin to single stalks at the desired spacing after the seedlings have become well established. This will allow for any reduction in seed germination and for loss of a few emerging seedlings to insects, birds and other garden pests. About a quarter pound of seed is sufficient for each 100 feet of row.

Successive plantings are recommended in order to provide a steady supply of fresh corn throughout the practical harvest season. Also, an early, followed by a full season variety, may be planted at the same time to give a prolonged harvest period.

Once the desired stand has been established, the area should be kept free of weeds by cultivation and hoeing.

Diseases: Diseases are generally not a serious threat to clean, well nourished sweet corn plantings. Those that do occur most frequently are seedling root rot, Stewart's bacterial wilt, and common corn smut.

Root rot is caused by rot-producing fungi in the soil. It is often associated with a damp, cold soil, and may be evident as a slight stunting and irregular plant growth. Seed treatment with a fungicide provides good protection for the seedling during its early growth. Most sweet corn seeds packaged and sold commercially today have been treated with a fungicide. Use treated seeds whenever practical.

Stewart's wilt may appear at any stage of growth, but is most noticeable when plants attain considerable size. It produces yellow to brown streaks up to an inch wide on the leaves, and may extend the entire length of the leaf. Brown discoloration and sunken cavities form in the stalk near the soil line. Plants that become infected early may wilt and die. Those infected later may be only stunted and have streaked leaves.

The disease tends to be more prevalent after mild winters and is known to be spread by corn flea beetle. There are no sprays or seed treatments effective for controlling this disease.

Where wilt is suspected of becoming a problem, resistant varieties should be planted. Most varieties developed in recent years are resistant to the disease, particularly among the full season maturity group.

Common smut is characterized by the presence of large, fleshy galls on the stalks, leaves, tassel and ears. At first the galls are silvery white and spongy. Later, they turn brown or black, rupture and release large masses of powdery black spores. Smut galls are unsightly and render the affected ears inedible. Smut is promoted by injuries to the plant during cultivation, by insects, or hail.

Again, there are no chemical treatments to control the disease.

The best means of control is to avoid injuring the plants, avoid areas where smut occurred the previous year, and remove and destroy smut galls before they break open. This last step will prevent the spores from being released to infect later plantings.

Corn earworm.

Insects: Many species of insects are known to attack and damage sweet corn at all stages of its growth. Those that attack the plants early are more apt to cause serious damage, and they need to be dealt with promptly. These include the southern corn rootworm, cutworms, white grubs, wireworms and flea beetles. You can get some protection against the rootworms, wireworms, and grubs by using seed treated with a combination fungicide-insecticide. Cutworms and flea beetles may require an application of insecticide for control.

Insects attacking sweet corn later in its growth are corn borers, armyworms, aphids and the corn earworm.

Several insecticides are available to control them. Recommendations for specific compounds to use and rates of application can usually be obtained from a reputable garden supply center.

Once the sweet corn becomes established and attains most of its plant growth it can withstand a surprising amount of insect feeding without drastic loss of yield. Earworms that begin feeding on the silks and burrow into the ear tips are difficult to control. Unless the infestation is extremely high and damaging, most gardeners choose to ignore the worm at the tip of the ear, merely clipping off the ear tip and any damaged kernels when the corn is husked.

Most State agricultural experiment stations publish current recommendations for controlling insects on sweet corn. These bulletins and circulars can be obtained by writing to your State Agricultural Extension Service.

Harvesting

Sweet corn should be harvested when the kernels are in the milk stage. At this stage the silks are brown and dry beyond the end of the husks and the ear has enlarged enough to fill the husks tightly to the tip. The kernels are about as large as they will become, but they are still soft, tender and filled with an opaque milky juice.

With some experience the optimum maturity for harvest can be recognized by sight and feel. The husks should never be disturbed to peek at the corn as this will permit insects and birds to invade the ear.

Another way to estimate harvest time is to note the date of silk emergence on the earliest plants in a row, then harvest those ears 17 to 24 days later. The number of days from silk emergence to prime harvest will vary according to weather conditions. If days and nights are exceptionally

warm, prime maturity may be reached 17 or 18 days after silking. If cooler weather prevails during this period, it may require 22 to 24 days. After picking a few ears, you usually can make an accurate determination about harvesting the remainder of the corn at its prime maturity.

Sweet corn passes through its prime maturity very quickly. With uniform hybrid varieties the harvest of a single planting will last only about 4 to 5 days. If harvest is delayed the kernels become tough, starchy and lose their sweet flavor.

Sweet corn also loses its quality rapidly after it has been picked from the plant. For best quality, the corn should be picked early in the morning and refrigerated immediately. The sooner it is prepared for serving the better, but it can be held in a refrigerator (35 to 40° F) for 2 to 3 days with only a moderate reduction in eating quality.

To harvest corn, break the ear shank as close to the ear as practicable without breaking the main stalk or tearing the entire shank from the stalk. Grasp the ear with one hand near its base and bend it sharply downward or to one side with a rotary motion of the wrist. The inexperienced may need to use both hands; hold the shank with one hand and use the other to snap the ear off. With practice and a strong grip, the ears of most varieties can be snapped off with one hand.

Many of the modern hybrids under optimum fertility and growing conditions will produce two nice ears per plant. The top ear will be the dominant one, and it will reach prime maturity a day or so ahead of the second ear. Under such conditions a 100-foot row should yield 100 to 120 nice ears.

Fully mature and well-developed ears of sweet corn.

Cucurbit Crops—Cucumbers, Gourds, Melons, Pumpkins, Squash—Have Uniform Needs

by Thomas W. Whitaker

Cucurbit crops should be staples for home gardeners from Maine to California. Cucumbers, gourds, muskmelons, pumpkins, squash, and watermelons will perform satisfactorily over a wide range of climate and soil conditions. The vine, or more properly, the cucurbit crops are extremely uniform in their environmental and cultural requirements. Thus, a set of procedures designed for the culture of cucumbers can be used equally well for raising squash, with perhaps some slight modification.

The wide adaptation of the cucurbit crops to culture in temperate zone areas is surprising because they are basically tropical or semi-tropical plants, annuals, extremely frost tender, and mostly incapable of functioning normally at temperatures below 60° F. For best seed germination, temperatures of 60° to 75° are required, and for maximum seasonal growth, average mean temperatures of 65° to 85° are needed.

Considering their tropical origins, the cucurbits should thrive during the long, hot, humid days and warm nights of summer in the north temperate zone—and they do. Under such circumstances, and with adequate soil moisture from rainfall or irrigation, the vines grow rapidly and respond by quickly producing fruit. Summer squash and pickling cucumbers will produce an edible product within 48 to 56 days from planting. Muskmelons require 130 to 140 days from planting. Some baking squash, pumpkins and gourds have best quality if the harvest is delayed until after the vines are senescent or have been killed by frost.

The cucurbit crops are a homogenous group, easily identified by their prostrate, sprawling vines, usually with tendrils. Each runner bears many large, lobed more or less palmate leaves (having the shape of a palm leaf). Except for the bottle gourd, the flowers are usually bright yellow, large and conspicuous. The bottle gourd has white flowers which open at night. They are pollinated by nocturnal insects.

Each vine bears two kinds of flowers; the large or pistillate (no anthers, female), and the smaller or staminate (no pistils, male). Commercial varieties of muskmelons have a variation on the basic pattern. In this group, perfect flowers (with both pistillate and staminate parts) are on the same plant with staminate flowers.

The botanical name for the fruit of a cucurbit is a pepo. A pepo is a fleshy, indehiscent (closed at maturity), berry-like structure, the product of an inferior ovary. Some of the fruits of cucurbits are among the largest in the plant kingdom. Squashes weighing 350 pounds have been reliably reported, and fruits of the bottle gourd are nearly as large.

The most obvious disadvantages of cucurbit crops from the viewpoint of the home gardener are their light and space requirements. They need maximum sunshine for best development. A few vigorous plants of pumpkin, watermelon or gourd can overwhelm the small garden. These disadvantages can be successfully overcome by careful site selection within the garden, by planting bush or dwarf varieties of squash, and by judicious use of a trellis, or using structures adjacent to the garden—

Thomas W. Whitaker is a Plant Geneticist (Collaborator) with the Agricultural Research Service, La Jolla, Calif.

Top, pumpkins and winter squash — banana, acorn, Hubbard, and butternut. Right, youngster displays cucumber grown in Children's Garden of Brooklyn Botanic Garden, New York. Above, easy does it! A little effort is worth the pies these pumpkins will make.

such as fences, garages, doghouses, etc.—as a substitute for a trellis.

Nutrients. Cucurbits are not consumed primarily for their nutritional value. They contain only a sprinkling of vitamins, minerals and protein, and except for baking squashes are low in calories. Since they are low in caloric content, they are frequently used in reducing diets.

The attraction of cucurbit fruits as food is mostly to the palate. Their aroma, flavor, texture, and juiciness are among the most attractive and delightful in the vegetable world. Muskmelons make a superb breakfast fruit or dessert; besides they are relatively high in vitamins A and C. The cool, crisp, juicy, refreshing taste of a watermelon on a warm summer day is an unforgettable experience. Cucumbers, fresh or pickled, are zesty ingredients of salads and sandwiches. Summer squash, boiled and seasoned, is an extremely tasty dish, and baked squash is comparable to sweet potatoes as a dietary staple. The dessert qualities of pumpkin pie are well-known.

Cucurbits are raised mostly for their fruits which are consumed in the immature stage (summer squash, pickles), or mature stage (muskmelons, watermelons, winter squash). Gourds are allowed to mature, and then can be used as ornamentals, planters, liquid containers, work baskets, rattles, drums, etc. As food, the cucurbits can be boiled, baked, stewed, dried, pickled, or eaten uncooked. In Latin America, the staminate flowers of squash are dipped in a batter, fried, and served as a fritter. Watermelon rinds are delicious pickled or candied.

There are reasons for thinking that squashes and pumpkins were originally domesticated for their tasty, nutritious seeds, rather than the fruit flesh. In Mexico, squash seeds, fried in oil and salted, are sold by street venders, much like peanuts are sold at baseball games in this country. Also, in Mexico, squashes have been selected for the number and quality of their seeds as food, while the flesh is ignored.

Soils. The cucurbits are not exacting in their soil requirements. They accept almost any good garden soil, well-drained, aerated, and enriched with a generous supply of plant compost or animal manures. Sandy loams which warm up quickly in the spring are preferred for an early maturing crop, but crops can be grown on heavier soils if they are properly managed. Heavier soils have greater water-holding capacity, hence they withstand droughty conditions much better than lighter soils.

One factor that places a definite limit on the culture of cucurbit crops is soil pH. They are uniformly sensitive to acidic soils, and they require a neutral (pH 7) or even better soil with a slightly alkaline reaction. For acidic soils, treatment with lime prior to planting is mandatory.

Nutrient Needs

While cucurbit crops do moderately well on most fertile soils, they benefit greatly from a generous supply of organic material in the form of green and animal manures. Well-composted animal manures worked into the soil and concentrated in the area where the seed is expected to be planted (hills) is the most efficient means of using these materials. In addition to manures, applying mineral fertilizers is usually needed for a satisfactory crop. Fertilizers act as a supplement to the manure and provide an added source of plant nutrients during the growing season.

It is difficult to be specific about fertilizer recommendations because of great variation in soil types, soil fertility, and other soil conditions. In general, cucurbit crops can be expected to respond to a complete fertilizer containing 4 to 6 pct nitrogen,

Spacing Distances, Planting Depths for Cucurbit Crops

Measurements are in Inches

Crop	Spacing Between plants in row	Spacing Between rows	Planting Depth
Cucumber	12 [1] 24–36 [2]	48–72	1
Muskmelon	12 [1] 24–36 [2]	60–84	1–1½
Pumpkin	36–40	72–96	2–3
Squash (bush)	24–30	36	2–3
Squash (vining)	36–40	72–96	2–3
Gourd	36–40	72–96	2–3
Watermelon	24–36 [1] 72 [2]	72–84	1–2

[1] Single plants.
[2] Hills.

8 to 10 pct phosphoric acid, and 5 to 10 pct potash. This translates into 1 to 2 tablespoons for each hill prior to planting. In light, sandy soils, that leach readily, one or two side dressings of ammonium sulphate may be needed during the season; perhaps a tablespoon per hill will suffice.

In the garden, cucurbit crops are normally planted in hills, specific spacing depending upon the crop. In commercial practice, however, cucumbers, muskmelons, bush squash, and watermelons are drilled in continuous rows, and thinned to stand.

Cucurbit seeds are relatively large, and should be covered to a depth of 1 to 3 inches. After covering, the soil is lightly tamped, but not so firmly as to create a crust. In light, sandy soils that tend to dry out rapidly, seeds should be planted at greater depth than in heavier soils.

Assuming normal germination (80 to 90 pct), 4 to 5 seeds are planted in each hill. Thin the seedlings when they have 2 to 3 leaves. Remove all but 1 or 2 large, healthy, well-spaced plants per hill. More than 2 plants per hill causes undesirable crowding, and competition for nutrients, water and light. Under such conditions, the final result is unthrifty plants and declining yields.

Irrigation. In the West and Southwest, cucurbit crops are totally dependent upon irrigation as a source of moisture, and even in the Midwest and East some form of supplementary irrigation may be desirable during drought periods. The cucurbits are moderately deep-rooted crops, filling the soil mass to a depth of three feet or more. This means the soil must be supplied with enough moisture to maintain a thoroughly moist condition to this depth.

Furrow irrigation is probably the most practical for home gardens, especially after the young plants have a dozen or more true leaves. Moisture on the leaves from whatever source encourages several foliar diseases difficult to suppress with fungicides. Hence, sprinkler irrigation is not recommended if alternative methods are available.

Mulching and weed control are not critical for growing cucurbits in the home garden. If the soil directly above the seed (the hills) is kept from crusting, no mulch is needed. A thick, hard crust will prevent emergence of the young seedlings. Therefore, after planting and firming, it is important to scatter a thin layer of loose soil over seed in the hill.

There is really no safe, effective chemical weed control for cucurbit crops. If planting is done in a well-prepared seed bed, weeds will seldom be a problem and can easily be controlled by hand or by hoe. The cucurbits are leafy, rapidly growing, vigorous plants. Consequently, as they grow older they tend to shade out competition from weeds. Usually weeds are not much of a problem until late in the season, but by this time the crop is mature and little harm will be done.

The cucurbits are subject to infection by several diseases and attack by insects that can damage or even destroy a potentially promising crop without much warning, and within a relatively short time period. Nearly all the important pests are destructive to cucurbits you are likely to plant in the home garden. There are exceptions, however, such as scab, a fungus damaging only to cucumbers, and anthracnose, also a fungus, commonly attacking only watermelon. Squash bug and squash vine borer are more of a problem on squashes and pumpkins than with other cucurbits.

Cucumber Beetles

Bacteria responsible for bacterial wilt are spread by cucumber beetles. This fact is the key to control. If the beetles are eliminated or reduced to low levels in the garden, bacterial wilt will not be a problem. The bacteria multiply rapidly and plug the water transportation system of the plant. This results in characteristic wilting of the vegetative parts. Older plants at first may have only one shoot affected, but later the entire plant will wilt and die. Younger plants die quickly. It helps to promptly remove infected plants from the garden.

Anthracnose, a disease of watermelon, flourishes in warm, moist weather, and is particularly troublesome in the Southeast. Under favorable conditions it also attacks cucumbers and muskmelons, but is an acute hazard only to watermelon production. The symptoms are small, round, water-soaked spots on the fruits. These spots later become enlarged, sunken, with dark centers, which may turn pinkish in moist, humid weather. Infected leaves have a scorched appearance, and the stems may be girdled. Vines with the disease may die. The fruits are worthless, often decaying before they are mature.

A few simple preventive measures often give satisfactory control if you are in an area where anthracnose is a problem. (a) Plant seed of varieties known to have a high level of resistance to the disease. (b) Plant seed that has been treated with a fundicide to remove spores of the fungus. (c) Practice garden sanitation, removing all the debris of cucurbit vines because the spores overwinter in this trash. (d) Select areas for planting that have not been used for cucurbit crops within the past 3 to 4 years.

Downy mildew fungus is a destructive disease of cucurbits, especially when meteorological conditions favor its rapid growth. Warm, moist conditions that occur at times during the growing season in the Atlantic and Gulf States are ideal for growth of downy mildew. However, low humidity, high temperatures and lack of free moisture on the leaves and stems immediately check the fungus growth. Spores of the fungus are produced on the underside of the leaves, and are spread by wind or splashing of raindrops.

Initial symptoms are small, yellowish spots, with irregular edges which appear on the leaves at about the time the vines commence to set fruit. Tissue at the center of each spot soon turns brown and dies. Later the spots become more numerous, coalesce, and the leaf shrivels and dies. The brown or blackish withered leaves curl upward, a characteristic that makes downy mildew easy to identify with certainty. The fruits are not attacked, but fruits from denuded vines are apt to lack flavor, be tasteless, and are practically inedible.

Whenever possible use varieties with some tolerance or even resistance to the disease. Generally the home gardener will have to lean heavily on the use of a suitable fungicide.

Powdery mildew fungus can be a devastating pest of nearly all cucurbits, except watermelons. The first symptoms are small, white patches on undersides of the older leaves. As the disease increases in intensity, leaves and stems become covered with the white, powdery spore masses. The foliage gradually dies, leaving the fruit exposed to the sun. In muskmelons, such fruits ripen prematurely, and are usually sunburned and of poor quality.

The fungus responsible for powdery mildew requires much sunshine and reasonably high temperatures for best growth. Rains and low light intensity tend to check its rampant increase.

Resistant Varieties

Resistant or tolerant varieties of most species are available, and should be used where powdery mildew is likely to be a problem. Several safe chemicals are effective against powdery mildew.

Scab is caused by a fungus which attacks the fruit, particularly cucumber, but it may occasionally damage young squash fruits. Sunken, dark brown, irregular spots appear on the fruits from which a gummy substance is extruded. The young fruits become malformed and cannot be used.

The disease spreads rapidly in cool, moist weather. It is most serious on cucumbers in the northern tier of States (Wisconsin, Michigan, Minnesota, New York and Maine). Good garden sanitation and use of resistant varieties should successfully solve this problem for the home gardener.

Mosaic caused by virus is one of the most widespread and serious diseases of cucurbits. Characteristic symptoms are light-green mottling of the leaves, and the younger leaves are malformed, dwarfed, and slightly curled. With late infection, the symptoms are mild and little harm is done to the crop. Vines infected in early stages of development normally are dwarfed, the leaves and flowers malformed, and they do not produce acceptable fruit.

There are roughly two classses of virus that attack cucurbits. (1) Watermelon mosaic viruses which are not seed-borne, and are spread by sucking insects—chiefly aphids. (2) Squash mosaic viruses, which are seed-borne and are spread by chewing insects—chiefly cucumber beetles.

The watermelon mosaic viruses are destructive pests of cucurbits because they are spread by aphids, and it is almost impossible to deny aphids access to home gardens. Apparently the virus is carried by several widely grown ornamentals, so sources of the virus are always present. Control is difficult, although there are resistant or tolerant varieties of cucumber, but resistance breeding programs in other species are not well-developed.

Control of the squash mosaic viruses can be established by planting virus-free seed. Also, controlling cucumber beetles and other chewing insects prevents dissemination of the virus.

Root knot disease is caused by minute eelworms or nematodes which enter the roots where they feed and breed. Feeding causes the root tissues to swell, producing nodules or galls on the roots. Some galls are small, others may be the size of a walnut. Plants become dwarfed, unthrifty, and often turn yellow and die. Soil heavily infested with nematodes should be fumigated, using one of the several nematocides that are effective when properly applied. Fumigation will reduce the population of nematodes to a point where good crops can be grown for at least 1 to 2 years. Some control can be obtained by crop rotation, that is, by not planting susceptible crops in the same soil for 3 to 5 years. Much research has been done, but resistant varieties are not yet available.

Cucumber beetles, both striped and 12-spotted, are common pests in most gardens. It is important to control them because the adults seriously damage or totally destroy the plants, and their activities also spread certain diseases of cucurbits (bacterial wilt, squash mosaic). Additionally, the larvae bore into roots and stems below the soil line, often causing the plants to suddenly wilt and die. There are satisfactory chemicals that will control these insects, but they must be applied at the first appearance of the beetles for acceptable control.

Aphids—small, fragile, soft-bodied insects with sucking mouth parts—can be a problem on cucurbits during the course of the growing season. Aphids come in several colors, such

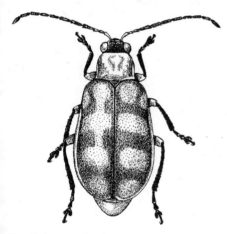

Banded cucumber beetle.

as black, green, yellow, or pink. They feed on the undersides of leaves, causing a curling or cupping. With severe infection, the leaves become sticky, lose color, and the plant dies. Winged females fly from plant to plant, establishing new colonies. Aphids can be controlled by chemicals, but be sure to use those insecticides that are least harmful to aphid predators.

The squash bug is a troublesome pest of squash and pumpkins, and occasionally attacks gourds. Adults and the immature forms (nymphs) suck the sap from leaves and stems, causing the plant to wilt and die. The adult bug is about a half inch long, and dirty brownish, or black. The nymphs are much smaller and steel-grey. The female deposits the relatively large and brownish eggs in regularly arranged masses on the underside of the leaves.

Good plant sanitation helps control this pest. Promptly remove and destroy debris from cucurbit vines and fruit. You can trap the bugs by placing a shingle or board at the base of the plants. The bugs collect here during the night, and can be destroyed the following morning before they become active. Also, searching out the egg masses and destroying them can be helpful. Insecticides are effective against heavy infestations.

The squash vine borer can damage squash and pumpkin in areas east of the Rocky Mountains. The borer is the larval stage of a day-flying moth that deposits her eggs on the stem of the plant, slightly above the soil line. The young borers penetrate the stem and burrow toward the base. When mature, the borer crawls from the plant to the soil where it pupates.

Garden sanitation, and fall tillage deep enough to destroy the cocoons, are effective control measures. Also, where only a few plants are involved, locate the wound and slit the stem with a razor blade or sharp knife, thus puncturing the borer. Then place moist soil around the stem to a height a little beyond the wound. New roots will develop, thus compensating for the injury.

Harvesting. The proper time to harvest is crucial for obtaining maximum quality of cucurbit fruits. Cucumbers for pickling should be harvested when the young fruits attain a length of 2 to 4 inches, roughly 3 to 4 days after the flower has opened. For dill and larger pickles, harvest is delayed until fruits are 6 inches or more in length. The slicing type is harvested when the fruit is 8 to 10 inches long.

Left, girl with squash she grew in Children's Garden at Brooklyn Botanic Garden, New York. Below, if vine separates from fruit easily, as with this cantaloup, melon is ripe and should be harvested. Note cantaloup is propped on flower pot to keep it off ground.

William E. Carnahan

William Aplin

For muskmelons, nature has provided an unmistakeable sign. At maturity, an abscission layer forms between stem and fruit. This layer appears as a crack, completely encircling the stem, at the point of attachment to the fruit. If the stem has to be forcibly separated from the fruit, the fruit is immature. The abscission layer, or "slip" as it is known in the trade, is characteristic of most muskmelon varieties except the casabas and honeydews. Maturity in the latter is judged by softening of the blossom end of the fruit, and subtle changes in the fruit's color.

The expertise required to select a watermelon at prime maturity is only acquired by experience. For the amateur, some of the guesswork can be eliminated by rapping the melon sharply with the knuckles. A crisp, metallic sound indicates immaturity, while a dull, flat sound suggests maturity. Also, at maturity the ground spot usually changes from white to light yellow.

Summer squash, which is consumed in the immature state, is best harvested when 3 to 6 inches long, about 3 to 4 days after the pistillate flower opens. At this stage summer squash is tender, crisp, and has a good flavor. It should be harvested 2 to 3 times per week.

Winter squash, or baking squash, normally is harvested when mature. Maturity can be roughly estimated by pressure from the thumbnail on the fruit exterior. If the skin is hard and impervious to scratching, the fruit is mature.

Fruits of pumpkins and gourds are customarily allowed to remain in the garden until frost destroys the vines, or they deteriorate.

Storage. Most cucurbit fruits are consumed fresh. Honeydew and casaba muskmelons can be stored as long as a month in a cool, dry environment. Pumpkins and the hard-shelled squashes can be stored for several months if properly cured. This means the fruit should be mature and carefully handled at harvest. After harvest, they should be placed in a room or other area with temperatures of 80° to 85° F for 10 days, then transferred to a cool, dry place, preferably with temperatures of 50° to 60°. In storage, the fruit must be well-ventilated, not piled on each other.

Since the cucurbits are large, vigorous plants, requiring relatively huge amounts of space for maximum development, the number of plants that can be accommodated in the averge garden is minimal.

Plant population should be enough to cover the needs of an average family. Six cucumber plants, if harvested regularly, will produce sufficient fruit for a family. Muskmelons produce 2 to 3 fruits per plant and needs should be based on this estimate.

Watermelons are not heavy producers. Most varieties produce 1 to 2 melons per vine. Some of the varieties with small fruits are more prolific.

Two to three well-grown plants of bush summer squash will produce an abundance of squash for the average family and the neighbors, if harvested at regular intervals. For the average garden, 4 to 6 vine type squash or pumpkin should be sufficient.

Gourds are terrifically prolific. Two or three vines on a trellis will produce 12 to 24 fruits depending upon the variety.

Thump test tells this young feller his icebox watermelon is ripe. But the most definite proof (and the most fun) comes from taste-test.

Photos by William Aplin

Asparagus Starts Up Slow But Goes On and On; Rhubarb Also Takes Its Own Sweet Time

ASPARAGUS
by Stephen A. Garrison and
J. Howard Ellison

The cultivated asparagus plant, *Asparagus officinalis*, is a perennial vegetable that can thrive in the home garden for 25 years or more when well cared for. However, plants may have to grow 3 years before they can be harvested.

The underground root system consists of an extensive network of fleshy storage roots with small feeder roots that absorb water and nutrients. The storage roots are about the diameter of a pencil and may be 5 to 10 feet long in mature plants, depending on the soil type in which the plants are growing.

These storage roots are attached to an underground stem called a rhizome. The storage roots and the rhizome are commonly referred to as an asparagus crown. When the soil is warm and the soil moisture favorable, buds arise from the rhizome and develop into edible spears, utilizing the carbohydrate and other nutrient reserves present in the storage roots.

The spears which are not harvested develop into attractive, green, fern-like stalks (brush). Through photosynthesis, the mature plant produces carbohydrates and synthesizes other essential nutrients that are translocated to the storage roots. The stored reserves supply the energy required to produce spears during the following growing season.

For this reason it is important to protect the fern-like foliage from insects, diseases, and other injury before natural senescence and cold weather terminate the functioning of the green foliage in the fall.

Spears should not be removed from the plants during the first two growing seasons in the permanent location. (See harvesting details later in chapter).

Unlike most plants that have both male and female parts on the same plant, the asparagus plant is dioecious. The male flowers that produce pollen are present on one plant and the female flowers that produce the berries and the seeds are on a separate plant. Bees transfer the pollen from the male to the female flowers.

Research has shown that female plants, which expend much energy in producing fruits and seeds, do not yield as well and are not as long-lived as male plants.

The genus *Asparagus*, a member of the Lily family, originated along the shores of the Mediterranean Sea and on its many islands. Asparagus (*Asparagus officinalis*) was considered a delicacy by the ancient Greeks. The elder Cato discussed its cultivation in 200 B.C., and 200 years later Pliny described the spear size of asparagus. A. W. Kidner, writing in England in 1959, said the spear size described by Pliny was very similar to that in England more than 19 centuries later. It is remarkable how little the cultivated asparagus has changed since the time of Christ.

Most asparagus strains grown in the United States today are seedling populations selected from the Martha and Mary Washington strains developed in the early 1900's by J. B. Norton of the U.S. Department of Agriculture.

All presently available asparagus strains produce plants with variable vigor, size and disease resistance.

Stephen A. Garrison and J. Howard Ellison are Associate Professor and Professor respectively in Vegetable Crops, Rutgers University, New Brunswick, N.J.

Emerging asparagus spear.

Plant breeders are developing more uniform plants by reproducing highly selected parent clones through test-tube tissue culture. In the near future, gardeners will be able to purchase these high-yielding uniform seeds or even highly selected, extremely productive clones propagated by tissue culture.

The Viking strain is suggested for gardeners in the northern United States. For the West Coast, California U.C. 157 is recommended. In the East and Midwest, Rutgers Beacon, Waltham Washington, and local selections of the "Washington" type are recommended to home gardeners and are available through the major seed distributors.

Asparagus is low in calories, but high in flavor. A serving of 4 spears of asparagus (60 grams) contains just 10 calories, 1 gram of protein, 2 grams of carbohydrates, and only traces of fat. When the nutrient content of vegetables is compared, asparagus is a good source of vitamin A and riboflavin, and a very good source of thiamin.

Environment Needs

Able to tolerate great variations in temperature, asparagus grows in places such as the Imperial Valley of southern California, where the temperature soars to 115° to 120° F in the summer, or Minnesota, where winter, minimums of 40° below zero occur.

However, asparagus grows best where the growing season is long and the days are sunny for maximum photosynthesis. Ideal day temperatures during the growing season are 75° to 85° F, and nighttime readings in the 60's to minimize respiration. These conditions favor maximum storage of carbohydrates in the root system for high yield and quality of spears the following season.

Asparagus can be a home garden vegetable in most parts of the country except the Deep South. Asparagus does not grow well in the Gulf Coast States, due to the moist, warm winters which may stimulate sporadic growth during winter.

Asparagus can be grown on a wide range of soil textures from loamy sands to clay loams as long as water drainage and aeration are good. In the more arid regions the heavy soils are satisfactorily aerated and produce excellent asparagus. However, in regions of moderate to high rainfall, asparagus grows best on deep, well-drained sandy loam soils. Asparagus plants lose vigor, become more susceptible to root rot, and die in poorly drained areas or following prolonged high rainfall.

The soil reaction (pH) should be maintained between 6.5 and 6.8. Medium fertility is best, to provide a balance between top growth and root growth, but the plant has a relatively high potassium requirement for maximum production.

Growing

In the past, most garden asparagus has been started by planting crowns.

Plant only healthy 1-year-old crowns and never use 2- or 3-year-old crowns. It is difficult to obtain commercial crowns free of Fusarium root rot (see discussion later in chapter). Fusarium carried on the seed will infect young seedlings and contaminate the garden permanently.

One way to avoid Fusarium is to plant asparagus seed that has been surface-sterilized. You can use this seed to grow your own crowns or seedling transplants.

Crowns are produced by seeding a nursery about 2 weeks before tomatoes are normally transplanted into the garden. Seed is sown 1 inch deep and 3 inches apart in the row with 2½ feet between rows in the nursery. Plant 3 to 4 seeds for each crown you want to plant in the permanent bed. The following spring (February-March), when the plants are still dormant and the ground has thawed, carefully dig the crowns to minimize damage to the root system. Immediately plant the crowns in the permanent bed as described later.

To grow transplants, sow the seed 12 to 14 weeks before transplanting seedlings to the garden in the spring. The sowing date will vary from late December in the Southwest to mid-February in Northern areas of the country. Use a commercial potting mixture of peat moss and vermiculite, (pH 5.5 to 6.0) and sow two seeds ¾ inch deep in small (2-inch diameter) pots, or seed in rows in flats. The rows should be 2 inches apart with seeds 2 inches apart in the row.

Maintain the temperature at 75° to 85° F while seeds are germinating. As seedlings emerge, they should grow at 70° to 75° during the day and 65° at night. Grow the plants in a greenhouse or window with full sunlight. Use supplementary fluorescent lights to extend the day length to 12 to 14 hours when plants are not grown in the greenhouse.

Apply a soluble complete fertilizer, such as 15-15-15, at half the recommended rate 4, 8, and 12 weeks after sowing the seed. Rinse the foliage lightly with water after fertilizing to avoid fertilizer injury to the tender growth. Excessive fertilization promotes large tender tops and small root systems with limited reserves in the storage roots. Make the last application just before transplanting to the garden, after danger of the last killing frost is past. If you don't have a suitable place to grow your own seedling transplants, you may be able to get seedlings from a commercial plant grower who specializes in bedding plants.

Asparagus should be planted with other perennial crops for convenience of tillage and management. It is preferable to plant on the north or east side of the garden so as not to shade other vegetables or low-growing fruits.

Asparagus can be planted along a fence, as long as there is plenty of sun. In fact, the beautiful green, fern-like foliage grows five to six feet high, and can be used as an ornamental summer screen. The female plants produce berries that become bright red in late summer and fall. The tops turn from green to an attractive yellow in fall and brown during winter.

Before planting, broadcast and turn under 1.2 pounds of 5-10-10 fertilizer (or equivalent) per 100 square feet of area. If lime is needed, turn it under along with the fertilizer. Keep a close check on soil pH, because asparagus does poorly at pH levels below 6.0.

Plant crowns with the buds up in the bottom of a 6-inch deep V-shaped furrow, and cover with 1 to 2 inches of soil. Plant seedlings 1 inch deep on small mounds in the bottom of a similar furrow. The seedlings will require some protection from water which may stand in the furrow and from soil which can wash into the furrow.

Spacing for crowns and transplants is 12 inches apart within the row and

4 to 5 feet between rows. If only one asparagus row is planted, allow at least 3 feet between the asparagus and the closest other vegetable crop.

As the asparagus grows, carefully fill in the furrow with soil so as to avoid covering any asparagus foliage. The furrows should be filled in by the end of the first growing season. Sidedress the plants with 1.2 pounds of 5-10-10 fertilizer (or equivalent) per 20 feet of row in late July or early August. Spread the fertilizer on either side of the asparagus and cultivate it lightly into the soil.

Adequate soil moisture is important during the first growing season. Don't let the plants suffer for lack of water during dry weather. Weekly applications of irrigation sufficient to wet the soil 8 inches deep should be adequate.

After the first growing season, asparagus plants do not require frequent irrigation because of the deep and extensive root system. Thorough watering (2 inches of water) slowly applied every 2 weeks during dry weather is sufficient.

During early spring of the year after planting remove the brush (old stalks) and any over-wintering weeds. Broadcast lime as needed to maintain the proper soil pH plus 1.2 pounds of 5-10-10 per 100 square feet of bed. Sidedress another 1.2 pounds of 5-10-10 per 100 square feet in late July or early August.

Remove brush during each succeeding spring before the asparagus emerges, and broadcast lime if needed. At the same time spread 3.4 pounds of 5-10-10 (or equivalent) fertilizer per 100 square feet of bed. Rake the fertilizer and lime 1 to 2 inches into the soil, taking care to avoid damage to the asparagus crowns.

Maintain good foliage growth for maximum photosynthesis. Tall weeds can shade asparagus and reduce photosynthesis. Even low weeds and grasses compete with the asparagus for water and nutrients. Thus, weed control is an important aspect of good asparagus culture. Chemical herbicides are toxic to many other vegetables and not recommended for the home garden.

Asparagus can be mulched with organic debris (leaves, grass clippings, etc.), but mulches can harbor pests, alter the soil pH, and change the crop's fertilizer needs. Clean cultivation of the asparagus is preferred.

Pests

The most serious disease of asparagus is Fusarium root rot, caused by the fungus *Fusarium oxysporum f. asparagi*. This organism is present at very low populations in most soils, where it grows slowly on organic matter in the soil.

Planting disease-free seeds or crowns of the host asparagus causes a slow build-up of the Fusarium population in the soil. However, planting infested seed or infected crowns leads to a rapid build-up that may adversely affect performance of the first asparagus planting, and permanently contaminate the garden in that location for asparagus in the future. The Fusarium infects the root system and kills the feeder roots. As a result, plant vigor declines, spear size decreases, and the weaker plants may die.

Fusarium can be identified by reddish-brown color of the feeder roots, reddish-brown spots and streaks on the storage roots, and large lesions on the base of the spears and stalks at or below the soil line.

The symptoms are much more severe when the plants are under stress due to excessively long harvests, poor drainage, competition from weeds, and damage from insects and diseases.

The best way to avoid Fusarium root rot in the garden is to plant surfaced-sterilized seed in disease-free soil. Although surface-sterilized seed is not generally available, you can treat the seed yourself.

Soak the seed for 2 minutes in a solution of 1 part laundry bleach (composed of approximately 5 percent sodium hypochlorite) plus 4 parts of water. Rinse the seeds for one minute with cool running water and plant in soil that has never grown asparagus previously, or start seedling transplants in a disease-free artificial soil mix.

Although research to control Fusarium by the use of fungicides is under way, no practical treatment is yet perfected. Progress is being made on development of Fusarium-resistant strains of asparagus at the New Jersey and California Agricultural Experiment Stations.

Rust (*Puccinia asparagi*) is another common fungus disease of asparagus in the East and Midwest, and certain valleys in California. Dew or other free water on the plant for 10 hours is enough for spore germination.

Ten days after germination, the fungus appears on the surface of stems and branches as small rust-colored spots, containing spores that cause spread of the disease. In the fall the rust colored areas produce black, over-wintering spores that can infect the plant the following year. Severe rust destroys much of the foliage, reducing reserves for the next year's crop.

Rust can be partially controlled by fungicides, and several asparagus strains selected in the East and Midwest have some rust resistance.

Several species of thrips are often a serious problem, primarily on small plants in a nursery or on transplanted seedlings in the garden. Thrips are small, white, flying insects which are very difficult to see. They suck plant juices and cause the green needles and stems on the young plants to turn dull gray-green, then brown. Insecticides can be used to control thrips.

The asparagus beetle (*Crioceris* species) is a serious pest every year, although the insect populations vary from season to season. Adult beetles, which look like slender lady-bird beetles, lay small black elongated eggs on end in rows on asparagus brush. The larvae (small, dark green worms) do the actual damage by eating the green epidermis from the fern.

Beetles can kill very young seedlings, and seriously damage fern growth of mature plants. They can be controlled by insecticides.

European asparagus aphids invaded the eastern United States in recent years. They are blue-green, with a metallic sheen (like aluminum) when in clusters on plants. The aphids feed on young growing shoots, and inhibit the elongation of internodes, producing a kind of rosette (bushy stunted growth). Heavy infestations of the insects can seriously decrease foliage area needed for photosynthesis.

Fortunately the European asparagus aphid has many natural enemies, and seems to be subsiding as a pest.

Harvesting: An important culture requirement of asparagus is that the crop must be grown for two full growing seasons before harvest begins. This is necessary to allow the plants to develop an adequate storage root system to produce spears during the first harvest season and beyond. Any harvesting or damage to the brush during the first two growing seasons dwarfs the plants and can reduce yield for the life of the bed.

When the first spears emerge in the spring, merely snap off spears 7 to 10 inches long, with tight heads, leaving the tough stub on the plant. The upper portion which snaps off should be "all green" and "all tender". Harvest all spears that come up during the harvest season.

A good general rule for length of harvest season for all areas except the cool central valleys of California is the 2-4-8 week sequence. Harvest

Asparagus spears harvested by snapping.

for 2 weeks the third year the plants are in the garden; 4 weeks the fourth year, and 8 weeks the fifth and following years. In the cool central valleys of California, a 4-8-12 week sequence is best.

When the harvest season is approximately half completed, 5 to 6 inches of soil may be carefully ridged over the row. This lowers the temperature around the crown and increases spear size. The ridge should be raked level right after the last harvest.

White asparagus, which has a distinctive flavor, can be produced by ridging 10 to 12 inches of soil over the row in the spring when the first spears emerge. When the tip of the spear breaks through the ridge of soil, carefully remove some soil from around the spear, and use a long knife or asparagus knife to cut the spear about 8 inches below the tip.

If the harvest from one day is not enough for a meal or if the asparagus is to be consumed later, wash the spears, place the cut ends in a shallow pan of water and immediately put them in the refrigerator. Good quality can be maintained for several days if the spears are kept at 35° to 40° F. A 40-foot long row of asparagus will yield approximately 10 to 25 pounds of spears during the average season.

RHUBARB
by Daniel Tompkins

Rhubarb, also known as pieplant, is a hardy perennial vegetable grown in many home gardens for its thick leaf stalks or petioles. It produces its crop early in spring, largely from food that has been stored in the large fleshy crowns and roots of the plant during the preceding year. It likes cool weather and grows best in the Northern States where the average summer temperatures are not much above 75° F. Rhubarb does not grow well in areas where the summers are quite warm.

Stalks can't be fully harvested until the third year of the planting.

A member of the buckwheat family, rhubarb is native to Central Asia. It was introduced into Great Britain, where it is grown extensively, in the 16th century and was probably brought to the United States from Italy late in the 18th century. It has long been grown in the Old World as a vegetable, an ornamental foliage plant, and for medicinal properties of the dried root which provides a strong purgative. Here in this country rhubarb is grown for its acid stalks which are stewed for pies and sauces, made into preserves, and sometimes used for making wine. It is also excellent baked.

Rhubarb leaves should never be eaten, since they contain levels of soluble oxalic acid that can make one quite ill or even cause death. The stalks are harmless since the oxalic acid is present in smaller amounts and mostly in an insoluble form.

Rhubarb contains vitamin C and calcium (largely insoluble). It also contains some vitamin A, iron, phosphorus, potassium and only about 60 calories per pound of stalks. It has been reported that rhubarb can pro-

---

Daniel Tompkins is a Horticulturist with the Cooperative State Research Service.

tect the teeth against acid erosion such as may be caused by excessive use of lemon juice or cola beverages. Rhubarb is one of the most acid of all vegetables; the juice has a pH of 3.1 to 3.2. The tender stalks are about 94 percent water.

For home use, rhubarb varieties or cultivars may be divided into two classes, those with red stalks and those with mainly green stalks when grown outdoors. The somewhat larger and more vigorous green stalk cultivars are Victoria, German Wine and Suttons' Seedless. These cultivars are commonly used by commercial growers for forcing where they produce stalks with a delicate pink-red color. The cultivars that produce red stalks when grown outdoors are Ruby, McDonald, Valentine, Canada Red, and Crimson Wine.

Rhubarb grows easily from seed but this is not recommended since many plants will not be like their parents.

Dividing Crowns

You usually propagate rhubarb by dividing the crowns in early spring. Dig the crowns and then split them into pieces with one large bud to each section of crown and root. Trim the pieces by removing all broken roots and shortening the long thin roots.

Crowns vary in size and number of buds produced due to cultivar, age, and growing conditions. Vigorous crowns will normally provide 5 to 10 pieces suitable for planting. Very old crowns may have only an outer fringe of buds suitable for dividing.

Protect the root pieces from excessive drying before planting.

Space the plants 2½ to 3 feet apart in rows 3½ to 4½ feet wide. Usually you plant in a furrow, placing the crown pieces at a depth so that the buds will not be more than two inches below the surface. Fill in soil around the pieces and firm well, but leave loose surface soil above and around the bud. The soil should be well fertilized and worked deeply and thoroughly before planting. Plant rhubarb as early in spring as the soil can be worked. For each person, about 3 to 4 plants should produce an ample supply. If well cared for, the new planting should last 5 to 7 years depending on cultivar and location.

A deep, rich, well-drained sandy loam soil is most desirable for production. However, rhubarb will grow well on any type soil from sand or peat to clay, provided it is well drained and has a good supply of moisture to encourage vigorous growth during hot summer months. Light sandy soils that warm up quickly provide earlier spring growth than the colder, heavier soils.

Rhubarb requires large amounts of plant food and abundant moisture during the growing season. If available, a heavy application of manure should be worked into the soil before planting to provide organic matter and nutrients for the growing plants. This should be followed by a manure mulch each fall.

Before planting, broadcast a complete fertilizer like 10-10-10 at the rate of 2 to 3 pounds per 100 square feet and thoroughly work it into the soil. In the following years a fertilizer like 10-10-10 should be broadcast or banded at the rate of 1½ pounds (sandy soil) to 2 pounds (clay soil) per 100 square feet before the new leaves begin to grow each spring. This fertilizer should be mixed 2 to 3 inches deep in the soil but not any closer than 10 inches from the plant.

After harvesting is completed, a sidedressing of ammonium nitrate (33.5 percent nitrogen) at the rate of 6 ounces per 100 square feet will stimulate summer growth and food storage in the roots.

If manure is not available, the rhubarb patch can be mulched with 1 to

3 inches of lawn clippings each year during late spring or early summer.

To promote good growth during the summer, water the plants whenever the soil begins to dry.

If the plants go dormant (leaves die) after harvest, little food is stored in the roots for the next year's crop.

Rhubarb is tolerant to soil acidity, and liming is seldom needed. It will grow well in soil as acid as pH 5.2, provided the essential nutrients of calcium, phosphorus, and magnesium are well supplied.

Weed control by hoeing and cultivation should be shallow and frequent enough to control emerging weeds. The most serious weed problems will usually occur early in spring before the newly planted root pieces start growing well.

When to Pick

Don't pick stalks during the first season or the year of planting. Food from the leaves is needed to enlarge the roots for the coming years' growth. During the second year stalks should be picked for only a short period (two weeks). Beginning with the third year the harvest period may extend as long as six weeks or until the stalks become small, indicating that food supplies in the roots are becoming depleted. Don't remove more than two-thirds of the developed stalks from the plant at any one time. Pull only the large stalks, leaving the young ones to grow.

Pick the stalks by pulling and not by cutting. Grasp the stalk near its base and pull it slightly to one side in the direction it grows. The stalks separate readily from the plant and are easily pulled. After the stalk is pulled, trim it by removing the leaf or leafblade.

If flower stalks appear, remove them at once so the plant's food will go into the roots for the next year's crop of stalks. Continued development of flower stalks will reduce rhubarb production during the following year.

Rhubarb is relatively free of insect and disease problems. But one insect that can cause problems is the rhubarb curculio. This snout beetle, common in the eastern half of the country, can puncture the stems—leaving black spots. The beetles average about a half-inch in length and are black. But they usually are so densely covered with a rusty powder that they appear reddish. Since the curculio as a rule feeds on curly dock weed, one control measure is to destroy these weeds growing near the rhubarb.

Phytophthora crown rot or foot rot is a disease that can affect rhubarb in the eastern half of the country. At the base of the stalk lesions develop rapidly to cause collapse of the whole stalk. In warm moist weather the stalks may continue to collapse until the plant is killed. There is no effective control at present.

Other diseases that may affect rhubarb are bacterial crown rot, pythium crown rot, rhizoctonia crown rot, gray mold or *Botrytis* (the most serious disease of commercial forcing operations), and ascochyta leaf spot. There also are a number of viruses that can reduce plant vigor and yields.

If you plan to propagate your own plants, identify the most vigorous 2- to 3-year-old plants with stakes during June. Leave the stakes by the plants (crown) until the following spring when the marked crowns are dug and split for planting stock. Replace plants when they start to produce fewer—and small size—stalks. Remove the old crowns and associated fleshy roots by digging, and make a new planting elsewhere in the garden.

Commercially, much of the rhubarb crop is produced by forcing during the winter months. This is a unique horticultural practice of producing an edible crop inside darkened, heated buildings. Rhubarb plants are grown in the field for 2 to 3 years during which the food materials produced in the leaves are stored in the thick fleshy roots. During winter months when the plant leaves are dead, the crown (roots) are brought from the field to the forcing structures. The large amount of food stored in fleshy roots of the crown enables it to produce many well-colored rhubarb stalks under the proper forcing conditions of darkness, water, and low even heat (50° to 55° F). After the crowns are forced they are discarded.

A crown can produce 4 to 12 pounds of stalks, depending on cultivar and crown size and vigor.

While it is a messy and time-consuming job, crowns may be forced in the home basement or cellar. About 6 to 10 good crowns should produce enough for the average family after you learn how to force rhubarb.

For Further Reading:

U. S. Department of Agriculture. *Rhubarb Production*, Leaflet 555, for sale by Superintendent of Documents, U.S. Government Printing Office, Washington, D.C. 20202. 25¢.

A Few Rows of Home Garden Potatoes Can Put Nutritious Food on Your Table

by Orrin C. Turnquist

The potato is probably the most important vegetable crop in the world today. None other is used as regularly and in such quantity in the average American home. Its culture is simple and it is a dependable and efficient food producer on any soil suitable for general garden crops.

Does the potato have a place in your home vegetable garden? The answer depends on the size of the garden.

On the average, a 100-foot row planted with 10 pounds of seed should yield between 1 and 2 bushels of potatoes. Obviously the 10 to 20 bushels a family of 5 might require for winter use would be difficult to produce in the small backyard garden. Yet, even in small gardens—after space has been provided for such vegetables as tomatoes, green beans, and leafy greens—the potato might be considered for planting on any remaining space.

Space-saving techniques such as intercropping can also be used in growing early potatoes in the small garden. Vining crops like cucumbers, melons and squash can be planted between rows of potatoes. After the early potatoes are harvested the area is free for the vine crops to spread and produce their crop.

The potato's home is in the mountainous regions of South America, although it is referred to as the Irish potato. It was cultivated rather extensively by the Inca Indians of Peru as far back as 200 A.D. Early explorers after Columbus introduced the potato to Europe between 1532 and 1550.

Orrin C. Turnquist is Professor and Extension Horticulturist, University of Minnesota, St. Paul.

Not until the potato was introduced into Ireland was it recognized for its great food value rather than as a curiosity, and by the 1600's it was cultivated extensively in that country. For approximately 250 years the potato was a major source of food in most of Europe. In fact the majority of the population in Ireland depended on this crop for its existence.

When the late blight disease came from America into Ireland (1845-1847) it caused a national disaster. Destruction of the vines and decay of the tubers caused a complete loss of the crop nationwide. The result was the Irish famine in which thousands starved to death.

A colony of Presbyterian Irish who settled in New Hampshire introduced the potato to our country in 1719. Soon after the Irish famine the potato gained in importance in the United States.

As late as 1771 only two varieties of potato were listed, but during the 19th century thousands of varieties were developed and introduced in America. Only a small number were accepted, however. Some varieties still prominent today originated during that period. They include Irish Cobbler (1875), Russet Burbank (1876), Green Mountain (1878), Red McClure (1880), and White Rose (1893).

The cultivated potato in North America and Europe is known botanically as Solanam tuberosum. It is a member of the nightshade family which includes such plants as tomato, egg plant, pepper, ground cherry, bittersweet, petunia and tobacco.

Although grown as an annual, it is often considered a perennial because of its ability to reproduce vegetatively by means of tubers that arise from

205

underground stems. In fact the tubers have all the characteristics of normal stems, including dormant true buds which are called eyes and rudimentary leaf scars that are called eyebrows.

The small dots on the tubers are identical to lenticels on a stem which facilitate the exchange of gases. These lenticels often become enlarged and objectionable when tubers develop in soils with excessive moisture and access of air is restricted.

Contrary to much common opinion, development of tubers does not depend upon flowering. Potato plants will form tubers without any flower development on top. The fruits or seed balls that develop from the flowers on some varieties are true fruits. These berries are not edible. They are not the result of cross pollination with the tomato, as many gardeners believe.

Although some of these fruits are seedless, normally they contain many small true seeds, no two of which are alike. The fact that these true seeds will not be the same as the variety from which they came is the reason we do not grow potatoes from true seed. Instead we propagate the potato by stem cuttings called seed pieces or seed eyes. The use of true seed is impractical for all except potato breeders who control the pollination and use the resulting seed in development of new varieties.

The tubers usually initiate at the tips of the stolons (underground stems) from 5 to 7 weeks after planting or when the plants are 6 to 8 inches tall. This varies with the variety and several environmental factors. As the plant grows the leaves make food for continuing growth. A point is reached, however, where a supply of food is made beyond what is needed for growth. This is when excess food is moved down into tubers for storage.

Environmental factors such as long

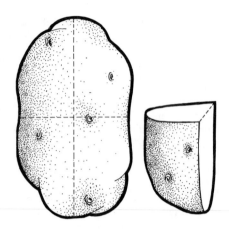

days, warm temperatures, high moisture and fertility tend to favor plant development whereas short days, cool temperatures, lower moisture, and less fertility promote tuber development. Don't forget that a good yield of potato tubers depends first of all upon a good healthy plant development.

Certified Seed

One of the most important steps to success in growing potatoes is the use of sound, healthy, certified seed. Potatoes are subject to several diseases whose symptoms are not easily recognized on either tubers or plants. Many are virus diseases that dwarf the plant and cut the yield in half. High-yielding seedstocks can be maintained only under carefully controlled conditions of isolation, disease control and storage.

Certified seed potatoes are grown mainly in the Northern States where lower growing temperatures favor the expression of virus disease symptoms so the infected plants can be eliminated. There are also fewer insects present that spread the diseases.

Potatoes are grown from "seed pieces" or "seed eyes" that are a quarter cut of the potato and include a couple of "eyes."

Home-produced potatoes may become infected in a single season, so there is no assurance that seed saved from a high yielding crop will perform satisfactorily the following year. It is best to buy new certified seed each year. Certified seed potatoes are usually identified by an official State Department of Agriculture tag on each bag. The higher cost of this seed is more than justified by the higher yield and better quality of the new crop.

Table stock potatoes that you buy at the food store should not be used for seed. They often have virus diseases present and may have been treated with a growth-inhibiting chemical to prevent them from sprouting in the market place.

Varieties

Many varieties of potatoes are certified in the United States each year. Some are more specific in their adaptability to certain regions than others. Furthermore, these varieties have a wide range of maturity from very early to late.

For the small garden, choose potato varieties that are early in maturity. This facilitates intercropping with some of the later vegetables that require more room further into the season after the potatoes have been harvested. For the larger garden the midseason and late varieties can be grown.

Potato varieties adapted to the home garden are:

Norland: A very early red variety with oblong, smooth tubers with shallow eyes. It has moderate resistance to common scab and good table quality. It is widely adapted and commonly available in the United States.

Irish Cobbler: An early maturity white variety with round to blocky tubers and deep eyes. It is a good producer and has excellent table quality. It is very susceptible to common scab.

Norgold: An early maturing russet variety with very smooth long to oblong tubers and shallow eyes. It is scab-resistant. Cooking quality is good, but under stressed growing conditions tubers could develop hollow heart.

Anoka: A newer early white variety with round to oval smooth uniform tubers. It is an all purpose potato well adapted to the home garden at 14-inch spacing. It has less tendency to darkening after cooking.

Superior. A midseason white variety with a rather tough skin. It is moderately resistant to scab and has high table quality.

Kennebec: A very popular late variety with white skin and shallow eyes. It has resistance to late blight disease. Cooking quality is excellent. Because of the thin skin the tubers are very susceptible to sunburn and greening.

Red Pontiac: A high yielding red variety with a midseason maturity. Tubers are oblong with medium deep eyes. Wih excessive moisture the tubers tend to become oversized. Cooking quality is only fair, but it is a good variety for winter storage.

Red LaSoda: A good red variety for gardens in the South. It is midseason in maturity with round to oblong tubers. Eyes are medium deep. It appears to have some tolerance to higher temperatures.

Katahdin: A very popular white variety with wide adaptability. Tubers are round to oblong with shallow eyes. It is midseason to late in maturity. Cooking quality is good.

Sebago: A late maturing white variety with resistance to late blight. It is popular in the South as well as in the East as a home garden variety. Tubers are smooth and nearly round with shallow eyes. Cooking quality is fair to good, but not as mealy as Kennebec.

Russet Burbank: Also known as Netted Gem and Idaho Russet. It is a late variety with long cylindrical to slightly flattened tubers with a heavily netted skin. Under conditions of moisture stress the tubers often develop growth cracks or knobs. It has some scab resistance. Table quality is excellent, especially for baking. Better size is obtained with a 14- to 16-inch spacing.

Consult the local extension service for new potato introductions with specific adaptation to your area. Potato breeders continue to develop varieties with improved market and culinary quality combined with high yield and multiple disease resistance.

Soil Needs

Like most vegetable crops, the potato is adapted to a wide variety of soils. It performs best, however, on a sandy loam well supplied with organic matter and plant food. High organic soils like peat or muck can be used if they are well drained. Heavy fine-textured soils are satisfactory if their structure is improved with organic matter.

Applying organic matter in the form of well-rotted manure, compost or similar materials will improve the structure for better air-holding capacity as well as water-holding capacity. Apply it to the top of the soil in early spring at the rate of 3 to 4 bushels per 100 square feet of area. With a rotary tiller, or spading fork, incorporate it thoroughly into the soil to make a uniform tilth or structure. This practice binds together coarse-textured sandy soils and breaks up fine-textured, heavy clay soils.

The best soil-acidity range for potatoes is between pH 4.8 to 5.4. On soils with a pH of 5.5 to 7.0 the potatoes usually are scabby. Lime should not be used unless a soil test indicates a pH below 4.8.

One of the most widely used commercial fertilizers for the home garden is 5-10-5. Another is 10-10-10. These analyses should provide sufficient plant food for a good potato crop. Either fertilizer can be spread at the rate of 15 pounds per 1,000 square feet at planting time.

Some gardeners prefer to place the fertilizer in bands below and slightly to the side of seed pieces. If the rows are 3 feet apart and 100 feet long, apply 6 to 8 pounds per row. A side dressing may be applied after the plants are 4 to 6 inches high. Use the same rate in an open trench 4 inches deep and 6 inches away from the plants.

Local agricultural extension offices will have more specific recommendations for your area; however, the suggestions given here should generally be adequate.

Manure is often used as a source of nutrients where scab is not a problem. This disease is more serious when fresh manure is used. Manure that is well-rotted or applied during fall or early winter is less apt to increase potato scab.

Rotate the location of potatoes in your garden each year. At least a 3-year rotation is suggested for all garden crops. In large gardens where space is not limited it helps to plant a small grain such as rye as a nurse crop seeded down to red clover the second year. After the hay is cut the third year, the plot can be plowed down and fallowed. Potatoes are then planted in the area the fourth year. Such a rotation will help control diseases, insects and weeds. Potato yield and quality will be improved.

Planting

The potato is considered a cool-season crop and can be planted as soon as frost is out of the ground and the soil dry enough to work. If soil temperature is below 40° F (5° C), however, there is a greater chance for seed piece decay, especially when cut seed is planted. Using whole seed or cut seed that has been suberized will help prevent the problem. This is a process of healing over the cut surface by holding the cut seed for a period of 7 to 10 days at a temperature of 60° to 70° F (16° to 21° C) and a relative humidity of 85 percent with good air circulation.

Planting dates will vary with the locality and soil type. In the North, potatoes usually are planted from April 15 to May 15, and in the South from November to February.

Plant early to get the highest yield. By planting early maturing varieties as soon as the soil is dry enough to prepare and warm enough to prevent seed decay, you can have potatoes of usable size by midsummer. These will not keep as long as potatoes that matured later in the growing season, however.

Proper size of the seed piece planted is important. Experience has shown that 1½- to 2-ounce seed pieces are best. This is about the size of an average hen's egg. The so-called "potato eyes" are generally too small to provide sufficient nourishment, and as a result weak plants develop. Potatoes can be planted either whole or cut. If cut, there should be at least one eye on each piece and the pieces should be uniform and blocky.

Plant the seed immediately after cutting, otherwise viability will be lowered by loss of moisture and entrance of rot organisms.

Some gardeners prefer to treat seed potatoes before planting. A good fungicide dust such as captan will help protect the cut seed pieces after planting and assure better emergence.

Seed can be cut and stored for 10 days or more if it is properly suberized.

If available, whole B-size seed is best. The value of planting small potatoes without cutting them has been known in Europe for many years. With the skin around the entire potato, it is protected from soil organisms and the result is less seed decay and better stands of plants.

If whole or B-size seed is used, it should be from certified stock. Virus-infected plants tend to produce smaller tubers, and when such seed is planted whole, weak and poor yielding plants result. However, small tubers from certified seed will produce as good a crop as large potatoes cut for seed.

Plant potatoes deep enough so the new tubers will develop sufficiently beneath the surface to avoid sunburning. Generally the seed should be planted in contact with moist soil, 3 to 4 inches deep. This depth will vary with the soil's temperature and moisture. A shallow covering of about 2 to 3 inches of soil over the seed will result in quicker emergence and less sprout damage from rhizoctonia and blackleg disease.

In the small garden a trench is usually opened with a hand hoe or cultivator. Distance between rows will depend on the type of cultivation to be used. A 36-inch row is quite common, but rows can be spaced 24 inches apart if cultivated by hand. Within the trench, place the seed pieces 12 to 18 inches apart, depending on variety. Oversize potatoes in varieties like Kennebec can be prevented by planting seed pieces as close as 6 to 8 inches. This also helps reduce the amount of hollow heart in tubers.

The amount of seed needed varies with the spacing and size of the seed piece. When cut to a 1½-ounce size and planted 12 inches apart in the trench, a 100-foot row would require about 9 pounds of seed potatoes.

Water when dry periods occur, but only if the soil needs it. Once watering is begun it should be continued until the soil is moistened to an 8- to 12-inch depth. Water thoroughly at weekly intervals when needed. Dry periods alternating with wet periods can cause potatoes to develop such abnormalties as hollow heart, growth cracks, and knobs.

Pest Control

Cultivate frequently from planting time on, to destroy weeds in the seedling stages. When the plants first emerge, rake lightly over the row as well as between rows. All cultivation should be shallow—2 inches or less—to prevent pruning the potato roots near the surface. Hill or ridge the soil over the rows only if the tubers become exposed. This will help prevent sunburn and frost damage. Cease all cultivation when the vines fill in between the rows.

A pre-emergence weed killer is registered for use in many vegetable crops, including potatoes. When used according to label directions, it is very good for control of annual broadleaved weeds and grasses. Another herbicide applied to the garden in the fall after harvest is an excellent treatment for quackgrass. It is not selective but will kill perennial grasses when applied as directed on the label. Plowing or rototilling can be deferred until spring with no residual effect on potatoes or other vegetables to be planted in the treated area.

Numerous insect and disease pests may attack potatoes. An ounce of prevention is worth a pound of cure. Use certified seed as a fundamental step in disease control. Practicing a 3- to 4-year rotation in the garden will help prevent both insect and disease problems.

Soil insects troublesome to potatoes are white grubs, wireworms and cutworms. Most garden soil insecticides give satisfactory control.

You can control leaf-feeding insects like the Colorado potato beetle and the flea beetle with recommended chemicals. Control leaf-sucking insects like leaf hoppers and aphids with a contact insecticide.

Prevent foliage diseases such as late blight and early blight by applying recommended fungicides every 7 to 10 days according to label directions. Several new materials are available for disease control. Soil-borne diseases such as scab and verticillium wilt may be reduced by growing resistant varieties and following a good rotation.

Any damage to potato plants—by insects, disease or other causes—will result in an abnormal crop of poor quality. The better the plant growth, the better the crop of potatoes.

Harvest, Storage

Potatoes may be harvested once the tubers are large enough. As long as the vines remain alive, the size of the potatoes and the yield will continue to increase. When the vines are completely dead and the skin ceases to slip from pressure by the thumb, the potatoes can be dug with a spading fork or plow.

Avoid bruises or other injuries such as cuts and fork holes in digging and handling. Do not expose the freshly-dug potatoes to sun or wind as they are very susceptible to scald and sunburn at that time. Potatoes have a sweating period the first two weeks after harvest. During this time they

Flea beetle that causes damage to many garden plants is not much larger than a flea and jumps like one when disturbed.

should be kept in a place where the temperature is about 65° F (18° C) and the relative humidity at 85% to 95%. This will help the healing-over of any injuries or wounds in harvest.

Sort over the potatoes and place only the best, sound tubers in bins or containers for winter storage. Store them dry in a room that can be kept at a temperature of 35° to 40° F (2° to 5° C) and a moderate humidity. Under these conditions well-matured tubers will keep in good condition for 7 to 8 months. Above 40° F (5° C) they may keep for 2 to 3 months but sprouting and shriveling may occur.

Sprout-inhibiting chemicals are available under various trade names from garden supply stores. A common type is applied to the vines according to label directions during the growing season. The harvested tubers are safe for table use and they will not sprout when kept at higher storage temperatures such as found in a modern basement room.

Potatoes are very sensitive to light, which causes green pigment to develop under the skin in the flesh of the tuber. This will make them bitter and unfit for table use. Tubers that have an excessive amount of greening should be discarded. If greening is only slight the affected area can be peeled away before use. Always keep potatoes in a place with total darkness to avoid greening in storage.

If the temperature in the storage room reaches 32° F (0° C), the potatoes often become sweet. Increasing the temperature for a few days will cause the sugar to revert to starch and good table quality will be restored.

Some people have developed the erroneous idea that potatoes are fattening, but studies show this is not true. Potatoes are less fattening pound for pound than most foods in the daily American diet. The potato alone is comparatively low in calories per pound, but when fried or served with a lot of butter or sour cream the caloric intake may be high. French fried potatoes have about five times the caloric value as the same weight of mashed potatoes.

Besides being a good source of food energy, the potato is also a source of iron, thiamin, niacin, and vitamin C.

A few rows of potatoes in the home garden properly planted and well cared for should provide satisfaction and achievement as well as some nutritious food to enjoy at your table.

Harvesting potatoes in the home garden — "it's a good feeling."

Sweet Potatoes—Buried Treasure

by John C. Bouwkamp

Since most of us will never savor the excitement of digging for buried treasure, digging sweet potatoes in your garden may be the next best thing. The day-by-day progress of your crop of tomatoes and beans is readily apparent but the reward for your efforts with sweet potatoes must await the day of harvest. Variety names such as Jewel, Goldrush, Nugget, Nemagold, Gem, and Maryland Golden allude to the feeling of buried treasure by their developers.

The sweet potato, like many of our vegetable crops, originated in the Tropics. The exact area of origin is subject to debate but was probably in tropical America or somewhere in the tropical South Pacific islands. Two of the ancient civilizations of tropical America, the Mayan and the Peruvian, grew and cultivated sweet potatoes, although maize was their staple crop.

Primitive cultures of the South Pacific islands give a central role to the sweet potato in their celebrations, suggesting that sweet potatoes were culivated for food in two widely separated parts of the world from ancient times. Whether the crop originated in the New World and was transported to the Polynesian islands or vice versa remains open to speculation.

Although Columbus noted the use of sweet potatoes by West Indian natives on his fourth voyage, there is no record of pre-Columbian cultivation of sweet potatoes by Indians in the continental United States. Sweet potatoes were grown in Virginia as early as 1648, most likely from roots obtained in the West Indies.

A frequent source of confusion is use of the terms sweet potato and yam. The true yam (Dioscorea sp.) is of African or Chinese origin and belongs to a different plant family than the sweet potato. It is only rarely grown in the United States.

The Blacks, when first brought to this country, mistook the sweet potato for a type of yam since the sweet potato was unknown to them and the two crops grew and were used in a similar manner. Their word *nyami* was shortened to yam and the sweet potato became known as a yam in many parts of the United States, particularly in the South.

The terminology was further confused since the dry-fleshed varieties grown in the Middle States and North (known as sweet potatoes) were noticeably different from the moist-fleshed varieties grown in the Deep South (known as yams). Thus many people came to believe sweet potatoes and yams were different vegetables.

At present, the same varieties are grown throughout the United States and a sweet potato by any other name is still the tasty, nutritious vegetable many people associate with holidays and special meals.

Almost everyone knows that sweet potatoes are a delicious addition to any meal, but not many know they also are very nutritious. An average sized boiled sweet potato (2-inch diameter, 5 inches long) will provide over half the recommended daily allowance (RDA) of Vitamin C and more than twice the RDA of Vitamin A for an adult male, 23 to 55 years old and weighing 154 pounds.

In addition, this root will provide nearly 5% of the protein, 6% of the calcium, 9% of the phosphorous, 11% of the iron, 10% of the thiamin, 5% of the niacin and nearly 6% of the riboflavin required by an adult male.

John C. Bouwkamp is an Associate Professor of Horticulture at the University of Maryland, College Park.

All these vitamins and minerals are provided with about 170 calories (6.4% of the RDA). Empty calories? Not sweet potatoes!

Growing Requirements

What will you need to know about growing sweet potatoes?

First, sweet potatoes are a tropical crop and need 4 to 5 frost-free months for growing. They thrive in hot weather. Probably little or no growth occurs when soil or air temperatures are below 60° F. Growth appears to be optimum when soil temperatures are near 70° and air temperatures near 85°.

Second, since the root is the part you will harvest, the soil should be loose and friable (crumbly)—allowing for unimpeded root enlargement and easy digging.

Meet these two requirements, and you should be a success as a grower.

Choose a well-drained site in the full sun. The pH should be in the range of 5.5 to 6.5. If your garden is moderately fertile, apply 3½ pounds of 5-10-20 (or a similar analysis) of fertilizer per 100 feet of row before ridge construction or transplanting. If your garden is very sandy and/or infertile, 5 pounds per 100 feet would be more appropriate. If it's very rich and fertile, 2 to 2½ pounds should suffice.

In many parts of the country, sweet potatoes are grown on ridges 8 to 15 inches high. These are especially important on heavy or poorly drained soils. If the soil remains water logged for several days, the roots may rot. Ridges are a good idea if this is likely to occur.

If your garden is a fine sandy loam and well drained, probably no ridge is necessary.

Should you wish to grow on ridges, construct them before transplanting. Rows should be 3 to 4 feet apart.

Plan to obtain disease-free plants, also called sprouts, from a reputable dealer. Your choice of varieties differs widely in various parts of the United States, and not all varieties will be available at any location. Your local Extension Office would be your best choice for help in choosing a variety.

You may also wish to produce your own sprouts after you have some experience with growing sweet potatoes. Set plants 2 to 3 inches deep, and 12 to 16 inches apart in the rows, one plant per hill. Don't transplant until soil temperature reaches 60° F. A good rule of thumb is to plant sweet potatoes 1 to 2 weeks after frost danger is past.

After transplanting you may wish to water the plants with ½ cup or more of water or ½ cup of a starter solution. Starter solution can be made by mixing ½ ounce of a soluble, high phosphorus fertilizer (15-30-15 or similar analysis) per gallon of water. Be careful not to overdo the starter solution or you may damage the plants.

Hoe and cultivate sweet potatoes frequently (once per week) after the vines have begun to run. This serves two purposes—to control the weeds and to keep the vines from rooting at the nodes. If these nodes are al-

Allan Stoner

Vigorous growth of young sweet potato vines.

lowed to root, some storage roots may develop and the main storage roots will not develop as quickly.

Three to four weeks after transplanting, make a second application of fertilizer at the same rate as before transplanting. Place the fertilizer near the rows and work it into the soil.

After 6 to 8 weeks the vines will become too large to cultivate and you must weed by hand.

Moisture Needs

Once established, sweet potatoes

Allan Stoner

Sweet potatoes at harvest.

may be considered drought-hardy, and will produce a fair crop even under quite dry conditions. But supply them with ample moisture for best yields. They need about ¾ inch of rain or irrigation per week while small, and up to 2 inches of water per week when growing vigorously in hot weather.

Occasional leaf wilting during the hot part of the day is no cause for concern. But if the plants do not revive by early evening, you may wish to irrigate.

Excessive moisture after a prolonged dry period may cause the roots to crack. These cracks usually heal over by harvest time, resulting in a less appealing appearance but with no loss to eating or storing quality.

Sweet potatoes do not ripen or mature, so the time for harvest is judged by root size. At about 110 to 120 days after transplanting (if you can wait that long) you may wish to check on root size. To do so, carefully dig away some of the soil from several hills near the center of the row—the end plants may develop faster. When a majority of the roots have reached a size you desire, they should be dug immediately. A light frost which only slightly damages the vines won't cause problems.

Dig roots carefully to avoid injury. A shovel usually works better than a fork. After digging, place the roots in containers. Keep separate any badly injured or cut roots for immediate use.

A yield of two pounds a plant is good.

Optimum conditions for curing and storing sweet potatoes are rather exact and should be maintained as nearly as possible. ("Curing" is a means of preparing sweet potatoes for storage.) Cure roots 6 to 8 days at about 85° F and 85% to 90% relative humidity. Temperatures should never exceed 90°. A high relative humidity can best be maintained by loosely wrapping the containers in polyethylene film. Don't cure in the sun as the roots may sunburn.

After curing, store the roots in a cool place (55° to 60° F) with a high relative humidity (85%). An unheated area of the basement or a root cellar might be okay.

If storage temperatures drop below 50° F for an extended time, chilling injury may develop and the roots may spoil. Temperatures above 65° will likely result in sprouting and pithiness. If the relative humidity is too low, shriveling will result. However, if proper conditions are maintained you may expect your sweet potatoes to keep well and provide good eating through winter and into spring.

If you wish to produce your own "seed" for the following year, begin when the plants have developed vines 1½ to 2 feet long. Cut a segment 8 to 10 inches long from the tip of a vine and transplant it into a row which has been fertilized as previously mentioned. Plants should be set about a foot apart and 2 to 3 inches deep, making sure to cover at least 1 node. Water with ½ cup of starter solution or water.

You can expect a 10-fold increase with most varieties. So if you want 100 plants next year, have 10 hills of vine cuttings. Fertilize and cultivate your cuttings as suggested for the regular crop.

If the majority of roots from the vine cuttings are 1 inch in diameter or larger, harvest them at the same time as your main crop. Roots produced for sprouts should be cured and stored the same way as roots for table use.

Producing Sprouts

If you have produced roots from vine cuttings and wish to produce your own sprouts, several methods may be used depending upon the facilities at hand. If a heated frame is available, place disease-free "seed" roots near each other but not touching, and cover them with 1½ to 2 inches of light sand 4 to 5 weeks before transplanting. Soil temperature should not exceed 85° F. Open the sash on warm sunny days to avoid excessive heat buildup.

When the sprouts are 6 inches tall (usually a week or two before transplanting) turn off the heat and leave the sash open, except for very cold nights. Water the bed as necessary to keep it damp but not waterlogged.

Alternatively, you may wish to bed roots in a cold frame. Put your cold frame in a sunny place protected from wind. Allow an extra week to produce sprouts without heat, but otherwise the procedure is about the same.

If neither a cold frame nor a heated bed is available, you can bed sweet potatoes in your garden. Choose a sunny spot protected from wind and make a raised bed 6 to 8 inches high and 1½ to 2 feet wide, 6 to 7 weeks before transplanting. Bed the roots as described and cover with 1½ to 2 inches of sand. Then cover the bed with clear polyethylene film, securely anchoring edges of the plastic with soil.

When the sprouts begin to emerge, ventilate the plastic by punching ¼-inch holes every 6 to 8 inches. After 2 to 3 weeks or when the sprouts are 3 to 4 inches tall, you may remove the plastic—but replace it if a frost is likely. The sprouts should be 6 to 8 inches tall by transplanting time.

Remove sprouts from the roots by grasping them firmly one at a time and pulling sharply from the soil. With a little practice you should be able to pull sprouts without disturbing the bedded root. Transplant the sprouts as soon as possible. The bedded roots will continue to produce sprouts which may be used for a second planting, or you may wish to dig up the roots and destroy them.

Fortunately for the home gardener, there are relatively few serious disease and insect pests of sweet potatoes. Fusarium wilt, also called blue stem and stem rot, attacks certain varieties resulting in both a loss of plants and a yield reduction. It can be best controlled by growing resistant varieties and a rotation of four or more years' duration.

Scurf, also called soil stain, causes a dark "freckled" appearance to the skin of storage roots. The fungus attacks only the skin and causes no loss to eating quality. Control it by using only vine cuttings to produce the "seed" roots.

Pox or soil rot results in deep corky pits on the root. Once again, although the appearance is greatly affected, eating quality is not. Control pox by reducing the soil pH to 5.0 to 5.5, resulting in greatly reduced incidence and severity.

Black rot is occasionally a problem both in the garden and in storage. Affected areas begin as small black, nearly round spots, but under favorable conditions they enlarge and may nearly cover the root and extend well into the flesh. Usually you can control it by using disease-free sprouts, not planting in the same area of your garden each year, and using clean containers for storage. Some fungicides are effective in controlling black rot.

Rhizopus rot or soft rot attacks the roots after harvest, usually entering through a wound. It results in total loss of the root. Control soft rot by carefully handling the roots during harvesting, thus avoiding as much as possible any injury to them, and by curing the roots properly as soon after harvest as possible in order to heal wounds.

Fusarium root rot or surface rot is usually not noticed until after some months of storage. It appears as a generally circular spot, black or dark brown, sometimes slightly sunken, and only extends about ⅛ inch into the root's flesh. If infected areas are pared away the remainder of the root may be eaten. There is no known means of control, but immediate and proper curing helps in reducing the incidence. The occurrence is sporadic and seldom reaches serious proportions.

Nematodes may cause serious losses to sweet potato yields. The infected roots become misshapen and cracked, and galls may be observed on feeder roots. At present the home gardener has no effective way to control nematodes. Many varieties are resistant to one or more races or species of nematodes. Your local Extension Office is the best source of information if nematodes become a problem.

Although several species of insects and soil grubs attack sweet potatoes, they rarely result in serious losses. Damage is usually confined to a small area of a root and may be pared away prior to cooking. If serious losses occur, consult your Extension Office.

I have tried to outline the how's and why's of sweet potato growing for the average gardener, but experience is the best teacher. You may find it advisable to change some of the recommendations.

Sweet potatoes are likely to produce a fair crop even if you make a few mistakes, so feel free to experiment with changes after you have a little experience. If you fail in your attempt to grow sweet potatoes, find out why and try again. After all, not everyone finds buried treasure on their first try.

Herbs for Flavor, Fragrances, Fun In Gardens, Pots, in Shade, in Sun

by Doris Thain Frost

Herbs give much pleasure and profit if you grow them yourself. Plant herbs in your garden, read books about them, and discover personal joy and an added dimension to your cooking.

First of all, a place is needed to plant the seeds or roots and this means productive soil. A grower with an outdoor plot is indeed fortunate. Herbs will grow well in any garden where vegetables thrive, in the garden rows or around the edges. Herbs will grow in flower beds, in borders, among ornamental shrubs and roses, just so there is good drainage and six or more hours of sun.

Most herbs prefer an alkaline soil, a pH of 6.5 to 7.5. If the soil test indicates acidity, work ground limestone into the soil. The amount will be indicated in the soil test analysis received from your county extension services.

If an outdoor plot is lacking, many herbs will grow in boxes, pots or hanging baskets if the same conditions—good soil, drainage and sun—exist.

When planning an herb garden, remember that herbs belong in different classes according to their life span. Annuals, tender and hardy, may be planted in the vegetable garden as they mature in one season. Biennials and tender and hardy perennials must be planted in locations that will not be disturbed by cultivation or rotation as they live several years.

Prepare the plot as for vegetables.

Animal manure and compost are good fertilizers, preferably applied in early spring. Use mulches to keep the herb foliage clean, for weed control, and to preserve soil moisture. Cocoa hulls, buckwheat hulls, leaves, straw and hay are popular mulches.

Herbs are propagated by seeds, cuttings, layering, and divisions. If you want only a few plants, buy them from commercial growers.

Seeds come in packets sold by established seed houses and by some commercial herb growers. Unless you have the equipment and space to start seeds indoors to transplant later, experience has proven that the average gardener had best buy seeds of annuals and plant them where they are to grow, and start perennials from cuttings, divisions and plants.

Parsley, the culinary biennial, can be started from seed if many plants are desired, or a few small plants purchased.

Cuttings can be rooted in water or in a medium of perlite, milled sphagnum moss, or in compressed peat pellets. The rooted cuttings may be transplanted into pots or into the garden if the season permits.

Divisions are made by digging up an older plant and pulling apart or cutting sections of the root and replanting each section individually.

Plants from commercial growers should be carefully examined for insects and disease, and if they are to grow outdoors, bought and planted when the ground has warmed and all danger of frost is past.

Bees, lady bugs, praying mantis and many other insects are friends in the herb garden as they pollinate plants and also destroy insect enemies. Herbs are peculiarly resistant to most insects and diseases.

Doris Thain Frost of Great Falls, Va., is a board member of the Herb Society of America, has taught herb classes at the National Arboretum, and is editor of the *Garden Bulletin* issued by the National Capital Area Federation of Garden Clubs.

Sometimes mints become mildewy early in the season. Either harvest early, or cut and destroy the affected stems. New growth for the second crop will be free of mildew.

Sometimes dill and fennel attract tomato worms. These can be removed by hand. Japanese beetles attack basil. Shake them off into a can of kerosene or into a bucket with warm water, salt and detergent added.

Do not use poisonous substances or powders on any herb to be used in food or beverages.

When to Harvest

The secret of a good harvest is timing, taking into account the readiness of the plant and the use to which it will be put. Just before the flowers fully open is said to be the time when the most oils and flavors are present and the richest fragrances prevail.

Successive harvests can be made of mints, comfrey, basils, parsley and others by cutting the stems early in the season, not too close to the ground so that new growth will start quickly. Cut again in late summer, and—with annuals—before frost when the entire plant may be harvested. If seeds are desired or self seeding is planned, a crop must be allowed to mature and ripen seeds.

Herbs for future use may be dried, frozen, the flavors preserved in vinegars and jellies, or kept fresh for a short time. To dry, cut the stems or stalks when the plant is ready, as I explained. Don't cut too close to the ground. Separate into small bunches, tie with string, and hang in a warm, dry, dark place such as an attic or vacant room until the leaves are crisp and brittle. In the summer this takes from three to ten days.

Strip the leaves, and buds or flowerettes if desired, and put as whole as possible into a jar with a tight lid. Check for a few days to be sure the herbs are perfectly dry, or mold, mildew or other problems will develop. The leaves are kept as whole as possible to preserve the flavor. They can be crumbled when used.

When only the leaves are dried, as with comfrey, gather the leaves, and spread thinly on newspaper in a warm, dry place until crispy dry. Then store in jars.

Basil, parsley and chives sometimes turn very dark if air dried. Stems of these can be laid on brown paper and put into an oven at 150° F or less. Leave the oven door open to allow moisture to escape. This method takes several hours.

The quickest, most modern way is to dry in a microwave oven. Place sprigs on a paper towel and cover with a paper towel. Put into the oven for one minute. Take out of oven and cool. If not completely dry put back into oven for a few seconds. When crumbly, store in jars. Basil, sage, parsley, mints and oregano, especially, retain beautiful, appetizing green colors when dried this way.

If only seeds are to be used, such as from dill, fennel or coriander, take care to cut the stems when ripe but before the seeds fall. A paper bag carefully put over a head or umbel and tied with string before the head is severed is then hung upside down (seed head down) to catch the seeds as they dry and fall.

Freezing

To freeze herbs, gather as for drying, wash if dusty, pat off excess water, place into plastic bags and put into the freezer immediately. When it's time to use them, snip or chop the herbs without thawing as they mince easily while frozen. Mint, tarragon, lovage, parsley, chives, sorrel, and sweet marjoram take kindly to this method.

Another good way to freeze is to put the chopped herbs into an ice cube tray, fill with water, and freeze. Then put the cubes into plastic bags and store in the freezer. The cubes can easily be popped into soups or stews when needed.

Herb flavors can be enjoyed in vinegars, jellies, and pickles.

Fresh herbs are probably the most desired as the flavor is at its best. Herbs can be kept in the vegetable bin of a refrigerator for a while, at least through the winter holidays. Or they may be planted in pots and kept in a sunny place or in a greenhouse.

Parsley, chives, sweet marjoram, thyme and basil are some of the easiest to grow in pots. Many herbs have deep or large root systems that require more space than is usually available in pots.

Herbs that are easy to grow and delightful to use are listed below. Do try to grow some of them in your home garden. Advanced herbalists will know many more.

BASIL—*Ocimum basilicum* (pronounced like dazzle). A tender annual. Plant seeds when all danger of frost is past, and cut the last harvest before cold winds turn the leaves black. Of the many varieties, lettuce leaf, dwarf bush, lemon, and the purple or opal basils are the ones used for flavoring food.

Harvest basil when the flower heads appear. If the leaves are to be kept growing, keep the flower heads pinched out. Use fresh in salads, salad dressing, soups, and vegetables. Basil's clove-like flavor has a special affinity for tomatoes, cottage cheese, and egg dishes. The leaves can be dried quickly in the oven, or made into a vinegar to which the red or opal variety gives a lovely ruby color.

CHIVES — *Allium schoenoprasum* is the most delicate tasting member of the onion family (see onion chapter for production details). The tender, hollow spears are cut and chopped finely to flavor a great variety of dishes.

The lavender flower heads of chives may be cut close to the ground and dried to go in winter arrangements, or chopped fresh and added to salad. A beautiful perennial, chives often are grown as garden borders.

Glenn M. Christiansen

Chives are best used fresh, and a fresh supply can be kept for winter by potting a few plants and bringing them indoors in fall. Or the snipped foliage may be frozen in ice cubes as described before.

Chives are good in herb butters, green salads, in sour cream for dressing potatoes, in fact in any dish where a mild onion flavor is desired.

DILL—*Anethum graveolens* is a hardy annual. Plant the seeds where they are to grow in the early spring, or in the autumn to get an early start. Make successive plantings from April to July. Dill reseeds very easily if a few plants are allowed to mature.

Both fresh foliage and seeds of dill are used in pickling, in vinegar, minced over salads, cottage cheese and potatoes, blended into sauces for veal and fish, or baked into dilly bread. Dill foliage is the dill weed found in the grocery store.

Fresh-picked dill grown indoors makes sandwich garnish.

Dill has such a refreshing flavor that it should be much more widely used. Green dill umbels are distinctive in flower arrangements.

EGYPTIAN ONION—*Allium cepa* var. *vivaparum* is a hardy perennial, a curious member of the onion family that forms its bulbs on the tips of its long green shoots rather than in the ground, as most of its relatives do. (See onion chapter for production details.)

Egyptian onion is a very ornamental plant in a garden border. The bulbs may be used in any way an ordinary onion is used. The fresh stalks may be chopped and used too.

GARLIC—*Allium sativum*. Garlic's health-giving qualities have been known since ancient times. It also serves as a bug repellent in the garden. (See onion chapter for production details.)

Garlic gives that extra touch to a salad, pickles, and vegetables. The taste for it develops and often it is used as a condiment at the table like salt and pepper. Garlic contains an important essential oil, allicin. Chewing a sprig of parsley or a whole clove is supposed to sweeten the breath after indulging in garlic-flavored food.

Garlic butter is excellent with sour dough or French bread, and garlic vinegar for flavoring potato salad, stews, and cooked greens.

LOVAGE—*Levisticum officinale* is a hardy perennial that often grows to six feet in height. Plant it in a permanent place as it lives many years. One or two plants are enough for a small garden. It grows from seed but the best way to begin is with a division from an old plant, or with a young plant from a nursery.

Lovage likes a bit of shade and moisture. The large celery-like leaves are the usable parts. They have a strong celery flavor. The leaves can be cut from the stems and put fresh in salads and soups, or like celery used with other vegetables. In the fall, the leaves can be spread out to dry on paper for winter use.

The dried leaves turn yellowish and do not keep their flavor much more than a year. All the foliage will die down in the fall but new shoots appear in early spring. Lovage is ornamental in a garden corner.

SWEET MARJORAM—*Marjorana hortensis* is a tender perennial, treated as an annual. It is sometimes called knotted marjoram because of the form of the flowers and seeds along the stem ends. It is a small, low plant, most easily started by buying plants and setting them into the garden when the ground is warm. Seeds may be sown indoors in early spring and transplanted later but this is for the advanced gardener.

Two harvests may be made. The first is when the plant starts to bloom. Cut back all the stems, leaving at least an inch of the stems above ground. The harvested marjoram may be dried in small bunches or spread on paper

M. S. Lowman

A pot marjoram.

in a warm, dry room. The plants can be cut again when they flower the second time.

Culinary uses are many, as marjoram is one of the most aromatic herbs. The delicious, spicy flavor remains when the leaves are dried. It can be used with fowl, lamb, herb butters, vegetables, and in herbal tea mixtures.

MINTS—*Mentha.* There are many varieties of these hardy perennials. The most popular mints are peppermint (of which curly mint is a variety), spearmint, and orange and apple mint. The best way to start is to find a friend who has a mint bed and get one or two root divisions, or buy them from a nursery.

Mint likes moist, fertile soil and doesn't mind some shade. It reproduces by sending long, lateral stolons (runners) under the ground. These may be divided to supply new plants.

Mints may be cut two or three times each year, leaving a few inches of stem to grow again. The leaves are stripped off and dried in a warm dry place, or oven, or hung to dry in small bunches and stripped later.

You can make mint tea, mint sauces, mint jelly, or mix mint with other herbs or citrus juices for teas and jellies. The fresh leaves may be used in green or fruit salads, with new peas, and in candy.

OREGANO—*Origanum* has many varieties. These hardy perennials are not clearly defined by herbal authorities. The oreganos and marjorams are closely related but the varieties are different in growth and flavor.

Greek oregano, *Origanum heracleoticum,* is thought by some to be the true oregano. This plant is treated as a tender perennial in the Washington (D.C.) area. It is slow growing and the flowers are white.

Others call *Origanum vulgare,* a very hardy perennial, the true oregano. It grows several feet high. The blooms are pink and purplish.

Oreganos are propagated by cuttings, divisions, and young plants from nurseries—often not labeled correctly.

As soon as the flowers appear the stems may be cut to dry in small bunches. The leaves may be used fresh all summer long.

Oregano goes well with tomato dishes, tomato juice, pizzas and other pastas, spaghetti, macaroni, and noodles. It also enhances lamb, beef, soups, and salads.

PARSLEY—*Petroselinum hortense* is a very hardy biennial. The seeds take about three weeks to germinate, unless soaked overnight in water before planting. An old saying is that the seeds must go to the devil seven times and back before they will come up.

Busy gardeners usually buy plants from a nursery every year and plant them in a permanent plot or border in order to always have a good supply. The foliage makes an attractive low border. Medicinal uses of parsley are ancient and numerous since the plant is a rich source of vitamins A and C and of calcium, niacin riboflavin and other properties. Parsley is often made into an infusion or tea and drunk alone or combined with other herbs to promote health. The leaves are used for flavoring in soups, stews, potato dishes, and as a breath sweetener with garlic-seasoned dishes.

Harvest parsley by cutting the stems an inch or two above the ground, and dry quickly on paper in a dry, shady place as the leaves turn dark very easily. Many think oven methods are best to preserve color and flavor.

Fresh parsley can be used most of the year as it is very hardy. It is also lovely in hanging baskets for indoor gardens.

ROSEMARY—*Rosmarinus officinalis* is a tender evergreen perennial, one of the most esteemed and decorative herbs. It needs a well drained alkaline soil, a sunny location, and protection during the winter until well established. It makes a beautiful large pot plant in a green house.

Rosemary is propagated by cuttings or layering. Young plants may be bought from nurseries. Late summer is the best time to take cuttings.

Rosemary has an assertive spicy flavor delicious with lamb, chicken, other meat dishes and stuffings. Rosemary butter is luscious on hot biscuits. The tips and leaves may be dried for future use, but since the plant is evergreen, fresh tips are always available.

SAGE—*Salvia officinalis* is a hardy evergreen perennial that becomes woody and sprawly after four or five years. Sage is most easily propagated from seeds or young plants from a nursery. It needs sun—and good drainage.

The fresh leaves may be used all year, but cuttings of sprigs may be dried in a warm shady place or by oven methods. Sage makes a good, healthful tea. Its strong, dominating flavor improves cheeses, poultry, dressings, sausages, pork, and wines.

SALAD BURNET—*Sanguisorba minor* is a hardy perennial easily grown from seed in full sun. As it grows 1 to 2 feet tall, it should be thinned to leave the plants 12 to 18 inches apart. The leaves are used fresh as the cucumbery taste and smell vanish when dried. It is a pretty border plant. The leaves are fern-like and usually evergreen. The flavor is good in fresh salads, vinegar, and wine punches.

SORREL, FRENCH—*Rumex scutatus* is a very hardy perennial, whose broad leaves add a nice sharp taste to spring greens, spinach, and herbal soups. It is used fresh. The leaves can be cut throughout the growing season. It is grown from seed or from root divisions. The flower stalks should be removed as they appear, so that the green leaves may be produced longer.

M.S. Lowman

SUMMER SAVORY—*Satureia hortensis* is a hardy annual whose seeds are sown in the garden in the spring. It grows fast and the plants should be hilled to keep them upright. Savory needs sun and plenty of moisture.

Savory can be cut and dried when the flowers open, or the tips of the plants pinched and used fresh throughout the summer. Savory makes a delicate tea. It is the Bohnenkraut of the Germans, excellent with green beans, butters, spreads, green salads, egg dishes, and all kinds of meat.

Winter Savory, *Satureia montana*, is a hardy perennial that forms a small bush with lavender flowers. It is very desirable as a border plant, but not as aromatic as summer savory—therefore less useful in cooking.

TARRAGON—*Artemisia dracunculus* is a tender perennial, unless the roots are somewhat protected with straw or mulch during the winter in the Northeastern areas. Avoid buying seeds as the true variety rarely sets seeds—you might find you have the Russian or Siberian variety which is very vigorous but lacks the aromatic scent and flavor of the true type.

Tarragon plant at harvest stage. Leaves and tops may be cut several times during season.

Propagate from a cutting or root division or buy young plants from a reputable source. Plant in a sunny place, especially well drained, with room for the shallow lateral roots. Stems should be harvested in early summer, leaving at least three inches of stem above the ground to furnish growth for one or two more harvests later in the year.

Dry the leaves quickly as they turn brown easily. Try oven methods. When dry, seal in dry tight containers. A better idea is to pot a plant or two to keep indoors and enjoy fresh. Fertilize regularly and keep on the dry side.

The culinary uses of tarragon are ancient. Tarragon vinegar is well known for flavoring sauces and salad dressings. Tarragon is especially delectable on fish, cauliflower, spinach, roast turkey, and egg dishes, and it makes sauce Bearnaise. The robust flavor is best used alone and not combined with other herbs.

THYME—*Thymus vulgaris* is a hardy perennial that can be started from seed, but best results are from divisions or plants purchased from a nursery. Plant in a sunny, well drained location. It is a low, bushy plant with lovely blooms that is attractive in a foreground. Of the many varieties, the so-called French and English thymes are best for culinary purposes.

One cutting, made when the flowers begin to open, is taken for drying. The next growth should be left to help the plant survive the winter. Dry on paper in a warm, dry room. When dry, rub off the leaves and discard the stems. Store in dry, tight jars.

Thyme makes a stimulating tea and can be used to flavor any meat, fish, or vegetable. It is good in most any food. Greek thyme honey is famous. It can be found in organic food stores, or a thyme syrup can be made from our native honey mixed with strong thyme tea.

Basic Herbal Recipes

Herb Butter
Soften one half stick butter (sweet, unsalted if possible)
Add one tablespoon finely minced fresh herb or one-half teaspoon dried herb
Cream together, adding a few drops lemon juice
Use on hot breads, vegetables, baked potatoes
Herbs to use: basil, tarragon, thyme, chives, dill, parsley, marjoram, rosemary

Herb Vinegar
Clean and dry wide-mouthed glass jars
Gather fresh herbs. If dusty rinse in cold water and pat dry (water clouds vinegar)
Fill jar lightly with herbs
Heat, do *not* boil, good cider or wine vinegar
Pour vinegar over herb, cover with a non-rust lid or just put waxed paper over mouth of jar
Set jar in room temperature location for two or three weeks
Strain through cheesecloth and bottle
Herbs to use: dill, basil, salad burnet, tarragon, mint

Herb Jelly
Two cups herb infusion
One fourth cup vinegar or apple cider
Four and one half cups sugar
Heat the above until sugar is dissolved (high heat)
When boiling add one half bottle liquid pectin
Rolling boil for one and one half minutes
Take off fire. Add one or two drops food coloring if desired
Fill sterilized jelly glasses and seal with melted paraffin.
Herbs to use: Sage, basil, thyme, parsley, marjoram, rosemary, mint
Infusion: 2½ cups boiling water over 1 cup fresh herb. Let cool and strain

For Further Reading:
A Primer for Herb Growing, The Herb Society of America, 300 Massachusetts Ave., Boston, Mass. 02115. 50¢.
Foster, Gertrude B., *Herbs for Every Garden*, E. P. Dutton and Company, Inc., 201 Park Ave. S., New York, N. Y. 10003. $5.95.

Okra Is Produced Primarily in the South As Main Dish Vegetable, and for Gumbos

by W. D. Kimbrough, L. G. Jones, and J. F. Fontenot

Okra (Abelmoschus esculentus (L) Moench) is a member of the mallow family, closely related to Chinese hibiscus and to cotton. Its beginnings are uncertain, but it probably originated in Africa or Asia and was brought to America by the Spanish. Okra is a warm weather plant. Under ideal conditions a perennial, it is grown in the United States as an annual, since cold usually kills it here.

Grown extensively in home gardens in the South, okra is commonly served as a main dish vegetable. It can be used fresh or may be frozen or canned.

The immature seed pods, produced over a relatively long period if harvested regularly, are the edible part of the okra plant. Okra is especially important to the lower South, where not many vegetables are productive from midsummer through early fall. Most home gardens of the South should have a place for okra. However, in small gardens there may not be room, as the plants get fairly large and occupy the space for a long time.

Okra is grown to some extent in Northern gardens, but due to the shorter growing season the yields will not be as large as in the South.

Composition of okra pods varies somewhat with growing conditions and stage of maturity. The more immature the pods, the less food value they have. They consist mainly of water and carbohydrates, like most vegetables. Several vitamins and minerals also are present.

Okra contains mucilaginous material that some people object to, as it makes certain okra dishes seem slimy. This material, however, is what makes okra so desirable in soups and gumbos.

Okra grows on a wide range of soil types and tolerates large variations in soil reaction (pH). Good drainage is essential. This does not mean just surface runoff, but that water will percolate through the soil.

Any good garden soil should be satisfactory, although a sandy loam soil with a porous clay subsoil is ideal. Adding manure or organic matter in some other form is usually helpful. Also, it is generally a good idea to apply a complete fertilizer, relatively high in phosphorus, before planting the seed.

From 1 to 1.5 pounds of 6-12-6 fertilizer per 25 feet of row may be used in a garden, or a similar amount of nutrients provided by another grade and rate. It should be worked in to a depth of about 4 inches.

As okra has a long growing season, application of additional readily available nitrogen during the season is often beneficial, especially on lighter soil types. If this is done, 2 moderate applications of nitrogen fertilizer 4 to 6 weeks apart would be preferable to 1 heavy application.

When fertilizing okra, give care to the rate and timing of nitrogen applications. This is because of the plant's tendency to become excessively vegetative and produce few pods if excessive nitrogen is available. Excess nitrogen can result from either applying too much fertilizer or breakdown of soil organic matter during the growing season.

Include only light to moderate rates of nitrogen in the preplanting fertilizer, especially on soil relatively high

---

W. D. Kimbrough is Professor Emeritus, and L. G. Jones and J. F. Fontenot are Professors in the Department of Horticulture, Louisiana State University, Baton Rouge.

Allan Stoner

in organic matter. Withhold the first side-dressing of nitrogen until after a few pods have set on each plant. Then you may make a moderate application of nitrogen fertilizer (⅛ to ¼ pound of ammonium nitrate or equivalent per 25 feet of row, depending on organic matter content of the soil). At this time, the plants usually will be about knee-high, depending on the variety.

As the season progresses and the plants reach a height of waist to shoulder, they may require a second moderate application of nitrogen fertilizer as side-dressing, especially if you intend to harvest pods late in the season.

Important Varieties

Okra varieties differ considerably in size of plants and shape of pods. Height of plant will vary from 5 to 10 feet after a few months of growth. Length of the internodes and the degree of lateral branching—both of which are influenced by the plant's genetic makeup—as well as growing conditions determine size of the plant. Dwarf varieties have shorter internodes.

Pod shape ranges from short to long and from nearly round to very ridged. Pod diameter also varies.

Okra varieties and strains differ greatly in plant growth, pod pigmentation, leaves, and stems. They also vary in leaf shape and flowers. Thus, okra may be enjoyed by the homeowner for ornamental purposes as well as for food.

Seed catalogs usually list about six varieties. A short description of the most important varieties follows:

Clemson Spineless is of medium plant height, about 5 feet; mature pods are 5 to 6 inches long, moderately ridged, straight, and green. The pods are very smooth and have few spines. The first fruit for harvest matures in about 55 days. It can be used fresh or processed.

Emerald is also of medium height, about 5 feet, with mature pods 8 inches long. The pods are straight, round, smooth, very slender and deep green. Production starts some 50 days from planting, and pods can be used fresh or processed.

Louisiana Green Velvet is medium-tall, 7 to 9 feet, with mature pods about 7 inches long, slender, round, straight, and green. Production starts about 60 days after planting. Pods can be used fresh or processed.

Perkins Mammoth Long Pod is very tall, 10 to 12 feet; mature pods are 7 to 8 inches long, ridged and green. The first fruit for harvest matures in about 60 days. It can be used fresh or processed.

Gold Coast is of medium plant height, about 4½ feet; mature pods are short and round, green, straight, and about 3 to 4 inches long. The first fruit is ready for harvest in some 55 days. It has a long shelf life, and is recommended for fresh use only.

Size of the garden plot may determine the variety to plant. Smaller growing plants may be preferable in gardens that aren't very big.

As okra is a warm weather vegetable, don't plant it until the soil

Okra plants showing hibiscus-type flower and young tender pods.

warms up in spring. If early okra is desired, place black plastic on the rows prepared for planting. Or grow plants in peat pots and transplant them to the garden when the soil is warm enough.

Soak Seed Overnight

Getting a good stand of okra is not as easy as with many other vegetables. Soaking seed overnight in water before planting usually results in quicker germination and a better plant stand. Seed generally is planted in the prepared row, or 3 or 4 seeds may be planted in hills at desired spacings. Plant seed at a depth of ½ to 1 inch, depending on soil type. The deeper planting is for sandier soils. Rows should be 3 to 5 feet apart.

After the seed germinates, thin the plants to a spacing of 1 to 2 feet, depending on the variety. Plant okra where it will get full sunlight.

Weeds and grass should be controlled. Since drought severe enough to affect the growth of okra plants occurs commonly in the South, irrigation may be needed. Too much water may cause excessive vegetative growth of plants and less pod production.

The okra plant and pods may have small spines which some people are allergic to. Plant breeders have developed varieties with fewer spines, especially on the pods. At harvest time wear gloves and a long sleeved shirt or blouse as skin protection. With large plants and hot weather, harvesting okra can be mighty uncomfortable.

Harvest Carefully

Cut the pods with a knife or shears. In any case, harvest very carefully so as not to injure the pods or the plant. Harvest pods when they are large enough to give a good yield but before they become fibrous. This is usually when the pods are about four inches long or between the fourth and seventh day from the time the bloom opens, depending on variety and weather conditions. Perference of the harvester as regards pod size is also a consideration.

The okra pod becomes fibrous from the tip down. To be edible a pod should allow a knife blade to pass through it without providing noticeable resistance. Okra pods that get fibrous or tough on the plant will slow down the plant's growth and decrease yield.

Periods between harvests should be short, usually not over two or three days. Even then there will be considerable variation in the length of pods harvested.

The harvested pods should be handled with care, as they bruise easily. The pods will wilt rather quickly when harvested during warm periods. Therefore they should be used or stored at 45° to 50° F and relatively high humidity as soon as possible after harvest. A home refrigerator provides adequate storage conditions.

Cutting Back

Sometimes when plants get too large they are cut back and allowed to sprout again near the soil surface. If this is done, apply a top-dressing of a readily available nitrogenous fertilizer.

A 25-foot row of okra should produce 25 to 75 pounds of edible pods, depending on variety, care, weather conditions, and length of harvest season.

You can grow your own seed if you raise only one variety of okra and have no near neighbor growing another variety. Late in the season allow a few plants to develop mature seed. After removing seed from the pod, dry thoroughly and store it in the refrigerator in a closed container until the next spring. Then you can be assured of having seed of the kind wanted at planting time.

Several pests can injure okra, with

nematodes probably the most serious. Gardens that have remained in the same location for several years usually are infested with these small round worms. The only remedy is to periodically treat the soil with materials that will kill most of them. Obtain directions for treating soil from your county agent.

With other conditions favorable, a fair crop of okra may be produced despite nematodes. But the harvest season is likely to be shortened.

Wilts may kill or injure okra plants. The organisms that cause wilt are soil-borne and require soil treatment.

Several insects also may cause damage to okra. These include the corn earworm, which eats into the pod, and stinkbugs and ants which are especially troublesome late in the season.

Recommendations for controlling pests vary as new products become available or old ones are taken off the market. Consult your county agent for current recommendations.

Stink bug, which damages okra, produces foul odor when annoyed.

Miscellany, Including Celeriac, Horseradish, Artichoke, Peanuts, Vegetable Soybeans

by Homer N. Metcalf and Milo Burnham

Plants discussed in this chapter will provide both experienced and beginning gardeners with an introduction to the fascination of growing uncommon vegetables. Once culture of the easier vegetables has been mastered, the natural tendency of gardeners is to extend their experience. Unusual vegetables which may require special production techniques provide a stimulating challenge.

Information is provided for celeriac, chayote, dasheen, globe antichoke, horseradish, husk tomato, martynia, mushrooms, peanuts, sunchoke, vegetable soybeans, and watercress. Of these, celeriac, chayote, husk tomato, martynia, peanuts and sunchoke occur naturally in some parts of the Western Hemisphere. The others are from Eurasia or Oceania. Irrespective of their ultimate origins, they all are fascinating to grow and delightful to eat.

Because these vegetables are, for the most part, of no or minor commercial importance in the United States, comparatively little research has been devoted to them. Hence, the cultural practices suggested are often less exact than would be the case for major vegetables. Experience will certainly improve the home gardener's skill in growing them, but lack of experience should not deter him from giving them a whirl.

Since soil conditions vary so widely over the country, no specific recommendations for fertilizer usage are included, other than for fresh manure. In general, most efficient utilization of applied fertilizers will be obtained where 5% to 6% soil organic matter can be maintained. Home gardeners are urged to follow locally applicable fertilizer recommendations based on the results of soil tests that can be obtained through their county Extension office or commercial soil testing laboratories.

Resistant varieties are the first line of defense against diseases, insects, nematodes and other troubles. Where resistant varieties are not available, the home gardener should consult his county Extension office for information about recommended pesticides.

Celeriac

Celeriac (*Apium graveolens* L. var. *rapaceum* DC.) is a botanical variety of celery, differing from that vegetable in producing at the base of the plant a large, turnip-like swelling that may be as much as 4 inches in diameter. This rather bulbous base is the commonly eaten part of the plant.

Celeriac—like celery, carrots, parsley, parsnips and dill—is a member of the Parsley Family (Umbelliferae or Apiaceae). The plant is a biennial, growing to about 36 inches in height, but in gardens it is treated as an annual. It is naturally distributed in marsh areas in the temperate zones of South America, South Africa, New Zealand and Eurasia.

Celeriac "bulbs" may be boiled like potatoes, and put to all the uses made of boiled potatoes. Boiling time for celeriac will be somewhat longer than for potatoes. Celeriac may also be grated raw as a salad topping. An intriguing use is serving sliced, boiled celeriac with French dressing as an hors d'oeuvre. Celeriac has the odor typical of celery due to the presence of a volatile oil. Celeriac is comparatively low in nutrients.

Homer N. Metcalf is Professor of Horticulture, Montana State University, Bozeman. Milo Burnham is Extension Horticulturist, Mississippi State University, Mississippi State.

Although several varieties of celeriac are grown in Europe, Giant Prague has been the only one readily available in the United States. A newer variety is Alabaster.

The culture of celeriac is very similar to that for its close relative, celery. It is a long-season vegetable, requiring nearly 6 months from seeding to harvest, and growing best where the monthly growing season mean temperature is 60° to 65° F (approximately 15.5° to 18.5° C). It is unlikely to be grown successfully in areas or at seasons where monthly mean growing season temperatures exceed 70° to 75° F or are lower than 45°. Since celeriac plants grow rather slowly, a freeze-free growing season of at least 110 days is desirable.

The home gardener has two options for production of celeriac. He may either purchase seedlings from a vegetable plant grower or raise them himself. The use of purchased transplants is suggested for short growing season areas, while gardeners living in milder climates may take either route. Gardeners raising their own transplants should be aware that celeriac must have 60° to 65° F night temperatures throughout the seedling and early transplant stages if premature seedstalk development is to be avoided.

When home-grown transplants are to be used, the gardener will soon discover that celeriac seeds are quite small (approximately 70,000 per ounce). They will remain viable for 4 to 5 years under cool, dry storage conditions. Thus, one need purchase only a small quantity of seed at a time.

Seeds may be sown in flats or individual peat pots. If sown in flats, a seeding rate of 8 seeds per inch of row is suggested. If seeded in individual peat pots, about 10 seeds per pot should be sown, with the extra seedlings later thinned to one per pot. Celeriac seeds characteristically have germination percentages of 50% to 70%, and germinate best at a 70° F minimum temperature. If maintained at this temperature, germination should be complete in about 10 days. Following germination, night temperatures of 60° to 65° are desirable, with daytime temperatures 10° warmer. If seeded directly in the garden, germination and emergence will be materially slower.

Flat-grown seedlings should be transplanted to individual pots as soon as the first true leaves develop. Weekly feedings with a complete liquid fertilizer will promote sturdy growth.

Plants will be ready to transplant to the garden 8 to 12 weeks after seed-sowing, or when they are 3 to 4 inches tall.

Celeriac needs a deep, fertile, well-manured soil which retains moisture well throughout the growing season. Fresh manure, if available, should be applied at the rate of 55 to 92 pounds per hundred square feet. The quantities of lime and commercial fertilizers used should be governed by the results of soil tests. Since celeriac is a gross feeder, two or more side-dressings with readily soluble nitrogenous fertilizers during the growing season are suggested. Regular irrigation will prove beneficial.

As a rule, wait until 5 days after the average last spring freeze-date to set out the transplants, at which time maples may be coming into leaf in many areas. In the garden, the transplants may be spaced, or seedlings thinned, to stand 6 to 7 inches apart in rows 24 to 36 inches apart. An alternative is to check-row the plants at 15 to 18 inches apart in rows 18 inches apart, a system more likely to produce exhibition-quality plants.

Higher quality celeriac will be produced if sideshoots and withered leaves are removed as they appear. Mulching the rows with lawngrass clippings is sometimes practiced, especially where no manure has been used and irrigation is not available. A further quality-producing practice is to draw soil up around the plants to the level of the leaves about two weeks before harvest. This will yield a whiter product, and is called "blanching".

Celeriac will generally be ready to harvest in October, or when the "bulbs" have attained diameters of 2 to 2.5 inches. The plants may be pulled or dug, depending on soil conditions. Tops are trimmed off and any branching basal roots removed. Yields of 200 bulbs per 100 feet of row are a reasonable expectation. The trimmed "bulbs" can be stored in moist sand in a cool root cellar, in a vegetable pit, or in colored plastic sacks in a refrigerator. Properly stored, they will keep about 6 months.

Celeriac troubles will be similar to those afflicting celery.

Chayote

Chayote (*Sechium edule* Swartz) is popularly known as mirliton and vegetable pear in South Louisiana where it is grown and relished. In the mild-winter regions along the Gulf Coast and in parts of California, chayote is grown for its light-green, pear-shaped fruits that are served with salad dressing or stuffed after boiling with ground meat or seafood.

The fruit is considered an excellent substitute for summer squash, but is of little nutritional value. Historically chayote dates back to the Aztecs before the Spanish conquest.

Chayote is related to all the cucumbers, pumpkins, squash, melons and gourds so popular with gardeners, but differs from them in having only one large seed. In Central America the plant is grown as a perennial and the large tuberous roots that develop over a 2- to 3-year period are eaten after roasting, boiling or frying and they are sometimes candied in sugar. The roots contain about 70 percent water and 20 percent starch. The young shoots are also sometimes cooked as a green vegetable.

In the southern United States, chayote is mostly grown as an annual since freezing kills the entire plant. Heavy mulching may protect the roots from freezing, and the plant will resprout in spring. In Northern States the short growing season will more than likely prevent fruiting since the plant is day-length sensitive and flowers only in late summer and early fall.

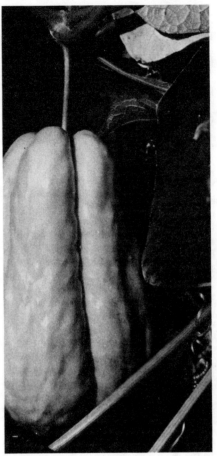

Milo Burnham

Chayote hanging on vine has pear shape and deeply furrowed surface.

Seed of chayote is difficult to locate since the whole fruit is used as a seed. It is not marketed through retail seed catalogs, and is often available only in localities where the plant is grown.

When all danger of frost and cold weather has passed in spring and the soil is warm, plant the whole fruit on a slant with the broad end down and the stem end slightly exposed. If the seed has sprouted before planting, which is often the case, cut the sprout back to a length of about 2 inches.

The vine grows rapidly when planted in a rich, well-drained soil with plenty of organic matter. Since the vine is large and vigorous, plant the seeds no closer than 10 feet apart and provide a trellis or some means of support. One plant may be sufficient for any garden since each vine produces 30 to 35 fruits. The plant should be supplied with generous amounts of water, and fertilizer rich in potash but low in nitrogen. An oversupply of nitrogen will result in excessive growth at the expense of fruit production.

Chayote is monoecious (male and female flower parts in separate flowers but on the same plant) and is dependent on flower-visiting insects for pollination. Bees swarm to the flowers for nectar. The fruits require about 30 days from pollination to mature sufficiently to harvest, and may weigh up to 2 pounds.

The type most commonly grown in the southern United States produces light-green, pear-shaped fruit.

Insect and disease problems of chayote are the same as for pumpkins, squash and other relatives. Among the insect pests are striped and spotted cucumber beetles, squash bug, squash vine borer and pickle worm. Disease problems include powdery and downy mildews.

To save seed, allow the fruits to reach full maturity on the vine but harvest them before they sprout. Wrap each fruit separately and store it in a cool ventilated place.

Dasheen

Dasheen (*Colocasia esculenta* Schott), also known as oriental taro, is a large perennial plant cultivated for its underground corms and tubers. In the United States its cultivation is limited to warm coastal regions. Closely related to ornamental elephant's ear, caladium, calla and the native jack-in-the-pulpit, dasheen differs from them in producing edible corms and tubers that contain practically no calcium oxalate, a harmful chemical. Dasheen varieties also differ from most taros in this respect.

Taro was first introduced into the southern United States with shipments of African slaves who used it for food. The origin of taro has been traced to India and following its dispersal it has served as a staple food crop of Pacific Island dwellers for thousands of years. In the early 1900's a superior type of oriental taro traced to China and known as dasheen was introduced into the United States. This type largely replaced the earlier introduced, acrid, coarse African types.

At one time dasheen was considered as a possible substitute crop on lands too wet to grow Irish potatoes. Nutritional properties of the tubers are similar to potatoes. The tubers can be prepared in any way that potatoes can and the flavor is described as delicate and nutty.

Dasheen requires a frost-free growing period of about seven months. It is therefore limited to the lowland Coastal Plains from South Carolina to Texas. In Hawaii, dasheen is a very common garden plant, used to make the popular poi.

Production of tubers is greatest in rich, loamy, well-drained soils with an abundance of moisture. Clay soils produce low quality dasheen, as do long droughts followed by regrowth and prolonged wet periods.

Plant whole tubers weighing 2 to 5 ounces, 2 to 3 inches deep, at 2-foot intervals in rows about 4 feet apart.

Begin planting about April 1 or earlier, up to 2 weeks before the average date of the last killing frost in spring. It is also possible to start plants indoors and set them in the garden when frost danger is past.

With adequate moisture and fertilizer the plants will reach 4 to 5 feet in height. Apply a preplant application of fertilizer and an equal amount before the plants reach 2 feet in height.

Dasheen is shallow rooted so a heavy mulch will help prevent loss of soil moisture in dry periods.

The corms and cormels (tubers) are mature enough to harvest when the tops have completely died down in fall (October-November). Dig the plants in dry weather if at all possible to avoid injury to the corms. The crop may be stored in the ground and dug as needed where the soil is well drained.

Each plant when dug should have at least one large central corm surrounded by smaller tubers with a combined weight of 2½ to 8 pounds.

The tubers are reported to be of better eating quality and will store longer than the large corm. Tubers will keep for several months at 50° F when they are provided with good air circulation.

Young unrolling leaves can be eaten as a table green. They are a rich source of vitamins A and C and when properly prepared are free of the harmful calcium oxylate. Leaves should be boiled with a large pinch of baking soda for 15 minutes and then boiled in fresh water till tender. The stored corms and tubers can be forced to sprout in the dark and the blanched shoots prepared and eaten.

A major disease of dasheen is root-knot nematodes. Plant only tubers free of evidence of nematodes. Storage rots occur if the tubers are dug before they are fully mature or if proper temperature and adequate ventilation are not provided in storage.

Magnifico globe artichoke with large terminal buds.

Globe Artichoke

Globe artichoke (Cynara scolymus L.) plants look like large thistles and may reach 3 to 4 feet in height and cover several square feet with their large, prickly deep-cut leaves. Artichokes are not for the gardener with limited space. Climate requirements restrict their culture even more.

Globe artichokes grow best in frost-free areas with cool, foggy summers. No wonder their commercial production in the United States is limited to one small area in coastal California! However, home gardeners with less than ideal growing conditions can successfully produce artichokes. A variety known as Creole grows in southern Louisiana, and a few artichokes have been grown in Michigan. The most familiar artichoke—and

Vincent E. Rubatzky

available in the seed trade—is Green Globe.

Historically, artichokes originated in southern Europe where they were cultivated since Roman times. They were brought to California by Spanish explorers. Artichokes are grown for the soft fleshy receptacle and thickened bases of the bracts of the flower heads. Each plant produces several stalks and each stalk bears several flower heads.

The plant is an herbaceous perennial that grows best in a rich, well-drained soil supplied with plenty of organic matter and having a pH of about 6.0. It lives for several years and increases in production provided it doesn't freeze

Seed of the Green Globe variety is advertised in several home garden seed and plant catalogs, but germination is apt to be low and the plants produced quite variable to type. Seed germination is improved by storing it for 2 weeks in the refrigerator in moist peat moss. Plant the seeds in individual cups or pots 4 to 6 weeks before you want to set the plants out. This will give you an early start and avoid the shock of transplanting.

Globe artichokes are best propagated by crown divisions or rooted suckers or sprouts from the base of the plant. Space the plants 4 to 6 feet apart and supply them with adequate fertilizer and water during the growing season. The plants grow best at temperatures from the mid-60's to the mid-70's (degrees Fahrenheit). At higher temperatures the buds open rapidly and the bracts become fibrous and tough.

Stalks and buds appear in late summer or early fall. In frost-free areas, flower bud production continues through winter into early spring. Cut artichokes while the buds are still tight. In overmature artichokes the green bracts loosen and point out and purple flowers show. The Creole variety is an exception since its bud bracts naturally point out. As each stalk is finished, remove it completely from the plant.

In northern gardens, most attempts to produce globe artichokes are unsuccessful. Frost and freezing temperatures kill the plants, and heavy mulches used for protection often result in the crowns rotting. In many instances the plants freeze before becoming large enough to flower.

Horseradish

Horseradish (*Armoracia rusticana* Gaertn., B. Mey. & Scherb.) is a member of the Mustard Family (Cruciferae or Brassicaceae). It is believed to be native in southeastern Europe, but is grown in cool temperate climates over much of the world and has frequently escaped from cultivation. The "radish" part of the common name derives from the latin, *radix*, for root, while the "horse" part may allude to the strong flavor of the root, or to the plant's coarse texture.

Although condiments derived from roots of the horseradish are now quite familiar, primary use of the plant before the 16th Century was for its alleged medicinal properties. However, both leaves and roots were eaten in Germany during medieval times. Today, peeled roots are either grated and prepared with diluted vinegar, or boiled, pureed, and used in preparing various sauces. The pungent flavor is due to the presence of the glucoside sinigrin.

Horseradish is a hardy perennial that produces a whorl of large, coarse-textured leaves. The seeds mature but rarely, and are not used in propagating the crop, which is raised from root cuttings.

A deep, rich, moist loamy soil is best for horseradish. It has also been grown successfully on organic soils. On hard, shallow, stony soils the roots tend to be malformed and yields are reduced. Unless the soil is already fertile and in good tilth, it should be manured the autumn prior to planting at the rate of 55 to 92 pounds of fresh manure per hundred square feet, and plowed or spaded to a depth of at least 10 inches. Where no manure is available, grow soil-improving crops for plowing down in the the autumn of the year preceding that in which the crop will be planted. In mild climates, a winter-grown soil-improving crop may be spring-plowed.

Phosphate and potash mineral fertilizers should be roto-tilled or spaded into the soil before planting. Manure should not be spring-applied in the year of planting, but nitrogenous fertilizers may be applied broadcast and the ground reworked at that season. Amounts of commercial fertilizers to be used should be guided by the results of soil tests.

Horseradish is best grown from root cuttings, sometimes called "sets". Sets are small or slender roots, 8 to 14 inches long, that are trimmed from the main roots at autumn harvest. As these cuttings are removed from the main root, it is wise to make a square cut at the top and a slanting cut at the bottom as an aid to subsequent proper planting procedure. The sets are cleaned, bundled, packaged and held under refrigeration or in a vegetable pit or root cellar until planting time the following spring.

An alternative procedure is to leave a few plants in the garden over winter for spring digging and taking of cuttings at or near planting time.

In spring, the fall-plowed soil should be well worked, including incorporation of any spring-applied mineral fertilizers, especially nitrogenous ones. It is a good idea to let the worked-up soil settle a few days before planting.

Horseradish is commonly grown in rows spaced 30 inches apart, with the plants spaced 24 inches apart in the rows. Yield estimates vary from 15 to 35 pounds of roots per 50 feet of row. Size of the horseradish plot will depend on the popularity of horseradish preparations with the family, but one or two dozen plants should be enough for the average family.

In planting, make furrows 3 to 5 inches deep. Plant the cuttings with the tops all in one direction in the row, dropping a cutting every 24 inches. As the cutting is dropped, draw a little soil over the lower end with your foot and tamp firmly. After all cuttings are dropped, they are covered with soil to slightly above ground level (to allow for soil settling), being sure that the soil is firmly in contact with the cutting.

Cultivation for weed control in horseradish (and other garden crops) is especially important early in the season when the plants are relatively small. If mechanical cultivation is practiced, it is best to cultivate in the same direction that the cuttings were dropped—toward the top end.

To grow high quality horseradish, remove all top and side roots, leaving only those at the bottom of the set. This is done twice during the growing season, first when the largest leaves are 8 to 10 inches long, and again about 6 weeks later.

To remove top and side roots from the sets, first carefully remove the soil around the top end of the main root, leaving roots at the lower end of the set undisturbed. Raise the crown and remove all but the best sprout or crown of leaves. Rub off any small roots that have started from the top or sides of the set, leaving only those at the bottom. Return the set to its original position and replace the soil. This procedure is called lifting and produces a relatively smooth root, free from side roots.

Horseradish makes its greatest growth during late summer and early autumn. For this reason, harvest usually is delayed until October or early November, or just before the ground freezes. In digging, it may prove wise to dig a trench 12 to 24 inches deep along one side of the row. Then, working from the opposite side of the row with a shovel or spading fork, dig the roots, using the tops as a handle for pulling laterally from the loosened soil. The tops should be trimmed from the roots to within one inch of the crown. Side and bottom roots are trimmed off, reserving the laterals for the succeeding season's crop.

If you wish to store horseradish roots for frequent fresh grinding, they may be cleaned, washed, and stored in plastic wrapping in a refrigerator, vegetable pit or root cellar. When stored in the refrigerator, protect horseradish roots from light to prevent their turning green. For this purpose, recycle the colored plastic bags in which potatoes often are marketed.

In relatively mild climates where frost penetration of the soil is not extensive, the roots may be stored in an 8- to 10-inch deep trench lined with clean straw. Place roots on the straw and cover with a 6-inch layer of clean straw. As the weather becomes colder, cover the straw with 6 inches or more of soil before the ground freezes. This will protect the roots from freezing injury.

Occasionally, horseradish may suffer from attacks of root rot. To avoid this, select only disease-free root cuttings for planting stock, and rotate the planting site so that horseradish is not grown on the same piece of ground more often than every 3 to 4 years.

Leafhoppers, flea beetles and grasshoppers may attack horseradish foliage. Leafhoppers spread the virus disease known in the inland Northwest as "curly top", which can have devastating effects on this and many other vegetables. There is no cure. Apply approved insecticides as soon as the insects appear. Consult your county Extension agent for current information on pesticides.

The most common way of preparing horseradish for table use is by peeling or scraping the roots and removing all defects. Then, grate the root directly into white wine vinegar or distilled vinegar. Avoid using cider vinegar, as it causes discoloration in the grated horseradish within a rather short time.

Depending on your preference, the vinegar may be slightly diluted before use. Bottle the horseradish and cap the containers as soon as possible after grating. Refrigerate the prepared product at all times to preserve the pungent flavor. It will keep for a few weeks. Then prepare a fresh supply.

Horseradish may also be dried, ground to a powder and put up in bottles in a dry form. So prepared, horseradish will keep much longer than the freshly grated product, but is not generally as high quality.

Husk Tomato

Husk tomato or ground cherry (*Physalis* spp.) is a member of the Nightshade Family (Solanaceae). The generic name is from the Greek for a bladder, in allusion to the charactistically inflated calyx ("husk"). Most *Physalis* species occur naturally in the Western Hemisphere. The forms of husk tomatoes in cultivation are usually ascribed to *Physalis pruinosa* L or *P. pubescens* L., but these species may be confused in gardens, and other species also may be involved.

Husk tomato plants are annuals growing 18 to 40 inches in height. They often inhabit sandy soils in nature. The fruit, which is the edible portion of the plant, is a berry completely enclosed in the thin, inflated calyx or husk.

The fruit may be eaten fresh-ripe, or prepared in a number of ways, including fried, baked, stewed, in meat dishes, soups or salads, or as dessert sauces and preserves. It is a common ingredient in Latin American cuisine.

Seeds are infrequently listed in

catalogs. An improved form, developed from Guatemalan material by the Iowa Agricultural Experiment Station, was introduced some years ago.

Plant growing and general culture are much the same as for the tomato. In cool climates, starting the plants indoors or in a greenhouse and transplanting 6-week-old seedlings to the garden about 5 days after the average date of the last spring freeze should help in attaining a good crop. Like tomatoes, husk tomatoes will respond to favorable levels of soil fertility and ample irrigation.

The fruits begin to mature from mid-summer to late summer, turning from green to yellow and becoming somewhat soft during ripening. They are not adapted for long-term storage and should be used or processed shortly after harvest. Yields as much as 2.5 pounds per plant have been achieved. Ten plants should produce enough husk tomatoes to supply the average family.

Troubles will be similar to those afflicting tomatoes.

Martynia

Martynia (*Proboscidea louisianica* [Miller] Thellung) is native to the Southwest but gardeners throughout the Nation who are interested in unusual plants grow it. The dried seed pod has an unusual appearance which accounts for the popular name "Unicorn Plant" and for the fact that several retail seed and plant catalogs offer seed. Dried pods are used in floral arrangements and as novelty items. Young immature seed pods can be pickled sweet like cucumbers.

Plant the seed ½ inch deep at 18- to 24-inch intervals in rows 3 feet apart. In Northern States start the plants indoors and set them out in the garden after frost danger is past. The plants grow about 18 inches tall and have a spread of some 30 inches. General cultural requirements are about the same as for okra.

Mushrooms

Edible mushrooms (*Agaricus bisporus*) are not easily produced in the home because of the exacting conditions required. Even so, mushroom spawn and culture kits are offered for sale by several retail plant and seed suppliers. However, a recently described method for small scale cultivation of mushrooms used for demonstration and class study provides a more certain way for the serious home gardener to grow mushrooms.

The reference to the article describing the method is: San Antonio, James P. 1975. "Commercial and small scale cultivation of the mushroom, *Agaricus bisporus* (Lange) Sing." HortScience. Vol. 10(5):451—458. Your library may have the article, or you can obtain a copy from the Vegetable Laboratory, Agricultural Research Center, Beltsville, Md. 20705.

Peanuts

Peanuts (*Arachis hypogaea* L.), a popular home garden crop in the Southeast and Southwest, are unique among garden plants. Showy yellow flowers are borne above ground but the ripened ovary and seeds (peanuts) develop below the ground.

The peanut originated in South America, was carried to Africa and Europe by Old World navigators and explorers, and was shipped to America as on-board food for slaves. Peanuts are now grown along the East Coast from Virginia to Florida, and along the Gulf Coast to Texas and in all adjoining inland States. Gardeners hold the peanut with the same high regard as Southern peas, okra and butter beans.

"Chock full" describes the nutritional and energy value of peanuts. They can be eaten raw, boiled, steamed or roasted. Raw, cured peanuts are rich in vegetable protein and oil and contain 564 calories per 100 grams.

Peanuts are divided into four general categories: Virginia, Runner, Spanish and Valencia. Virginia and Runner types are large-seeded and contain 2 seeds per pod. Spanish and Valencia are small-seeded with the Spanish having 2 to 3 seeds and the Valencia 3 to 6 seeds per pod.

Peanuts require a long, warm growing season (110 to 120 days). They flower 6 to 8 weeks after planting. Following self-pollination and wilting of the flower, the ovary (peg) emerges and grows downward until it enters the soil and the nut begins to form.

Best soils for peanuts are coarse textured (sandy loams) adequately supplied with calcium and with a pH of 5.8 to 6.2. Add lime to soils with a pH below 5.8. Spanish types grow in both fine and coarse textured soils, but the Virginia types should be limited to coarse soils.

Plant Spanish types with a spacing of 4 to 6 inches in rows 24 inches apart. Virginia types need more room so plant them 6 to 8 inches apart in rows 36 inches apart. Plant only shelled seed. One-half pound of seed will plant 100 feet of row. Plant the seed 1½ to 2 inches deep in coarse soils but only 1 inch deep in fine soils. Planting can begin about 2 weeks after the average date of the last killing frost in spring.

Prepare the garden soil completely before planting. All crop residues should be turned under in fall to permit decomposition. Do not plant peanuts in the same location 2 years in a row, to prevent build-up of diseases.

A soil test is the best means of determining fertilizer needs. Where the garden has been heavily fertilized for previous crops, you may not need to apply additional fertilizer since peanuts are good foragers. The young peanut plant is sensitive to fertilizer burn, so spread any fertilizer applied over the entire planting area rather than putting it in the row.

A shortage of water when the plants are flowering vigorously and when the pegs are entering the soil will reduce the yield of peanuts. As harvest draws near, do not water peanuts. Any excess water at this time may break dormancy and cause the mature peanuts to sprout.

To prevent development of "pops" (empty pods) the soil must have a good supply of available calcium. On soils known to be low in calcium, sprinkle about 2½ pounds of gypsum per 100 feet of row over the plants when they begin to bloom.

Cultivate the soil to control weeds and to keep the soil loose so the pegs can penetrate the surface. Once the pods are developing in the soil, cultivation without damaging the plants is almost impossible. Do not throw or pull soil to the plants while cultivating. Peanut plants are low growing; covering branches and leaves with soil kills leaves and interferes with flowering.

As peanuts mature the leaves will begin to turn yellow. Since flowers appear for several weeks, all the peanuts do not mature at the same time. If you delay harvest until the last formed pods are mature, the first-formed pods may rot or sprout or be left in the ground when the plants are dug.

Dig the entire plant and turn it over in the row with the peanuts facing up. Pull peanuts for boiling at digging time when they contain 40 to 50 percent water (the peanut inside of the shell will rattle). After several days under good drying conditions, moisture content of the exposed pods drops to about 15 percent and the plants can be moved to a warm, airy place and stacked for 2 to 3 weeks to complete curing before the peanuts are stripped from the plants. Some gardeners stack the plants around poles out in the open until the peanuts are cured.

Several insects and diseases attack

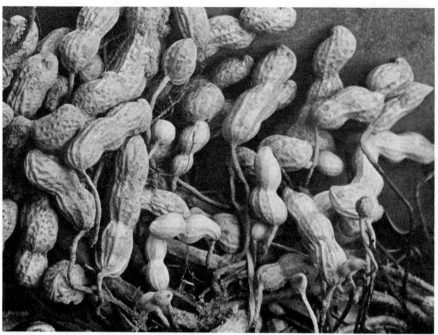

Milo Burnham

peanuts and reduce yields or kill the plants. Corn earworm, cutworms, fall armyworms and velvetbean caterpillars feed on the foliage while the whitefringed beetle feeds on underground plant parts. Leafspots and southern stem blight are among the most serious diseases.

Leaves infected with leafspots drop from the plants and result in "false" maturity and low yields and poor quality. Leafspot diseases can be controlled by spraying with recommended fungicides and changing the location of peanuts in the garden every year.

Southern stem blight (stem rot) attacks stems, roots, pods and pod stems. This disease is best controlled by turning under plant residues in fall so they have time to decompose, and by moving the location of peanuts in the garden every year.

Sunchoke (Jerusalem Artichoke)

The frequently used common name, Jerusalem artichoke, for *Helianthus tuberosus* L. is really a misnomer, and the plant might better be known as sunchoke. The plant has no biological association with Biblical lands, being native in North America east of the 20-inch precipitation line, ranging from Kansas and Minnesota east and north to Nova Scotia.

The "Jerusalem" part of the common name is thought to be a corruption of "girasole", the Italian name for the sunflower (*Helianthus annuus* L.), to which the sunchoke is closely related. The literal meaning of "girasole" is "turning to the sun". Further, the sunchoke is not really an artichoke, since that common name should be reserved for *Cynara scolymus* L., the globe artichoke. All these plants are members of the Composite Family (Compositae or Asteraceae).

Mature and immature peanut pods, showing fruiting habit of the plant.

238

Early adventurers and colonists found the sunchoke being used as a food crop by Amerinds along the Atlantic coast. It was taken to Europe early in the 17th Century. The sunchoke is now cultivated and naturalized extensively on well-drained soils throughout the cool-temperature climatic regions of the world.

Sunchokes are grown for the edible tubers produced on the ends and branches of underground stems. The tubers may reach 4 inches in length and 2 to 2.75 inches in diameter. They are of special interest because the principal storage carbohydrate in them is inulin, a substance reputed to be of value in the diet of diabetics as a substitute for ordinary starch. Sunchoke tubers may be prepared for the table in the same ways that potatoes are used.

The sunchoke plant is a rather coarse, rough-surfaced perennial that grows 6 to 9 feet tall. The leaves are large, 4 to 8 inches long, oblong and toothed. At the base of the plant they may appear to be opposite each other, but in the upper part of the plant they

Thin-skinned, often knobby subterranean stem tubers are edible portions of sunchoke plant (Jerusalem artichoke).

may be alternate. The flowering heads terminate the branched stems, looking much like small sunflowers, with yellow ray and disk florets.

In choosing a site for sunchokes in the garden, keep in mind their potential height and vigor. Although sunchokes will grow on soils too dry or infertile for potatoes or beets, they will respond readily to better growing conditions. If they can be planted on sandy or loamy soils, the task of digging the tubers will be much easier. It is wise to plant them where they will be more or less out of the way and not shade other sun-loving plants.

Sunchokes are sensitive to borate herbicides, so areas that have been treated with them should be avoided.

Cultural practices for sunchoke are generally similar to those for the potato. Although sunchoke is a perennial plant, it is usually treated as an

D. H. Fritts

annual in the garden. Planting may be done either in autumn at harvest time, or in spring as soon as the soil can be worked readily. Both white- and red-skinned forms are known, but the red is quite rare.

For planting stock, either whole tubers or tuber-pieces of about 2-ounce weight are suggested. These should be planted about 4 inches deep and spaced about 24 inches apart in the row, the rows being spaced 36 to 40 inches apart. If the soil is very fertile, spacing may be increased.

Very little information about fertilizer requirements for sunchokes is available, but European experiments suggest the potash requirement may be high. Hence it is probable that fertilizer and irrigation regimes producing good yields of potatoes will give favorable results with sunchokes. Established plants require relatively little care beyond weed control. The average gardeners may expect to harvest about 3 bushels of tubers per 100 feet of row.

Since sunchoke tubers are hardy, harvest may be delayed until the tops have frozen down in autumn or even be deferred until early spring. Spring-dug tubers will taste somewhat sweeter than autumn-dug ones.

Digging will be easier if the tops are first removed. Using a spading fork or potato hook, a thorough search for tubers should be undertaken, extending some distance from the plant. Missed tubers can lead to a weedy growth of sunchokes the succeeding year.

Because sunchoke tubers are thin-skinned and do not store nearly as well as potato tubers, dig them only as needed so long as the soil remains workable in autumn. When freeze-up is at hand, enough tubers to supply winter needs should be dug, cleaned, washed, and prepared for storage.

Sunchoke tubers store best at 32° F and 90% to 95% relative humidity. If these conditions can be provided, tubers may be stored successfully up to 5 months. It should be possible to store an adequate supply in damp sand in a root cellar or vegetable pit, or in colored plastic bags in a refrigerator.

Sunchokes are seldom bothered by insects or diseases.

Vegetable Soybeans

Soybeans are members of the genus *Glycine* L., which consists of 10 species of mostly viny perennial legumes native primarily in tropical and warm temperate parts of Africa and Asia. The cultivated soybean, *Glyine max* (L.) Merrill, is the only member of the genus having an erect bushy plant with an annual growth habit. Not known in the wild state, it is thought to be derived, at least in part, from the viny North Asiatic species, *Glycine ussuriensis* Regel & Maack.

Soybeans first appeared as a cultivated crop in northern China about 3,000 years ago. Although apparently imported to North America at various times during the Colonial era and the early days of the Republic, they did not become a major crop in the U.S. much before World War II.

For human consumption, soybeans may be eaten either in the immature or mature stages of growth, or in various processed forms.

Adding a dry soybean product to small grain cereals substantially improves protein utilization over the components consumed individually.

Since fresh, immature soybeans are seldom found in either canned or frozen forms on supermarket shelves, they are an excellent vegetable for home gardeners, who may expect yields of 2 bushels of green pods per 100 feet or row. Soybeans have a relatively high protein content for a vegetable and are a good source of vitamin A.

Vegetable soybeans grow best where nights are warm and days not too long. Only very early varieties

Toasted soybeans make a tasty snack.

should be attempted at higher latitudes. They are unlikely to succeed in areas having frost-free growing seasons of less than 130 to 135 days.

Seeds of vegetable soybeans are usually larger than those grown as a field crop, and only a limited number of varieties, such as Fiskeby V and Kanrich, are offered currently by seedsmen. Most vegetable soybeans have yellow seeds, but other colors are known, such as green, black and green, and black and yellow.

Land on which edible soybeans are to be grown should be well prepared before planting. Soybeans do not thrive on strongly acid soils, and liming may be desirable if indicated by soil tests. Because they are legumes, nitrogenous fertilizers are seldom used, but on many soils they will benefit from application of phosphorus and potassium fertilizers (again the gardener should be guided by soil tests).

If soybeans have never been grown on the soil, it may prove wise to inoculate the seeds with nitrogen-fixing bacteria. *Rhizobium japonicum* is said to be specific for soybeans, and should be available in commercial preparations.

Weed control will be more convenient if the plants are spaced 4 to 6 inches apart, or in hills spaced about 8 inches apart with the rows 30 inches apart. When seeds of varieties differing in maturity are available, better results will ensue if a single planting of these is made—rather than successive plantings of a single variety.

Under favorable conditions, edible soybeans will be ready for harvest as immature beans from early varieties about 2 months after planting, while 100 or more days of favorable weather will be needed to mature dry beans.

Soybeans are self-fertile and have

mostly self-pollinated flowers. The beans are borne in pods that are produced in clusters of 3 to 15. The pods are slightly curved and hairy, and will average 2 to 3 seeds per pod. In the Orient, the immature pods and seed are eaten together, but this has seldom been done in the United States. When eaten in the immature stage, vegetable soybeans are harvested at about the same maturity as immature lima beans.

If vegetable soybeans are to be eaten as green beans, the pods will shell much easier if they are plunged into boiling water for about 2 minutes, after which the beans can be squeezed from the pods without any difficulty.

If dry mature soybeans are desired, the plants should be cut when the pods are turning brown and windrowed or placed on a rack under shelter until fully matured, when the seeds may be beaten out. This prevents loss through shattering in the garden. Following shelling, dry the seeds thoroughly before storage.

Another way of using edible soybeans is as sprouts (in the same manner as the sprouts of mung beans). Soybeans can be sprouted in any container that has holes in the bottom for drainage and can be covered.

In preparing the sprouts, soak the soybeans overnight and then place them in a container large enough for the beans to swell at least six times

Cabbage looper, a major vegetable pest, feeds on soybean leaf.

Fred Farout

their original bulk as they sprout. Cover container to keep out light.

Moisten the beans at least 3 times a day in summer and twice in winter. In winter add warm water and keep the beans in a warm place.

Time to maturity for soybean sprouts is 3 to 5 days in summer and 10 to 15 days in winter. The sprouts are fully grown and ready to be used when 2 to 3 inches long. Once harvested, sprouts should be kept in a cool, humid place.

At least 25 parasitic diseases are common on soybeans in various parts of the United States, variously caused by bacteria, fungi and viruses. In most cases, the best defense is the planting of resistant varieties, or in the case of seed-borne viruses, of virus-free seeds.

In various parts of the United States, soybeans may be attacked by the green clover worm, the thistle caterpillar, the army worm, by leafhoppers, mites, grasshoppers and blister beetles.

Watercress

Watercress (*Rorippa nasturtium-aquaticm* [L.] Schinz & Thell.) is a popular cool-season salad vegetable. Like collards, broccoli, turnips and their relatives, watercress is a member of the Mustard Family (Cruciferae or Brassicaceae). The generic name is thought to be derived from *rorippen*, the ancient Saxon common name for watercress. This perennial plant has its natural distribution in Europe and western Asia, but has become naturalized extensively throughout the cool-temperate climatic areas of the world.

Portions of the plant commonly eaten are the upper 4 to 6 inches of the vegetative stems and associated leaves. Watercress gives a pungent flavor to salads, makes a novelty sandwich filling, is sometimes used as a flavoring in soups, and serves as an attractive garnish.

Watercress grows naturally in clear, cold, shallow, slow-moving creeks. It may grow either as a floating plant, become rooted in the bottom, or creep along wet stream margins.

Watercress grows well on rich, slightly acid to slightly alkaline garden soils (pH not lower than 6.0), for which ample irrigation is available. Plantings can be established readily by means of stem cuttings, or by raising plants from seeds.

Home gardeners may adopt either the trench or surface culture systems for watercress.

For those particularly fond of high quality watercress, the trench system of culture may be well worth the effort involved. In this system a trench is dug 2 feet deep and 2 to 3 feet wide. A 9-inch layer of well-rotted compost or manure is placed in the bottom of the trench and allowed to settle for about 2 weeks. Irrigate daily at the rate of 3 to 4 gallons of water per yard of trench length. At the end of the period, put 4 inches of good topsoil over the organic mass and press it down firmly.

The prepared trench may be planted either with seeds or stem cuttings. If seeds are used, the gardener will find them to be quite small—about 150,000 per ounce—and to have a rather low germination standard, about 40% to 50%. The trench should be marked off in a grid of 8-inch squares, and several seeds planted at each intersection in the grid.

Cover the seeds with about 1/16th inch of fine sand. Mist the planting frequently, so that it never dries out, and keep the trench dark by covering with some material that will exclude light, laid over a supporting framework.

After germination is complete, remove the opaque covering and thin the seedlings to 1 per intersection. Give the trench a good flooding after thinning.

If stem cuttings are to be used as

planting stock, they may be collected from the wild. Or bunched watercress from the supermarket may be used. Plant one cutting at each intersection on the 8 x 8 inch grid, and water thoroughly.

With either starting method, keep the soil free from weeds by hand weeding until the watercress plants grow large enough to provide strong competition for them. Never allow the soil to dry out. The plants will benefit from being irrigated daily with a fine mist nozzle, except in rainy weather.

As soon as the plants have reached about 6 inches in height, pinch the leading shoot to encourage branching. The plants should not be permitted to flower. As soon as signs of flower buds are observed, cut the plants back.

It takes 60 to 70 days for watercress to reach harvest maturity from seeds, somewhat less from cuttings. Hence, in mild climatic areas it may be a good plan to start a succession of trench cultures about a month apart to assure continuity of quality harvests.

In the surface culture system, prepare the soil with a high level of organic matter from manure or compost. A bed of several short rows will be easier to handle than a single long row. Rows in the bed may be placed 12 to 18 inches apart. Sow the seeds very shallowly at the rate of 0.5 ounce per 100 feet of row. Even distribution of the seeds will be readily achieved if they are first mixed with a very fine-textured dry sand.

The planting should be kept wet throughout germination, emergence, and seedling establishment. As soon as the first true leaves appear, the seedlings may be thinned to stand 8 to 10 inches apart in the row. Maintain constantly high soil moisture levels for best results.

An alternative to direct seeding in the garden is starting the seed in small peat pots under a mist propagation system, provided you have a greenhouse equipped for this. Sow several seeds in each pot, thinning to one seedling after the first true leaves have appeared.

Well-started seedlings may be transplanted to the prepared bed in the garden, pot and all, or possibly maintained under mist in the greenhouse for nearly year-around harvest. Due to leaching effects of the mist, occasional application of liquid fertilizer is advised to maintain vigorous growth.

Apart from the above, cultural practices for surface culture are similar to those for the trench system.

In harvesting watercress, take a sharp knife and cut about 6 inches of the leading shoots or side shoots. Tie the cut pieces into bunches and trim the butt ends so the finished bunch is about 4 inches long. The harvested bunches may be kept in water, or possibly in plastic wrapping in the refrigerator for limited periods. The home gardener can expect to harvest one bunch of watercress per foot of row.

Watercress in the United States has not been found to be damaged seriously by diseases. Aphids, leaf beetles, leafhoppers and sowbugs sometimes attack watercress. For currently recommended control measures, consult your county Extension agent.

PART 3

Fruits and Nuts

Growing Apples, Pears, and Quinces; Pest Control, Air Drainage Important

by Roger D. Way

Growing fruit in the home garden is a good, profitable hobby, but it also can be challenging because of pest control problems. This chapter gives instructions on how to select, plant, and maintain an apple, pear or quince orchard and how to harvest, store and use the fruit.

Home-grown apples are excellent for eating fresh or for apple sauce and baking. Pears can be home canned as halves or eaten fresh. Quinces are used for jelly making or preserves. Fruit trees can do double-duty in producing fruit and also in landscaping.

Climate is more important than soil in determining where apples will grow successfully. Apples do not grow well in central and southern Florida or southern California because winters are not sufficiently cold to satisfy the necessary chilling requirements. Some new varieties with low chilling requirements, such as Anna, will grow farther south than most varieties.

Conversely, severe cold (−45° or colder) will kill most apple trees, making it impractical to grow them in northern North Dakota and in other very cold locations. They do well between Georgia and New Mexico and Maine, Wisconsin and the southern parts of Canada.

Before planting, survey your area and determine what varieties grow best.

Air drainage can be very important to the successful growing of fruit trees. Trees planted on sloping land (5 to 10 percent slope) will sometimes escape late spring frosts which could kill blossoms. Windswept hill tops or low valleys where frosts settle should be avoided.

Apples grow on a wide range of soil types. An ideal soil is a well drained, fertile, sandy loam at least four feet deep. Good drainage is more important than good fertility. Soils that remain wet late into the spring are not suitable. Apple and pear trees tolerate a wide range of soil acidity. Liming before plowing may not be necessary unless the pH is below 5.5.

There are at least 6,000 apple varieties. They can be classified according to their time of harvest. In the Northeast, the very earliest summer varieties are harvested in mid-July and the latest in late October at the time winter freezes begin.

In their approximate order of harvest, some of the best mid-July to mid-August varieties are: Vista Bella, Julyred, Jerseymac, Viking, Tydeman Early; mid-August to late September: Gravenstein, Paulared, Prima, McIntosh, Cortland, Macoun, Spartan, Jonathan, Rhode Island Greening, Empire; late September to late October spur type Delicious, Priscilla, Jonagold, Golden Delicious, Spigold, Northern Spy, Stayman Winesap, Idared, Red Rome, Mutsu, and Melrose. There are also many old apple varieties still available from nurseries. Contact your county Extension office for a list of recommended varieties.

Scab-resistant varieties include Prima, Priscilla, Macfree, Nova Easygro, Priam, and Sir Prize. Unfortunately, no insect-resistant varieties are yet available.

Satisfactory pear varieties are Bartlett, Spartlett, Moonglow, Seckel, Clapps Favorite, Aurora, Gorham,

---

Roger D. Way is Professor of Pomology, New York State Agricultural Experiment Station, Cornell University, Geneva.

Magness, Highland and Bosc. Magness and Moonglow have some resistance to fire blight, a severe disease on pears, and are recommended for the South where this disease is a special problem. Orange is the most popular quince variety.

Variety selection can be greatly aided by studying nursery catalogs which give good descriptions of varietal attributes, but often fail to point out their weak features.

Pollination is essential for the setting of flowers to initiate fruit development. Apple varieties cannot be fertilized by their own pollen. However, pollen from almost any other apple variety will cause fruits to set. Therefore, in order to obtain fruit set, you need to provide for cross-pollination by having more than one variety within 100 feet. Bouquets of another variety may be brought in and placed in a pail of water beside the tree.

The pollen source variety has no effect on fruit characteristics of the variety being pollinated.

Some varieties, although they bear heavy crops when pollinated by another pollen-producing variety, do not themselves produce good pollen. Examples are Gravenstein, Rhode Island Greening, Mutsu, and Jonagold. When these are grown, it is necessary to have three varieties in order to provide cross-pollination.

Blooming Times

Varieties differ in their time of blooming. Some early bloomers are Vista Bella, McIntosh, and Idared. Late bloomers follow about a week later. These include Macoun, Rome Beauty, and Golden Delicious. In most years, early and late bloom will overlap and result in good cross-pollination but in some cool springs, the overlap may be insufficient. It may be desirable to plant two early bloomers or two late bloomers.

Pears will not pollinate apples nor vice versa. Pears also need two varieties to cause effective pollination and fruit set. Bartlett and Seckel are cross-incompatible. Most other pear varieties are cross-compatible. Magness and Alexander Lucus do not produce good pollen.

Bees carry pollen from one variety to the other. Bees fly at temperatures above 65° F. In some springs, the temperature during bloom may never rise above 65°. Due to bee inactivity, little cross-pollination will occur. Thus, not much fruit will be set.

Dwarf Trees

Dwarfing rootstocks are a beneficial innovation for the modern home orchardist, as well as for the commercial apple grower. Although dwarf trees are somewhat more expensive to buy, they are easier to prune, spray and harvest. They also begin to bear crops of fruit at a younger age than full-sized trees.

The fruit of Delicious, or any other variety, which is borne on a dwarf tree is just as large and otherwise identical with the fruits of Delicious borne on a full-sized tree. Dwarf rootstocks do not shorten tree life.

Dwarfing in apple and pear trees is caused by specific dwarfing rootstocks onto which common varieties are budded. Tree size at maturity depends on which rootstock is used. True dwarf trees grow to a height of about 10 feet when fully mature at 15 to 20 years of age.

The most dwarfed trees are propagated on Malling 9 or M.27 roots; semi-dwarf on M. 26 (12 feet), M. 7 (15 feet), or M.9/Malling-Merton 106 interstems; semi-vigorous on MM.106 or MM.111; and vigorous (30 feet high) on seedling roots.

An interstem tree, such as M.9/MM.106, is one with strong growing, well-anchored MM.106 roots. By double budding, it has a 6-inch trunk section of the very dwarfing M.9 to produce a semidwarf tree. Finally, the variety is budded on the top.

Pear trees are dwarfed by growing them on quince roots with an interstem of Old Home to overcome the graft incompatibility which exists between Bartlett and quince.

Tree size also is influenced by the inherent varietal vigor, soil fertility, severity of pruning, and several other factors. For example, on a given rootstock, Cortland grows into a larger tree than Golden Delicious or Rome Beauty. Also, nonspur Delicious and McIntosh will grow into bigger trees than spur type Delicious and McIntosh. Spur varieties are mutations which grow into compact trees that are smaller than normal and usually more desirable for the home orchard.

Buying the proper tree from a reliable, local nursery is one of the most critical decisions in the successful growing of fruit in the home orchard. Nursery catalogs contain much information about varieties and planting tips; study them carefully before buying. Medium-sized, 1-year-old trees are preferred to 2- or 3-year-old trees.

Do not attempt to grow fruit trees from seed; they do not come true to variety. Trees from seeds produce very small, poor quality fruits.

Trees with five varieties on one tree can be grown, but they are not recommended because the different varieties will grow unequally, making tree shaping difficult.

Order from the nursery the precise variety/rootstock combination you want. Do not buy a tree which is called "dwarf"; the specific rootstock should be identified.

Spacing between trees in the orchard will depend on such factors as how much land is available, vigor of the variety, rootstock vigor, soil fertility, and drainage.

Photos by Roger Way

Dwarf apple trees provide easy access to the fruit. They also bear fruit much quicker than standard trees.

Apple Tree Spacing in the Home Garden

Varietal vigor	Rootstock	Soil fertility	Tree spacing (feet)	
			Between trees	Between rows
Semi-vigorous (Golden Delicious, Rome Beauty, spur varieties, etc.)	Dwarf (M.9, M.27)	Medium	6	12
		High	8	16
	Semidwarf (M.7, M.26, M.9/MM.106 interstems)	Medium	8	16
		High	10	18
	Semivigorous (MM.106, 111)	Medium	14	22
		High	16	24
Vigorous (Cortland, nonspur Delicious, McIntosh, etc.)	Dwarf	Medium	10	18
		High	12	20
	Semidwarf	Medium	12	20
		High	14	22
	Semivigorous	Medium	18	26
		High	20	28

Planting of fruit trees in the northeastern United States should be done as early as possible—as soon as the land is dry enough to work in the very early spring. Fall planting is risky because it may result in winter injury to the trees, but in warmer regions, fall planting is practical.

As soon as the trees are received, they should be unwrapped. The roots should be kept moist and above freezing. If the planting site is not ready, dig a temporary hole in the garden and heel-in the roots in moist soil in a shady spot. Trees should be planted while still dormant, or at the latest, before much leaf growth occurs.

Fruit trees require full sunlight and should not be planted in the shade of a building or large tree. The orchard area should be plowed and the soil disked before planting. Prune off damaged, broken, diseased or dead roots. Cut off the tips of excessively long roots so they are no more than 15 inches long. Usually very little root pruning is needed. Dig a large hole a foot or more deep and wide enough to contain the roots without crowding when they are extended in their natural position. Do not put fresh manure in the hole.

Depth of the hole is important and must be adjusted according to special needs of each tree. If the roots have been budded onto size-controlling rootstocks such as M.9, it is essential that the scion be above the surface of the ground. The bud union is the point at which the scion variety bud had been inserted into the rootstock and there usually is a small crook at this point. Dwarf trees are budded high in the nursery (14 inches) so they can be planted deep for good anchorage.

Put the topsoil into the bottom of the hole. Do not use grass sod to fill the hole. Bring in good soil from another part of the garden if necessary. Tramp hard with the heel of your boot to firmly pack the soil. Pour on a pail of water just after planting. No fertilizer is applied at planting time, nor during the first summer.

Any labels attached to the tree

must be removed at planting, as the wire or string will girdle the trunk after growth begins.

Just after planting, cut off the top of the tree at a height of 30 inches. If there are several side branches, remove half of them to balance the root loss.

You need to stake dwarf apple trees because they have poor anchorage due to their brittle roots; heavy fruit crops will topple them. They need support throughout their lifetimes. Semidwarf trees may also require staking. Semivigorous and vigorous trees will stand alone without staking.

Just after planting a dwarf tree, a 2 x 2 inch stake, 4 feet long, is driven 2 feet into the ground at a distance of about 6 inches from the tree trunk. The tree is supported by tying it to the top of the stake with a strip of cloth. As the tree grows, the tie must be loosened so that it does not girdle the tree.

Pruning

Pruning fruit trees is an art. A strict rule is not to prune too much, especially in the early years. The tree should be trained in the first year or two so that it will begin to develop into its proper shape. Select branches with wide crotch angles. In the third to sixth years, very little pruning may be needed. Overpuning will delay bearing.

Prune in late winter or early spring. Summer pruning of small twigs is also a good practice and helps to keep tree size down. Each spring cut out all dead and broken branches. During the first two years, select branches spaced at intervals along the main stem. Cut off all branches lower than 20 inches. A short trunk and closely spaced lateral branches will aid in developing a small, compact tree.

Train the tree so that it has a central leader which is taller than the other side branches. Shape it like a Christmas tree. A 5-year-old tree should have 5 to 7 side branches, well spaced around the main central trunk.

You may need to partially shorten the leader and to remove some of the high central branches to prevent the tree from growing too high. Heavy cropping tends to deter too much branch growth.

Trees on dwarf rootstocks have less excessively vigorous growth than trees on seedling roots. Pruning itself has a dwarfing effect on the tree. Sometimes, dwarf trees need very little pruning.

Cut out branches which cross each other, as well as vigorous upright suckers in the middle of the tree. Thin out parts of the tree which are too thick and which hamper penetration of chemical sprays and sunlight.

Make pruning cuts flush with the main limb, without leaving stubs. On young trees no healing paint is necessary.

Crotch angles of some varieties, especially spur type Delicious, tend to be vary narrow. The branches grow almost straight up, close to the center of the tree. These should be spread out in the first, second, and third years and later if necessary.

Spreaders should be placed before you make pruning cuts. Cut a thin lath board about 18 inches long with V cuts on both ends. The V on one end is braced against the central leader and the other against the branch to force it outward to an angle of about 45°. Branches can also be forced outward by loosely tying their tips to a stake driven firmly into the ground. Clothespins or No. 9 wire 6 to 16 inches long and sharpened on both ends also are very effective in spreading small branches. Spreading of branches discourages too much vigorous tree growth and also induces early fruit production.

Dwarf trees can also be trained on a wire trellis similar to grapevines.

Fertilizer

Beginning with the second year, apply fertilizer annually about two weeks before bloom. It must not be applied in mid to late summer because this stimulates late summer growth which will be too tender and result in winter kill during very cold winters.

Ammonium nitrate at ¼ pound per tree multiplied by the number of years the tree has been set, but never more than 2½ pounds per tree, is applied to moderately fertile soils. Very fertile soils need less.

The quantity of fertilizer is also adjusted according to the tree's vigor. If shoot growth the previous year was more than 12 inches long, less fertilizer will be needed.

Too much nitrogen causes excessive branch growth, inhibits fruit set, causes poor fruit color and flavor, delays ripening, and encourages fire blight disease. Conversely, pale green or yellowish leaves in the summer and short shoot growth may indicate the need for more nitrogen fertilizer the following spring.

Phosphorous fertilizer generally neither benefits nor harms apple trees. If the soil is low in potassium, apply some.

A 10-10-10 fertilizer at three times the above rates of ammonium nitrate is an equally good substitute.

Fertilizer is scattered under the outer parts of the branches. Since ammonium nitrate dissolves easily, you don't need to dig holes in the soil.

Stable manure can be used instead of chemical fertilizers, but it is usually more expensive and less available.

Mulches of any plant material, such as straw, grass or sawdust, suppress weed growth, hold soil moisture during a dry summer, maintain favorable soil temperatures, and add organic matter to the soil. Mulches should be six inches or more deep and extend to the tips of the branches. Woody materials, such as sawdust, wood chips or coarse hay, will require extra nitrogen fertilizer to aid decomposition. Mulches may harbor harmful mice in winter.

Thorough irrigation benefits fruit trees in midsummer when little rainfall occurs. Watering is especially important at planting time and during the first summer. In Western States, of course, fruit trees are almost always irrigated every summer.

Frost (28° F or lower) occurring after bloom kills all the blooms and young fruit and there will be no crop. Frost injury to blossoms occurs most frequently in the low parts of a valley, because cool air is heavier than warm and it drains to the low areas. On a sloping hill there may be good air drainage and less damage will occur. Trees near houses in suburban areas will suffer less frost damage than those in rural areas. Late blooming varieties, such as Golden Delicious or Rome Beauty, sometimes escape late spring frosts.

Frequently there is little the home orchardist can do to protect against frost.

Pest control is one of the most difficult aspects of growing your own fruit. It is not possible to produce usable apples without applying chemical sprays. Pears do better. When pears are grown without insect control, it is often possible to use about half of the fruits.

Insects which can be really serious problems on apples include the codling moth, apple maggot, red-banded leaf roller, tent caterpillar, aphids, mites, and apple tree borers. Serious diseases include apple scab, powdery mildew, and fire blight. The home gardener usually must accept less than complete control or apply more spray than is actually needed.

Mow the orchard weekly so that tall grass and weeds do not compete with the trees for soil moisture and nutrients. Fertilizers do not reduce the need for grass and weed control, but mulches and herbicides can be useful in helping control weeds. Modern herbicides can kill all grass without damage to the trees. If weeds are controlled, no soil cultivation will be needed.

Meadow mice and rabbits can chew off the trunk bark. If the bark is chewed completely around the trunk, the tree will die unless it is bridge grafted. Crushed stone packed around the base of the trunk, and wire guards of ¼-inch mesh screen made into a tightly closed cylinder 1½ feet high and 6 inches in diameter wrapped around the base of the trunk, can help control mice. Rodenticides are sometimes used.

If the orchard is near a wooded area, deer may chew off the growing tips of young shoots. Tankage, a pulverized animal slaughter by-product, in a small cloth bag hung in the tree sometimes helps to repel deer.

Birds may peck into early summer apples which ripen from mid-July to mid-August, and can cause serious damage. Late ripening varieties are not injured by birds.

Thinning is necessary when too heavy a crop of fruit is set. It results in larger, better colored and higher quality fruits.

Prevent too early cropping by removing all fruits just after bloom in the spring of the first and second years. This encourages maximum early tree growth. Fruit removal from the leader will encourage an upright leader.

Up to five or six years of age, apple trees usually do not overset. But after six years, thinning may be needed. Reduce the crop to a fruit per spur, spaced 4 to 6 inches apart.

By thinning in the years of excessive fruit set, alternate cropping varieties such as Baldwin and Wealthy can be forced into a more consistently annual cropping behavior.

Harvest when fruits begin to drop and soften or become fully colored, and have developed good eating quality. Early summer apple varieties tend to ripen unevenly, and several pickings over a 2-week period may be needed, but the fruits on individual trees of late varieties all ripen at once.

Certain varieties such as McIntosh begin to drop even before they ripen. Others such as Cortland will not drop, even long after they have become overripe.

Harvest Bartlett pears before they begin to turn yellow. Ripen them at room temperature off the tree in the basket.

Yields of fruit trees will vary, depending on such factors as pests, rootstocks and variety. Fruit buds for the 1979 crop develop on the tree beginning in June, 1978, and they require adequate foliage for proper flower development.

Some varieties such as Golden Delicious crop at a young age, often 3 years, but others, such as Northern Spy, are much less precocious, often beginning about the 8th year.

Mature apple trees on dwarfing rootstocks usually produce 1 to 2 bushels per tree. On seedling roots, 15-year-old trees may bear 5 to 15 bushels.

Pear trees may produce half the volume of fruit borne by apple trees.

Storage of early ripening summer apple varieties is generally not practical, but late October varieties store well. Such apples store best at 31° F at high humidity and will stay tree-fresh through the winter. A home fruit storage can be made from a large garbage can or a large discarded home refrigerator buried in the ground with its door at the surface. In the case of a refrigerator, the lock must be removed to prevent children from accidentally being trapped inside.

Peaches, Nectarines, Plums, Apricots, Cherries...
Climate Puts Limits on What You Can Raise

by John H. Weinberger and Harold W. Fogle

Growing peaches or other stone fruits in your home garden can reward you with luscious, tree-ripened fruit. But you must give your trees the care they require. Here are three specifics:

—Regular sprays for insect and disease control are absolute necessities to growing unblemished fruits.

—Birds and bees will take their share unless you protect ripening fruits.

—You must select varieties adapted to your climate to be assured of regular crops.

Despite such requirements, the opportunity to have fruits of various flavors, tastes and textures for up to six months in your home garden is a real inducement to plant stone fruits.

Peaches, nectarines, plums, apricots and cherries are called stone fruits because they have a hard, stony pit. They can be eaten fresh, or saved for future enjoyment by canning, preserving, freezing, or drying. Sour cherries are most often used in pies.

The climate where you live limits your selection of the kinds of stone fruits you can grow. Individual varieties must be adapted also. One or more of the stone fruits can be grown in every State except Alaska.

Low winter temperatures hamper the growing of stone fruits in Northern States. Some fruit buds of peaches, nectarines, and Japanese plums are usually killed by temperatures below 0° F and a reduced crop results. Lower temperatures damage or may kill the trees. European and native plums, cherries and apricots are hardier in fruit bud and wood than peaches or Japanese plums.

Along the southern border of the country, winter temperatures may be too high to break the rest period of the buds of many varieties. Only varieties with a low chilling requirement succeed there. In the vast area of the country between the marginal areas most stone fruits can be grown successfully.

Nectarines do best in a climate where rain rarely falls in the three weeks before ripening. They are very susceptible to brown rot disease.

Japanese plums, except for a few varieties, are not adapted to the humid climate of the Southeast. Diseases affect the trees and the fruits of most varieties.

Apricots bloom early in spring. The blossoms are usually killed by frost or freezes each year in all but the most favorable locations.

Sweet cherries are not adaptable to the extreme North or South. Everywhere birds will get a good share of the fruit before the home gardener is ready to harvest unless the tree is protected by netting or otherwise.

Peaches should receive first consideration by the home gardener for their wide adaptability, long ripening period, and ease of growing. Nectarines, where adaptable, are equally as good.

European plums need more care than peaches. They bloom later than Japanese plums and may escape frost.

Japanese plums, where adapted, produce large and attractive fruits with a minimum of care.

Apricots in the home garden should be tried only in commercial apricot-growing regions.

John H. Weinberger is a Collaborator with Horticultural Crops Production, Agricultural Research Service, Fresno, Calif. Harold W. Fogle is a Research Horticulturist, Agricultural Research Service, Beltsville, Md.

Sweet cherry trees reach large size, which adds to the problem of growing them in a backyard.

Contact your county Extension office for recommended variety lists and cultural practices.

Location, Spacing

Stone fruit trees should not be planted in a low or frosty location, where frost damage to blossoms and young fruits is probable. Moderately elevated ground or a slope will provide the necessary air drainage. Temperatures below 30° F will kill most fruits.

The soil should be reasonably fertile, with a pH of 5.5 to 6.5. Poorly drained soils are not suitable for stone fruits. Avoid planting trees in the permanent sod part of the lawn. Plant them in border plots or edges of the lawn. Fruit trees need full sun. Do not plant them in the shade of larger trees.

Peach, nectarine, plum, and sour cherry trees need the least space for maximum production; 18 to 24 feet is adequate. Apricot and sweet cherries need 25 to 30 feet. Peach and plum trees can be kept small by pruning and maintained in a 10- to 12-foot spacing if necessary.

All fruit can be picked from the ground if trees are kept low by pruning. Training trees on a wall or wire trellis is practical where space is limited.

A single tree can have fruit ripening over several months if three to six early, medium and late ripening varieties are budded into one tree. Budding is best done in late August while the bark still slips. The buds remain dormant until spring, when they are forced by cutting off the branch just above the bud. The ordinary "T" bud is the simplest type to use.

Sweet cherries and some Japanese plum varieties require cross-pollination in order to set fruit. A tree of another variety capable of cross-fertilization must be planted nearby. For best results select two varieties to plant which are known to be cross-fertile. An alternative is to bud or graft a branch of the pollinator variety in the desired tree.

Nearly all peach, nectarine and apricot varieties set fruit with their own pollen. Avoid self-sterile varieties of these fruits.

Select only varieties which do well in your locality. The fruits should have good flavor and smooth texture to make your efforts worthwhile. Extreme firmness and slow softening are not necessary since the fruit will not be shipped. On the other hand, rapid softening makes handling difficult.

Fruits of most commercial varieties will fill these requirements satisfactorily when picked at their peak of perfection. Do not let nostalgia for old varieties overly influence your choice. Peaches, nectarines, and plums have been much improved in recent years.

Hundreds of peach varieties are available. Freestone peaches are preferred for fresh use and for freezing. Both freestone and clingstone peaches may be canned. Varieties grown in the humid region east of the Rocky Mountains are usually different from those grown in dry irrigated areas west of the Rockies.

For the eastern part of the country, a succession of varieties in time of ripening from early to late would be Springold, Candor, Early Redhaven, Dixired, Harbrite, Redhaven, Redglobe, Loring, Redskin, and Monroe.

In States from Texas to Maryland where bacterial leaf spot disease is a problem, give special consideration to resistant varieties such as Sentinel, Ranger, and Dixiland.

Special varieties having a low chilling requirement are needed where winters are too warm for the above varieties. These include Maygold, Junegold, and Suwanee. Desertgold and Flordasun, which require even less chilling, are suitable for central Florida and the Rio Grande Valley.

A succession of peach varieties for the dry, irrigated areas west of the Rocky Mountains are Springold, Springcrest, Royal May, Flavorcrest, Regina, Redtop, Suncrest, Fayette, Summerset, and Fairtime. They ripen from mid-April to mid-September. Firm-fleshed clingstone peaches for canning are Loadel, Andross, and Halford. Junegold, Sunnyside, and Fairway varieties are freestones adapted to warmer areas of the region. Desertgold can be grown where winters are short.

Some of these peach varieties may not be available in your area. You might visit a local fruitstand where you can select a locally-grown, adapted variety suiting your needs.

Nectarines are beautiful fruits. In recent years some non-patented varieties have been developed which are available to the home gardener. The earliest is Firebrite, followed in order by Independence, Flavortop, Fantasia, Late Le Grand, Flamekist, and Fairlane. Fairlane ripens about September 1 in California. Remember that in humid climates, nectarines are harder to grow than peaches.

Plum Varieties

European plums can be grown in most States, including some of those too cold for peaches. Suggested varieties are the self-fruitful Fellenberg (Italian Prune), Stanley, and Shropshire. In the Far West, Tragedy and President can be grown and they pollinate each other. French Prune is used for drying. It is self-fertile and can be planted alone.

Varieties developed from native American species of plums are available for areas with severe winters.

Japanese plums ripen from May to September in California. A succession of varieties in season of ripening is Burmosa, Santa Rosa, El Dorado, Laroda, Friar, and Casselman. Santa Rosa, and Casselman are partially self-fertile and will pollinate the other varieties. In the Southeast, Frontier and Ozark Premier can be grown. Frontier needs cross-pollination. Methley and Santa Rosa are useful farther north.

Frost-protected locations are best for Japanese plums because of their early blossoming.

Blenheim (Royal), Tilton, and Castleton are suitable apricot varieties in California. In other areas Wenatchee (Moorpark), Goldrich, and Early Golden may be used. Apricots bloom earlier than Japanese plums.

Robert Bjork

A sour cherry variety.

The home gardener who wishes to challenge the birds for his crop of sweet cherries might plant Bing, Rainier or Van. They ripen in June and July. Two or more varieties are needed for pollination.

Sour cherry varieties available are Montmourency, English Morello, and Early Richmond.

Planting

A comercial nursery is the most convenient source of trees for the home gardener. Trees are graded by height in feet, or trunk caliper in inches. A medium-sized tree (4 to 6 feet in height or ½ to ⅝ inch in diameter) often gives best survival and growth.

Trees should be dormant when

255

planted. Spring planting is satisfactory in most areas providing the trees are kept dormant before planting. Fall or winter planting also is satisfactory and sometimes preferred in southern areas. If conditions are not suitable for planting at time of purchase, store in moist cellar or "heel-in" outdoors in a trench. Keep the roots moist and cool but avoid freezing.

Remove broken or diseased roots. If the roots have dried out in handling, soak them for several hours or overnight. Avoid planting when roots might be exposed to freezing. Plant the tree 1 or 2 inches deeper than it was growing in the nursery. Fill around the roots with topsoil, and tamp the soil. If the soil is dry, add 1 or 2 gallons of water to the hole. Fill the hole with soil and round off slightly.

Trees usually bear their first appreciable crop the third or fourth year after planting.

At planting time, the nursery tree usually has a single upright stem which should be cut back to 24 to 36 inches. This cut should be just above a mature bud. If there are wide-angled, strong lateral branches, select 2 to 4 which are separated up to 6 inches and spaced around the trunk. Tip these slightly. Remove or severely stub remaining laterals.

Select 3 or 4 scaffold branches during the first dormant period. These selected branches become the tree's primary framework.

Pruning should be minimal until the tree bears fruit. Unpruned trees tend to bear younger than pruned ones. However, branches which cross or interfere with good exposure of the scaffolds should be removed. An open-centered vase-type tree will give good exposure. Some secondary branches may be kept for early fruit and protection of the trunk and crotches from sunscald, but they should not interfere with the tree's basic framework.

In mature trees, keep vigorous current-season wood coming along to bear next year's crop of peaches and nectarines. The other stone fruits produce spurs which bear part of the fruit. Cut the tops back to reasonable picking height but change the height of cuts slightly each year. Prune to renew the bearing wood annually.

Fertilization

Most of the 12 nutrient elements essential for growth are available in nearly all soils. The tree's growth and production can tell you which ones are deficient. Leaf sample tests may be useful in diagnosing deficiencies.

Nitrogen is most often needed. A tree deficient in nitrogen will have light green to yellowish foliage and reduced shoot growth. In severe cases of N deficiency, small leaves, red specks on leaves and sometimes on fruit, misshapen and insipid fruit, and greatly reduced growth are com-

Peach tree with well-spaced, strong scaffold limbs and open center for maximum exposure to light on the fruit-bearing surface.

mon symptoms. Excessive nitrogen causes rank growth, poor fruit color and flavor, and may subject the tree to winter damage.

The dormant period is a convenient time to apply fertilizers. For a tree growing in sod, you need extra fertilizer to satisfy requirements of both tree and sod.

To remove competiton for nutrients and water, keep a bare area under the spread of the tree by culivation, herbicides or mulches.

Apply a complete fertilizer (10-10-10 or similar mixture) after the newly planted tree starts to put out leaves.

Each subsequent year apply fertilizer in amounts judged necessary for the individual tree based on appearance or leaf analysis. An approximate amount to apply is ⅛ pound of actual nitrogen per year of tree age up to 1 pound per tree. Spread the fertilizer evenly in a circle slightly larger than the tree spread. Applications may be split, with part applied during the growing season. Avoid nitrogen applications after late July in northern areas.

Regular irrigation in arid areas is needed, and supplemental irrigation in natural rainfall areas is desirable. A temporary drought, particularly during the month before harvest, may severely reduce fruit size and quality, even though the annual rainfall is adequate. Sod or shallow-rooted plants will show water stress before the tree suffers from lack of moisture.

Avoid frequent light irrigation. Instead soak the soil thoroughly to root depth and wait for signs of moisture stress in the indicator plant before irrigating again. Too heavy or too frequent irrigation may damage roots. Cherries are particularly susceptible to excessive moisture.

Heavier soils—particularly those with considerable clay—require less frequent irrigation than light, sandy soils and are subject to slower loss of nutrients to the subsoil.

Frost Protection

Protecting fruit trees from frost is difficult in the backyard. Anti-smog restrictions prohibit use of smudge pots or similar protection.

Choose the most frost-free site available before planting. Record the minimum temperatures in available sites for at least a year in critical areas. Avoid planting in draws or basins where cold air settles. Higher elevations are usually best, but windswept knolls should be avoided.

Covering trees with tarpaulins or other material to prevent radiation cooling is one way of protecting them. However, some framework is usually necessary to avoid tree damage. It is cumbersome to cover large trees, and the cover must be left on until air temperature is safely above freezing and then removed before damaging heat is built up. Hence, covering is usually impractical except for small trees.

Low volume sprinkling can be used for frost control. Pruning must be altered to give a heavy, stiff framework to hold the ice load from all-night sprinkling. Protection depends on a continuous film of unfrozen water which releases heat for bud protection. Sprinkling must be continued until air temperature is well above freezing or the night's effort may be lost.

Don't try growing stone fruits unless you provide for adequate pest control. You need spray equipment capable of reaching the tops of mature trees, or you need to be able to hire a custom spraying service when required.

Obtain and follow carefully the pest control calendar from your county Extension office. Timing of sprays is extremely important. Use only currently recomended materials at the rates specified.

Brown rot, caused by *Monilinia* spp., destroys more ripening fruit than any other pest. This is particularly serious in areas where it rains during and just before harvest. Removal of rotting fruit and "mummies" from the trees will help control spread of the fungus.

Scab, leaf curl, and cherry leafspot usually are not troublesome if trees are sprayed regularly. Bacterial leaf spot is not adequately controlled by spraying in extremely sandy soils of the Eastern United States—resistant varieties should be planted.

William E. Carnahan

Several virus and virus-like diseases can spread unchecked unless diseased trees are recognized and removed. They will not recover, and endanger nearby trees.

Precise timing of sprays should give adequate control of insects. However, missed sprays can result in wormy fruit, dead "flags" in the terminal growth, girdling of trunks by borers, or leaf damage by aphids and mites.

Protecting trunks from mice and rabbits with wire screens or plastic wrap-arounds may be necessary on trees growing near forests. Covering ripening fruit with netting is often the only way to protect it from birds and squirrels.

Fruit Thinning

Adequate dormant pruning removes a large number of flower buds. Pruning is the only practical thinning method for cherries, and can do a partial thinning in the other stone fruits. Heavy pruning may reduce the number of buds too drastically if later frosts kill additional fruit buds.

Additional thinning usually is needed after fruits have started development. Trees overloaded with fruit must have the crop thinned out to produce fruit of adequate size and good quality, and to prevent limb breakage.

Peaches, nectarines, plums, and apricots should be spaced 6 to 8 inches apart. Early ripening varieties need the greater spacing, and must be thinned early to give large fruit. Later varieties can be thinned at the pit-hardening stage without much loss in final size.

The advantage of homegrown fruit is that the best quality possible can be attained by ripening it on the tree. Most fruit for commercial use must

It's tempting to want to leave all the fruit on a peach tree. By thinning the peaches, those left on tree will be larger.

be picked three to seven days before soft ripeness to withstand handling and shipping.

Ripeness can be estimated by the disappearance of green and the development of yellow undercolor. Pressing the pads of your fingers against a fruit in your cupped hands will indicate softening of the fruit without damaging it. The fruit should be harvested by this same method, adding a slight twist of the wrist to loosen the fruit from its stem.

Pick the fruit into shallow containers to keep bruising at a minimum. Handle the fruit gently in moving and transporting it.

Fruit which will be used within a short time need not be refrigerated. It will attain its best quality in relatively warm storage.

Most varieties of fruit can be held in refrigerated storage for two to three weeks without excessive loss of quality. Longer storage usually results in internal breakdown of the flesh.

Stored fruit should be checked regularly for rotting or internal breakdown. Use the fruit as close to its prime quality as possible.

For Further Reading:

Barden, J. A., R. E. Byers, F. R. Dreiling, and others. *Production Management Practices for Apples, Peaches and Nectarines*, Virginia Agri. Exp. Station Publ. 595, Virginia Polytechnic Institute and State Univ., Blacksburg, Va. 24061. 1974. 54¢.

Benner, B., *Fruit and Vegetable Facts and Pointers; Peaches*, United Fresh Fruit and Vegetable Association, 777 14th St., N.W., Washington, D.C. 20005. 1963. $2.

Bobb, M. L., *Insect and Mite Pests of Apple and Peach in Virginia*, Virginia Agri. Exp. Sta. Publ. 566, Virginia Polytechnic Institute and State Univ., Blacksburg, Va. 24061. 1973. 26¢.

Childers, N. F., *The Peach* (Third Edition), Childers Horticultural Publications, New Brunswick, N. J. 08903. 1975. $13.95.

Drake, C. R., *Diseases of Stone Fruits and Their Control in Virginia*, Virginia Agri. Exp. Sta. Publ. 475, Virginia Polytechnic Institute and State Univ., Blacksburg, Va. 24061. 1972. 85¢.

Fogle, H. W., J. C. Snyder, H. Baker, and others. *Sweet Cherries: Production, Marketing, and Processing*, U.S. Dept. Agri. Handbook 442, on sale by Superintendent of Documents, U.S. Government Printing Office, Washington, D.C. 20402. 1973. $1.40.

———, L. C. Cochran and H. L. Keil, *Growing Sour Cherries*, U.S. Dept. of Agri. Handbook 451, on sale by Superintendent of Documents, U.S. Government Printing Office, Washington, D.C. 20402. 1974. 40¢.

———, H. L. Keil, W. L. Smith, and others. *Peach Production*, U.S. Dept. of Agri. Handbook 463, on sale by Superintendent of Documents, U.S. Government Printing Office, Washington, D.C. 20402. 1974. $1.35.

Gerdts, M., and J. H. LaRue, *Growing Shipping Peaches and Nectarines in California*, Calif. Agri. Exp. Sta. Leaflet 2851, Univ. of Calif., Davis, Calif. 95616. 1976. Free.

Harvey, J. M., W. L. Smith, and J. Kaufman, *Market Diseases of Stone Fruits; Cherries, Peaches, Nectarines, Apricots, and Plums*, U.S. Department of Agri. Handbook 414, on sale by Superintendent of Documents, U.S. Government Printing Office, Washington, D.C. 20402. 1972. 65¢.

Rizzi, A. D., and J. A. Beutel, *Care of Standard Fruit Trees*, Calif. Agri. Exp. Sta. Leaflet 2759, Univ. of Calif., Davis, Calif. 95616. 1975. Free.

Savage, E. F., and V. E. Prince, *Performance of Peach Cultivars in Georgia*, Georgia Agri. Exp. Sta. Res. Bul. 114, Univ. of Georgia, Athens, Ga. 30601. 1972. Free.

Schwartz, P. H., *Insects on Deciduous Fruits and Tree Nuts in the Home Orchard*, U.S. Dept. of Agri. H&G Bul. 190, on sale by Superintendent of Documents, U.S. Government Printing Office, Washington, D.C. 20402. 1972. 40¢.

Seelig, R. A. *Fruit and Vegetable Facts and Pointers; Plums-Prunes*, United Fresh Fruit & Vegetables Assn., 777 14th St., N.W., Washington, D.C. 20005, 1969. $1.90.

Grapes Are Great But You May Have to Wait; Buying Rooted Vines Can Save You a Year

by J. R. McGrew

Grapes may be used as fresh or stored table fruit, made into jellies or juice, or fermented into wine. There is a wide range of flavors among the many varieties. Grapes can be one of the easiest home-garden fruits to grow and one of the most rewarding.

There are several types, each suited to particular climates, areas and use. Trying to grow types not adapted to your area can be a frustrating experience.

In an article this brief, there is no way to cover all varieties or all the methods of growing grapes. Nor can all the possible mistakes, hazards or pests be discussed. What follows are general statements.

Before deciding to try to grow grapes, you should consider the basic requirements for success:

—A growing season of at least 140 frost-free days.

—A site with full sunshine and good air drainage (not frosty).

—Soils that are neither waterlogged nor shallow, at least 3 feet deep.

—Willingness to spray at least three times per year to control insects and diseases.

—Patience to wait three to four years for vines to reach maturity before cropping.

—Annual pruning of vines.

—Readiness to defend the fruit against birds by netting the vines or bagging clusters.

A few vines may be planted along an existing fence, or a fence or arbor may be built in an esthetically pleasant place. Vines form an excellent summer privacy screen, but after leaf fall and pruning there is little left.

Purchase of rooted vines from a nursery or garden store saves a year over propagating your own vines from cuttings. If muscadines or grafted vines are to be grown, the purchase of plants is preferable. Spacing of vines is not critical. Six to 10 feet between vines gives room for each vine, makes pruning easier, and is a more economical use of the space.

Planting will be easier if the soil is spaded or tilled beforehand. Grapevine roots rapidly grow out several feet in the first two years, so working compost or fertilizer into the planting hole will be of little value.

For at least the first 2 years, an area one to two feet around each vine should be kept free of weeds by hoeing, or with a heavy mulch of grass clippings or black plastic. Fertilize young plants only on very poor soils.

Varieties

The choice of grape varieties is both important and complicated. Advice from neighbors, your county Extension office or from State Agricultural Experiment Station bulletins can be most helpful.

For California and parts of the Southwest, there are many excellent varieties of Old World grapes (Vitis vinifera). There are seedless table varieties, muscats and many wine varieties, each best adapted to certain areas.

For the Southeast (from Tidewater Virginia, through the central areas of the Carolinas, south through Florida, and west through the southern part of Texas) Pierce's disease kills or shortens the life expectancy of many popular grape varieties.

J. R. McGrew is a Research Plant Pathologist with the Fruit Laboratory, Plant Genetics and Germplasm Institute, Agricultural Research Service, Beltsville, Md.

In these areas the kinds of grapes that may be expected to give the best results are the muscadines, like Scuppernong or modern self-fertile varieties, and a few tolerant varieties introduced from the Florida Experiment Station at Leesburg; Stover, and Lake Emerald, and a few older varieties such as Champanel, Herbemont and Lukfata. Other varieties may survive to produce a crop or two, but have not proven successful over a longer period.

For the rest of the country, where the climate is humid enough to permit wild grapes to survive, the problem of variety selection is complicated by the several diseases and insects that attack cultivated grapes. The American and French-American varieties are somewhat tolerant of these problems and therefore less risky to grow.

In the shorter season areas (140 to 160 frost-free days), you can grow early ripening varieties such as Beta (blue) for juice and jelly; Foch (blue), Cascade (blue), and in better sites Aurore (white), for wine. Light cropping of vines may be useful in short-season areas because it can advance ripening of the fruit by about two weeks.

In the medium season areas (160 to 200 frost-free days), Concord (blue) and Niagara (white) are two of the most popular and easily grown varieties for table use and for juice and jelly. There are several semi-seedless varieties, like Himrod (white) and Suffolk Red, table grapes such as Seneca (white), Alden (blue) and Steuben (blue), and many French-American wine grapes that are satisfactory. Chardonnay and White Reisling, representatives of vinifera wine grapes, may survive if sprayed carefully and frequently.

For growing seasons longer than

Top, table variety, Steuben. Right, semi-seedless Suffolk Red.

200 days, late ripening varieties are preferred. Concord and Niagara are suitable for juice or jelly. White wine varieties include Villard blanc and Vidal 256, for red wine-Chambourcin and Villard noir. A muscat flavored grape of interest is Golden Muscat.

If you have a protected site, in cold areas, and if you are willing to take a chance on occasional crop loss and especially if you are willing to take extra effort to protect vines against pests, you may succeed with varieties that might otherwise fail.

Vines should be planted at about the same depth they were grown in the nursery. If vines are grafted, the graft union should be about 2 inches above ground level.

Roots should be spread out in all directions in the planting hole. They may be trimmed to about 2 inches if you choose to plant the vines in a narrow hole made with a post-hole digger.

The top should be cut back to leave two or three buds. When the new shoots begin to grow, remove all except the one or two shoots that are the most vigorous and straight. Tie these loosely to a light stake. Several times during the first season remove lateral shoots that develop at the point of attachment of each leaf. This allows the main shoot to grow more rapidly and a full year may be gained in establishing the vine.

Failure to remove these lateral shoots and the sprouts that appear from the base of the vine throughout the season will result in a bushy vine which seldom has any shoots long enough to reach the trellis.

Leave about four lateral shoots just below any horizontal wires along which you want the vine to grow. When the shoot or shoots reach the highest point of the trellis or arbor, tie them there, pinch off the tip and allow several of the lateral shoots to grow.

If for any reason a vine fails to make good growth during the first growing season, cut the top back to two buds and treat it as a newly planted vine. It will generally grow more vigorously during the second season.

Training places the crop in a convenient location for vineyard operations and harvest. Pruning controls the size of the crop to a level that can be ripened successfully.

Structures

Structures on which the vines may be trained range from two or more posts set in the ground and strung with two or three horizontal wires (a trellis) to decorative arbors. Bracing should be sufficient to carry the weight of vines and crop under the sort of wind conditions experienced in the area. Trellis posts should not be more than 20 feet apart and arbor posts not more than 10 feet apart.

Wires (11- or 12-gauge smooth galvanized) should be spaced about 2 feet apart up the posts or along the top of an arbor. Closer spacing causes excessive shading. To permit weed control under the vine and to keep the fruit up, the lowest wire should be 30 to 36 inches above the ground.

Train a permanent trunk to the top wire of a trellis or to the top edge of an arbor.

During the dormant season when vines are pruned, fruiting canes (see below) should be trained outward along each wire on the trellis or along an arbor's top edges.

Each bud on the fruiting canes grows into a shoot from 4 to 20 feet long. These are tied along trellis wires as they grow, or on an arbor are spaced out across the top wires to give even exposure to sunlight.

Fruiting canes can be readily identified if we look at a vine in the spring before growth begins. They are the one-year-old shoots (wood of the previous season), with bark that is smooth and brown. At each place where a leaf grew the previous season, there is a conical swelling, or bud.

During the growing season, each bud grows into a shoot which bears leaves and generally three clusters of grapes. The more buds that are left after pruning, the more clusters will appear on the vine.

An unpruned grape vine will set far more fruit than it can ripen successfully. Fruit from overcropped vines is low in sugar, sour, and has poor color. Excessive over-cropping can severely damage the vine.

Obviously the cluster size must be considered in calculating size of a crop. With very large clustered varieties, such as Thompson Seedless, as few as 10 clusters per vine (8-foot spacing) should be left. Perhaps 50 clusters of Concord can ripen and as many as 100 of small clustered varieties such as Beta or Foch.

The commercial grower controls crop size by leaving exactly the right number of buds. The home gardener can achieve a far more accurate control of crop size, and do it despite variations in weather or fruitset, by leaving an excess number of buds, two or three times as many as needed, and removing clusters until the right number remain. Removal of excess clusters can be done any time from before bloom until mid-season.

Pest Control

Most county Extension offices have spray schedules for the home gardener and in those areas where grapes are grown, appropriate sprays for diseases and insects of grapes are included. You may be able to get an occasional crop without spraying, but both diseases and insects tend to become progressively more severe from year to year.

Control of weeds for a foot or two around young vines is worth the effort in the improvement of growth you can expect. Once established, the vine will shade out some weed growth.

Some types of weedkiller should not be used near grapes as they are extremely sensitive. Do not use the combination of fertilizer plus weedkiller on lawn areas within 15 feet of a grape vine. The weedkiller may be picked up by the grape roots that extend out this far and the vine can be damaged.

In many areas birds can be a major problem. Netting, which can be used earlier in the season for strawberries and blueberries, is available and if placed carefully over the vines will protect the fruit.

Hornets and wasps on ripe fruit are a common complaint. They are able to attack the fruit only if it has been damaged by insects, diseases or birds, or if it is overripe.

An acceptable taste is the main criterion for table use. On a vine that is not overcropped, the berries of blue varieties will lose their red color and white varieties will change from green to golden yellow. Ripe berries will soften and seeds become brown.

Black rot fungus on an American bunch grape leaf.

As the berries ripen, sugar content rises while the acid level decreases. Both these changes are reflected in improved taste.

Determining the harvest of wine grapes requires either experience or a means of measuring both sugar and acid levels.

The yields of a grapevine greatly affect fruit quality. If you permit vines spaced at 8 feet to produce over 30 pounds of fruit each, the quality will almost surely be low. Only under ideal circumstances and climates can this size crop be ripened successfully.

It is better, especially on young vines, to leave a smaller crop than optimum, say 5 to 10 pounds of fruit, until you find out how much fruit can be ripened successfully in your particular situation.

There are several sources of information at all levels of complexity for the home grape-grower.

The U.S. Department of Agriculture and State Agricultural Experiment Station bulletins and leaflets cover general grape growing, variety recommendations, descriptions of diseases and insects, pests, and recommended spray programs. States which have an established grape industry tend to have more complete and extensive publications.

For Further Reading:

Banta, E. S., G. A. Cahoon, and R. G. Hill, *Grape Growing*, Bul. 509, Coop. Ext. Serv., Ohio State University, Columbus, Ohio 43210. 1969. Free.

Control of Grape Diseases and Insects in the Eastern United States, U.S. Department of Agriculture F 1893, on sale by Superintendent of Documents, U.S. Government Printing Office, Washington, D.C. 20402. 1972. 45¢.

Cultural Practices for New York Vineyards, Cornell Ext. Bul. 805, Publication Office, New York Agri. Exp. Sta., Ithaca, N.Y. 14850. 50¢.

Grape Growing in Virginia, Bul. 175, Virginia Polytechnic Institute and State University, Blacksburg, Va. 24061. 14¢.

Growing American Bunch Grapes, U.S. Department of Agriculture F 2123, on sale by Superintendent of Documents, U.S. Government Printing Office, Washington, D.C. 20402. 1974. 40¢.

Muscadine Grapes, A Fruit for the South, U.S. Department of Agriculture F 2157, on sale by Superintendent of Documents, U.S. Government Printing Office, Washington, D.C. 20402. 1971. 30¢.

Wagner, P. M., *A Wine-Grower's Guide*, Alfred A. Knopf, New York, N.Y. 11022. $7.95.

Weaver, R. J., *Grape Growing*, Wiley & Sons, New York, N.Y. 10016. 1976. $16.50.

Winkler, A. J., *General Viticulture* (Second Edition). Univ. of California Press, Berkeley, Calif. 94720. 1974. $27.50.

Strawberries Like Full Sun — and a Good Deal of Attention

by Robert G. Hill, Jr., James D. Utzinger, and Elden J. Stang

Success in growing strawberries depends on close attention to cultural details. Small well-cared for plantings are generally more rewarding than larger plantings which receive less care.

You can expect nearly a quart of berries from each plant you set if you follow good cultural practices. Well established and cared for plantings can produce berries up to three years.

Besides being a versatile dessert fruit, strawberries are highly nutritious. One cup of fresh strawberries supplies more than the recommended daily Vitamin C requirement. Strawberries also make a welcome addition to the home freezer or can be used for jellies and jams and other preserves.

Strawberries do best when planted where they receive full sun most of the day. They grow and produce well in a wide range of soils—from sandy to heavy loams, but sandy loams are preferred. Strawberries are not particularly sensitive to soil acidity or alkalinity. However, they produce best on acid soils with a pH of 5.8 to 6.5.

Key factors in site selection are soil drainage and freedom from frost. Don't expect good production without adequate soil drainage during the entire year. Strawberries can't tolerate standing water. Since strawberries bloom very early in spring, don't plant them in a frost pocket. Frost pockets are low-lying areas into which cold air drains. The crop in such areas is likely to be lost to late spring frosts, which destroy the flowers.

Avoid planting strawberries on steep slopes. Heavy rains are apt to bury some plants and wash others out of the soil. If you must use a sloping site, rows should run across the slope or on the contour.

Avoid areas used recently to grow tomatoes, potatoes, or sod. These sites are likely to contain disease and insect pests that may attack strawberries. Likewise, avoid sites heavily infested with quackgrass, Johnson grass or thistle, or else treat the site well before planting to destroy those weed pests.

Performance of a strawberry variety is markedly influenced by local soil and climatic conditions. A variety highly successful in one area may be of little value in another. So select varieties on the basis of area conditions. Ask your county Extension office what varieties are best adapted for your area. No cultural practice can overcome a handicap imposed by poor selection of varieties.

Commercially available strawberry varieties are self-fruitful (they don't require cross pollination), and will produce good crops when only one variety is planted. However, most home gardeners prefer to plant several varieties to extend the harvest.

Two types of strawberries are available. June bearing strawberries produce a single crop each year. Everbearing strawberries produce one crop during the normal season and a second crop during fall of the same year. Probably the normal June bearing type is the most popular.

Other factors to consider are quality of the berries, their suitability for freezing, and the degree of disease resistance.

The authors are all members of the Department of Horticulture at the Ohio State University and the Ohio Agricultural Research and Development Center, Wooster. Robert G. Hill, Jr., is Professor and Associate Chairman. James D. Utzinger is Associate Professor. Elden J. Stang is Assistant Professor.

Buying Plants

Buy from a reputable nursery to be sure of getting quality plants true to name. To get plants of the desired varieties, order as early as possible and indicate the desired delivery date.

The best kinds of plants to buy are "virus-free". They can yield 50 to 75 percent more fruit than plants from ordinary planting stock.

There is no apparent visual difference between virus-free and ordinary planting stock. The only way to be certain the plants are virus-free is to purchase *registered* plants. These have been grown under State supervision, and the word *registered* on the bundle label indicates the plants are substantially virus-free, the best that can be obtained.

Virus-free plants of many varieties are available. Use them whenever possible.

Another class of plants, *certified*, also is grown under State supervision. Certification indicates the plants are free of most noxious diseases and insects; however, they may carry virus. Certified plants are the best available of some varieties.

Dormant strawberry plants are best for spring planting. Plants dug early and held dormant in storage, if properly stored, are as good as freshly dug plants. In some cases, stored plants are superior.

When plants arrive, check the bundles. If necessary, moisten the roots, but do not soak them. Plants which cannot be set immediately may be stored in a refrigerator for several weeks, or until planting conditions are satisfactory. Hold plants as close to 32° F as possible in the plastic bags they are shipped in. Be sure the bags are closed by folding only, and not tightly closed.

If no storage facilities are available and planting can't be done in a few days, carefully unpack plants and heel them in. To heel-in, pick a sheltered and well-drained area and dig a shallow trench deep enough to accommodate the roots. Open the bundles and place a single layer of plants against one side of the trench so the crowns are partially above the soil line. Cover the roots with soil, moist peat moss, or sawdust, and firm carefully.

Plants so handled can be held for several weeks, if not allowed to dry out. But don't leave plants heeled-in any longer than necessary.

It is wiser and cheaper, in the long run, to purchase nursery stock than to secure planting stock from your own or a neighbor's plantings. The better the planting stock, the better yields you can expect.

Site Preparation

It is best to begin preparing the strawberry site the year before planting. Use the proposed site to grow a cultivated crop during the season prior to planting. Chronic weeds can be controlled and soil fertility levels adjusted during this period. Use soil test results as a guide in adjusting the fertility level. If necessary to plant an area that has been in sod, turn or spade the sod over during the fall before planting. Weed problems in plantings set into newly-turned sod can be overwhelming.

Ideally, work the site during late summer. Seeding the area to ordinary rye in early September helps control erosion. Usually, 2 to 3 pounds of rye seed per 1,000 square feet will give the desired results.

If animal manures are available, they may be applied in fall. A suitable application is 50 to 75 pounds of strawy manure per 100 square feet.

Prepare the site for planting as early as possible in spring, during late March or early April, before the rye gets too tall. Work the soil until it is near seedbed condition.

Apply fertilizer and work it in as you prepare the soil. Adjust rates of fertilizer application based on your experience with the site, or results of a soil test. On most sites, 1 pound of a 5-10-10 fertilizer per 100 square feet will be beneficial. For convenience, you may mix and spread the recommended soil insecticides with this fertilizer.

Don't work the soil when it is wet.

Making the Planting

The training system to be followed in strawberry planting determines the distance between rows and between plants in the row. Most home gardeners use the *matted row*. No effort is made to limit the number of runner plants, but they are kept within a row 18 to 24 inches wide. Plants of most varieties are usually set every 18 inches in rows 48 inches apart.

The other system is the *spaced row*. The number and location of runner plants is predetermined. The spaced row system requires much more labor than the matted row. It also requires setting more plants per unit area.

Early spring planting is best. Set plants as soon as the soil can be prepared, normally during late March and early April. Don't attempt to plant until the soil is dry enough to work. If plants can be maintained in a dormant condition and irrigation is available, planting can be delayed.

Before planting, remove all but two or three of the most vigorous leaves and prune away about a third of the roots. Place the plants in the soil so the roots are spread out. Cover the roots until the crown (where the leaves arise) is just above the soil surface. If the crowns are covered with soil, or the roots exposed, plants will do poorly and may die.

Strawberries planted in matted row system used by most home gardeners.

Hand planting can be done by a two-person team. One person forces a spade or long-handled shovel about 6 inches into the soil and pushes it forward to open the hole. The other inserts the plant to the proper depth and holds it against the side of the hole while the spade is removed. The one with the spade then closes the hole by inserting the spade in front of the hole and pushing forward on the handle. Soil about the roots is then firmed with the foot.

After setting, if the soil is dry, give each plant at least a cup of water. Regardless of the planting methods, make every effort to prevent the plants from drying out.

First Season Care

Remove flower stalks as they appear. If berries are allowed to develop, they will reduce plant growth, runnering, and next year's crop.

Carefully cultivate and hand hoe

the planting throughout the season to control weeds. Frequent, shallow cultivation in one direction only is best. Infrequent, deep cultivation can damage strawberry plants.

Cultivation helps keep runner plants within the allotted row area and permits easier rooting of runner plants. Don't allow the rows to get wider than 18 to 24 inches. Chemical herbicides may be used as an aid in weed control during this period. But for most garden plantings their use is not recommended because of problems in application.

Runner plants produced after August 15 are relatively unproductive and should be removed, unless the desired matted row has not been obtained.

Fertilizer applications are seldom needed during the growing season. But if the new plants appear light green and don't grow well, sidedress with nitrogen fertilizer about a month after planting. Apply 1½ to 2 pounds of ammonium nitrate per 100 feet of row. When applying this fertilizer, select a dry day and brush all fertilizer off the leaves to protect them against fertilizer burn.

If the plants continue to have light green leaves, a similar application may be spread over the rows about August 1. Avoid getting fertilizer on the leaves.

Fertilizer applications in the spring of fruiting years are apt to cause soft berries and reduce yields.

Since too much nitrogen may cause excessive growth and reduce yields, exercise care in application rates.

Use irrigation and pest control practices as needed.

Winter Care

Mulch strawberry plantings in colder regions as a winter safeguard. A mulch protects plants from severe cold and against soil heaving caused by alternate freezing and thawing of the soil.

The best mulching materials are clean, seed-free wheat or rye straw. Tree leaves and oat straw tend to pack and smother the plants. Coarser materials offer little protection. Sawdust may be used, but straw is preferred. Expose the straw to weather by placing it near the planting in early fall. Most of the grain and weed seeds will germinate before the mulch application, thus reducing a serious spring weed problem.

Apply mulch 3 to 4 inches deep over the plant rows. Do this only after the planting has experienced several sharp freezes—in the lower 20's. It is easier to apply mulch when the ground is frozen. When mulch is applied before growth stops, damage to the crowns could occur. If mulching is delayed, low temperatures could damage the crowns.

Mice sometimes will damage strawberry plants under the mulch. Check with your county Extension office for control methods.

In spring when new leaves begin to develop, fork the mulch off the plants, placing it between the rows. So placed, the mulch controls weed

Removal of flower clusters from new leaf set plants will increase future yields.

Don Normark

growth, conserves moisture, and helps keep the berries clean. Remove only enough mulch from the rows to let the plants develop.

Winter mulch may be re-used to protect flowers from frost. If frost is predicted, mulch can be spread evenly back over the rows. The mulch acts as an insulation barrier which traps radiant heat from the soil and holds it around the plants. Mulch can be left in place for several days, if necessary.

Harvesting

Harvest berries when they are fully colored; those with white areas are not ripe. Pick the berries with the caps and stems attached. To do

Left, mulching strawberries helps keep the fruit clean and conserves moisture. Right, strawberry bloom at frost-hazardous stage. If frost is predicted, remulch the planting until hazard passes.

this, snap the stem, using the thumbnail. Avoid bruising the berries. Keep harvested berries out of the sun and refrigerate as soon as possible. The first harvest generally can be made about 30 days after first bloom.

One advantage of growing your own strawberries is that they may be eaten at peak of quality.

If possible, avoid picking berries when the plants are wet. Harvest as often as necessary, about every other day. The harvest season of a given variety extends over 5 to 7 pickings. Harvesting is made easier by the use of "carriers" that hold 4 or 6 quart baskets.

Pick and remove berries damaged by birds, and any rotted berries.

Renewing Plantings

Strawberries may be fruited more than 1 year. Yield and size of berries are progressively smaller the second and third years. As a rule, it is unwise to attempt more than 3 crops from a single planting.

Only good plantings should be maintained and renewed. Destroy weak, weedy or diseased plantings right after harvest.

Renewal of a planting should be done shortly after harvest. Start by mowing off the tops as close to the ground as possible without damage to the crowns. Then with a spade or a rotary type tiller, narrow the row width to a strip of plants 8 to 10 inches wide. This can be done by destroying plants on both sides of the row or one side only. The latter is preferred.

As you narrow the row, work the mulch and other organic material into the soil. Fertilize the row as indicated for summer fertilization of a newly set planting. Handle the renewed planting the same way as a first-year planting. Generally a planting should not be renewed for more than 2 seasons.

Irrigation

Strawberries can be grown on a limited basis without irrigation. However, in many seasons full production can't be realized unless the plantings are irrigated. During these seasons, supplemental water assures formation of a good row and helps assure good berry size.

Apply enough water during the growing season to supplement rainfall and to provide an average of 1 inch of water per week. Irrigate when a water shortage is apparent, even before the plants show drought symptoms.

Strawberries are relatively free from disease and insect problems. Normally they produce satisfactory crops in home gardens without spraying. Full production of high-quality berries, however, requires that you follow a careful pest control program.

You can avoid many headaches by selecting sites free of disease and insect problems, getting suitable planting stock, and following good cultural practices. As with other fruit crops, good pest control practices are based on preventing problems rather than overcoming them. Specific problems and control practices will vary from region to region. Contact your

William E. Carnahan

county Extension office for details.

Birds may be a problem during the fruiting season, damaging the ripe fruit. Bird control netting, available commercially, is the best solution. Stretch it over the beds.

Everbearing Berries

Everbearing strawberries are grown primarily for the fall crop. They will produce satisfactorily if grown under the spaced-plant system of culture. Successful production of this type strawberry requires much labor, so planting should be of limited size. They will not do well when grown in matted rows.

The site for Everbearing varieties should be prepared and the plants set with the same considerations as June bearing varieties.

Stretch netting over strawberry plants, so as to protect them against the birds.

These plantings should be maintained under either a sawdust mulch or a black plastic mulch. With a sawdust mulch, care for the planting as if it were a regular planting until early June when the runners appear, then stop cultivation. Fertilize each plant with 2 tablespoons of a 16 percent nitrogen fertilizer or equivalent, spreading the fertilizer uniformly over the soil around each plant.

Ripening cluster of strawberries.

Then cover the entire area of the planting with 1 inch of either hardwood or softwood sawdust. It may be fresh or weathered. Don't apply excessive amounts. Further weed control must be done by hand, since hoeing and cultivation will mix the sawdust with the soil, thus destroying the mulch benefits.

After applying the mulch, start training the runner plants, locating each in the desired position. Force the plants gently but firmly through the sawdust so their roots contact the soil. The distance between runner plants varies from season to season but will be about 8 to 10 inches.

After the desired number of runner plants has been established, remove all others as they develop through the remainder of the season.

Removal of flowers should continue until the first to the middle of July. The exact date for discontinuing blossom removal depends on the planting. The more vigorous it is, the earlier blossom removal can be stopped.

Harvesting will begin about 30 days after first blossoms appear. The first berries will ripen in August and harvesting should continue twice a week until frost.

Black plastic offers advantages over sawdust as a mulch for everbearing strawberries. Its use minimizes problems of weed control and helps keep the berries cleaner. With black plastic, only a slight variation in cultural practice is needed. The mulch may be spread over the row area and the plants set through it at desired locations. Cover edges of the mulch with soil.

Establish runner plants where needed by cutting a slit in the plastic and placing the plant firmly into the soil. Blossom and runner removal are the same under both mulch systems.

Trickle type irrigation lines installed under the plastic mulch can prove helpful during drought periods. Take care not to over water.

June bearing strawberries may be grown according to the spaced plant systems, too, but benefits do not justify the added efforts.

Cane and Bush Fruits Are the Berries; Often It's Grow Them or Go Without

by John P. Tomkins

Homeowners frequently overlook the possibilities of growing raspberries, blackberries, gooseberries, currants, and similar berries. During recent years these fruits in local markets have been scarce and rather expensive. The homeowner might have to grow them or do without.

These cane and bush fruits are easier to grow and much more practical as compared with the tree fruits in home gardens. Cane fruits require less work, occupy a smaller area, need relatively few sprays for pest control, and will produce fruit within a year or two of planting. A small area devoted to berries will give rich dividends in fresh fruit for the home or to be passed along to friends.

Raspberries, currants, and gooseberries are excellent sources of vitamin C. Raspberries and blackberries may be used fresh, canned, frozen, or in pies, jellies, jams or preserves. Gooseberries are used mainly in sauces and pies. Currants may be used alone or mixed with berries to make a very tart and tasty jelly.

These fruits vary greatly in hardiness to low winter temperature. Currants and gooseberries are the hardiest and can withstand —40° F and still be productive. Red raspberry is next in hardiness; some varieties will withstand —35°, although others may be injured at 0°. Black and purple raspberries are next in hardiness and may withstand —25°. Blackberries may be injured around —15° while thornless blackberries, boysenberries, and youngberries may be injured when the temperature falls close to zero.

All cane fruits may be injured at

John P. Tomkins is Associate Professor of Pomology at Cornell University, Ithaca, N.Y.

somewhat higher temperatures than indicated if they receive poor cultural care during the growing season. However, they may withstand temperatures 5° to 10° F lower than indicated if properly hardened or given some winter protection.

Success of a cane fruit planting depends largely on selection of the proper varieties for your area. Varieties differ greatly in berry quality and size, season of ripening, hardiness to low winter temperature, and disease susceptibility. Growth and yield are influenced by length of growing season, temperature, rainfall and humidity. A good variety in New York may be very poor in Maryland, Maine, California or Washington.

The most reliable information on varieties for a given area may be obtained in fruit publications available at your county Extension office.

Fall Bearers

Much good work is being done by the berry breeders in the U. S. Department of Agriculture and at various state Agricultural Experiment Stations. One of the most interesting developments has been the production of fall-bearing raspberries which have a summer crop on floricanes (canes produced the previous year) and a fall crop on primocanes (canes produced during the current season).

The fall crop is very successful in areas with a frost-free growing season of 165 days or longer.

The best fall-bearing raspberry variety is Heritage. Another promising fall-bearing variety is Augustred from New Hampshire. In New York it has a fall crop that ripens 35 days earlier than Heritage. Augustred is worth a try where the frost-free growing season is 130 days or less.

During recent years, breeders have developed some varieties of thornless blackberries which are very vigorous and productive. The weakness of these varieties is a tendency to injury during winter by temperatures near 0° F. However, the home gardener can cover these canes with mulch during winter and the buds and canes will survive at temperatures much lower than 0°.

The most important soil factor for growing raspberries is good drainage to a depth of 3 to 4 feet. Plants in full leaf will not tolerate standing water or a high water table for 2 to 3 days without root injury, subsequent decreased yields, and perhaps death of the plant. Raspberries tolerate a wide range of soil types from clay to sand if drainage is good.

Currants and gooseberries tolerate a heavier soil and poorer drainage than cane fruits.

The most suitable soil is a sandy loam, rich in organic residues, with a good moisture-holding capacity and a pH of 5.5 to 6.8.

Select a site for planting in which perennial weeds have been eliminated and where tomatoes, eggplants, peppers and potatoes have not been grown within the previous two years. These crops build up a fungus disease known as verticillium wilt which damages or kills many cane fruit varieties. A location in full sunlight is desirable, although cane fruits will thrive in areas shaded for part of the day.

Buy plants from a reliable nursery. This is a key to success in bramble fruit production. The disadvantages of poor stock can never be overcome by a good site or even superior cultural knowledge. Be sure plants come from certified or inspected stock.

Prepare the site by growing cultivated crops at least a year before planting cane fruits. Work 1 inch or more of organic residues such as lawn clippings, rotten leaves, or well rotted manure into the top 4 to 6 inches of soil. Before planting, mix about 1 pound of 10-10-10 or equivalent fertilizer per 100 square feet of soil. If the soil reaction (pH) is under 5.5, add limestone or hydrated lime as indicated by a soil test. Your county Extension office or a reputable garden store can suggest how to go about having a test made.

Plants are usually obtained and planted in early spring. If they arrive before soil preparation or when the soil is too wet for planting, store the plants—if well wrapped—in a cool place. If unpacked, heel them into the ground in a shallow trench in a cool shady area so the roots do not dry out.

Set plants in rows with 30 inches between plants in the row for raspberries and blackberries, 3 to 4 feet between plants for currants and gooseberries, and 5 to 8 feet for trailing and thornless blackberries.

Space between rows will vary from 6 to 10 feet depending on cultivation equipment. Nine to 10 feet between rows is usually adequate and helps to prevent spread of fungus diseases which tend to be prevalent with the plantings in closer rows.

Set plants in the ground to a depth of 5 to 6 inches, or at least 1 inch deeper than the plants were grown in a nursery row.

Purple or black raspberries are grown in hills. Black raspberry never develops new canes from root suckers; purple raspberry may develop a few.

Red raspberry develops many new plants from root suckers, and is usually grown in a hedgerow system. New suckers developing along the row are controlled either by cultivation or timely mowing to keep the row of new canes 6 to 18 inches wide.

Cane fruits have an unusual growth habit. The canes are biennial and the roots perennial. The new canes are known as primocanes. The next year these canes are known as floricanes.

Buds on floricanes develop shoots with leaves and terminate in flowers. After the canes have fruited they die. Remove them at this time or when pruning in winter.

Obviously new primocanes are being formed each year. Primocanes on thornless blackberries, dewberries, and boysenberries tend to grow along the ground. The next year these canes are tied along the trellis where they fruit. After the canes have fruited they may be removed and destroyed.

Pruning

Black and purple raspberries usually are not trellised. When the primocanes get 18 to 24 inches long—generally during May or June—prune off the tip of each cane. This forces 3 to 7 buds nearest the severed tip to start growth, and eventually they form laterals 3 to 7 feet long. The following winter prune back the laterals to about 8 inches from each main cane. All buds on the laterals and main canes are potential fruit buds.

If you live in a snow belt, prune black and purple raspberries in early November before heavy snows. Melting snows in early spring may damage unpruned laterals in areas where snow is 3 to 7 feet deep.

You prune red raspberry quite differently from black and purple raspberry. Tips of the primocanes are not pruned the first year. Red raspberries may be grown without a trellis, but a better crop will be obtained with less damage from fungus diseases if the raspberries are confined to a row 6 to 9 inches wide. The canes are secured to a vertical trellis with the bottom wire at 36 inches and the top wire at 60.

Red raspberry normally is pruned in late winter. Remove old fruiting canes and new canes which are damaged or weak. Leave only 2 to 4 robust canes, preferably 2 canes per foot of row. Tie these canes securely to each wire.

Sometimes red raspberry is grown in a hedgerow 18 to 24 inches wide. A horizontal trellis may be used with wires 36 to 42 inches above the ground. Frequently a trellis is not used and the fruiting canes are pruned to a height of 36 to 42 inches so they stand upright when loaded with fruit. This is a cheap and simple system but the plants are subject to fungus diseases.

Boysenberry, dewberry, thornless blackberry and youngberry are usually trellised. Fruiting canes are separately wrapped around the wire and tied along the bottom and/or top wire of the trellis. Allow new canes to grow along the row on the ground during the first year. Remove old canes after fruiting.

Currants and gooseberries are perennial bushes. Usually you remove canes or branches after the fourth year. A mature bush might consist of 12 to 15 branches under 4 years of age. You usually prune currants and gooseberries in early spring before growth starts.

All these berries may be grown under clean cultivating, sod, or a permanent mulch. The author prefers either clean cultivating or the mulch. The average homeowner has many organic residues around the home such as lawn clippings, leaves, or shredded vegetation. A raspberry planting is an ideal place to use these materials.

If you grow raspberries under clean cultivation, the area between rows is cultivated to a depth of 1 to 2 inches at intervals of 2 weeks from early spring to end of the harvest season. This controls weeds and raspberry suckers in the row. If you use sod culture, mow the area between the row like a lawn throughout summer to control growth of weeds, grasses and suckers. Where a permanent mulch is used, mow at timely intervals to control raspberry suckers between the rows.

Highest yields will be obtained with permanent mulch. Clean cultivation is the next highest in yield. Sod usually results in the lowest yield, but in some cases is the easiest system for a homeowner to maintain.

Two Cautions

A key to cane fruit is a weed-free planting given tender loving care the first half of the season and then somewhat neglected the second half after the berries are harvested. Serious mistakes are using too much nitrogen fertilizer, and cultivating after the last of August.

Cane fruits respond to timely irrigation during periods of drought. Water may be applied by sprinklers, soaker hoses, or trickle irrigation.

Sprinklers are easier to operate and faster, but if the foliage is wet for extended periods the plants are more susceptible to certain fungus diseases. It is suggested the homeowner irrigate by soaker type hoses or trickle irrigation.

Probably the best system is trickle irrigation, which applies a few gallons of water a day. It requires very little water at very low pressure. With trickle irrigation the water is applied along the row at intervals of 18 to 24 inches. It does not wet the entire surface.

Most people recognize the importance of irrigation between bloom and harvest for the fruiting canes. But irrigation may be needed from May to September for adequate growth and development of the primocanes which are next year's fruiting canes. A permanent mulch controls weeds and reduces water loss by evaporation from the soil's surface.

A raspberry planting usually responds to 1 to 2 pounds of 10-10-10 fertilizer or equivalent applied per 20 feet of row in early spring before start of growth. If you use less nitrogen, cease cultivation by mid-July and allow weeds, grass or cover crop to grow between the rows. The plants then are less likely to be susceptible to low winter temperatures.

In a very cold area you can protect canes and buds from winter injury by bending the canes to the ground in late fall before the ground freezes, covering the tips with soil. This keeps the canes below the snow line. Or you can cover canes with mulch. The author was able to grow Thornfree blackberries at Ithaca, N.Y., and had a beautiful crop after a winter of −18° F during the coldest night.

When canes are given winter protection by covering with mulch, how do you know when to uncover them in spring? Wait until warm weather —but uncover them before the buds start growth.

Harvest red raspberries when the fruit is ripe, usually about the end of strawberry season. There are early, mid-season and late varieties of raspberries. A given variety will produce fruit over a 17- to 25-day period. Blackberries ripen after raspberries.

If you laid out a good variety planting of red, black or purple raspberries; blackberries and thornless blackberries, and fall-bearing raspberries, in some States you can have fresh berries for your table almost daily from strawberry season until early November.

Currants and gooseberries are ripe when soft, well colored, and tasty. In making jelly, some people like tart berries which are not fully ripe. Others prefer berries that are fully ripe and high in sugars. Few people are aware that immature gooseberries, "hard as marbles," make a tastier pie than ripe berries.

Yields per plant will differ greatly depending on variety and cultural

Photos by Paris Trail

Top left, Jewel black raspberry. Top right, Brandywine purple raspberry. Bottom, Heritage fall-bearing red raspberry.

Left, Poorman gooseberry at time of harvest. Below, White Imperial currant.

care. Raspberries and blackberries should average a quart or more of fruit per plant. Thornless blackberries and boysenberries may produce somewhat higher yields. A mature currant or gooseberry plant should yield 4 to 6 quarts a year.

A raspberry planting may be productive for 7 to 12 years. Currants and gooseberries should be fruitful and productive for 20 years, and in some home gardens the same bush has been productive for 50 years.

Remove any plants that become unproductive. Buy new plants and set them in a different location.

Ripe fruit is prized by birds, mice, and raccoons. If birds are a serious problem, obtain netting with ¾- to 1-inch mesh and cover the planting. An electric fence with wires at a height of 5 and 10 inches is effective in keeping raccoons and similar animals away. Mice are likely to be a problem with a permanent mulch.

Mice and bird problems can be solved in part by owning a cat which travels through the garden. Presence of the cat tends to keep birds away. If mice are a serious problem, apply a prepared mouse bait under the mulch.

It is likely that a few birds and a few mice will have to be tolerated. You might try to grow a little more fruit and share with the birds. Birds often are helpful in controlling certain insects.

Many insects and diseases damage cane fruits. Only a few of these problems are likely to occur in a given area. You might attempt to grow these fruits without pest control treatments until a problem arises, then

George A. Robinson

have the problem identified by the county Extension office or a good garden store.

You can avoid many pest problems by: (1) planting only quality nursery stock; (2) keeping plants well spaced with a narrow wall of foliage well exposed to light; (3) use nitrogen fertilizer at a light to moderate rate; (4) removing diseased or sick plants and all cane after they have fruited, either burning them or removing them; (5) allowing a few birds to live in the vicinity of the garden; (6) replanting with quality stock every 5 to 7 years; and (7) being cautious about accepting a plant from a friend because it may be diseased.

For Further Reading:
Blackberries, Currants and Gooseberries, Ext. Bul. 97, Cornell University, Ithaca, N. Y. 14850. 30¢.
Controlling Diseases of Raspberries and Blackberries, U.S. Department of Agriculture Farmer's Bul. No. 2208, on sale by Superintendent of Documents, U.S. Government Printing Office, Washington, D.C. 20402. 35¢.
Growing Blackberries, U.S. Department of Agriculture Farmer's Bul. No. 2160, on sale by Superintendent of Documents, U.S. Government Printing Office, Washington, D.C. 20402. 35¢.
Growing Raspberries, U.S. Department of Agriculture Farmer's Bul. 2165, on sale by Superintendent of Documents, U.S. Government Printing Office, Washington, D.C. 20402. 25¢.
Shoemaker, James S., Small Fruit Culture, 4th Ed., AVI Publishing Co., Westport, Conn. 06880. $19.
Thornless Blackberries for the Home Garden, U. S. Department of Agriculture H&G Bul. No. 207, on sale by Superintendent of Documents, U.S. Government Printing Office, Washington, D.C. 20402. 30¢.

Japanese beetles are a common garden pest. Top, adult beetle feeds on blackberry leaf. Bottom, Japanese beetles in grub stage.

Just About Any Home Garden Can Produce Blueberries

by G. J. Galletta and A. D. Draper

Nearly everyone knows how great blueberries are in pies, muffins, pancakes, or as a fruit topping for breakfast cereals. However, most people have not experienced the delightful flavor and texture or subtle and delicate aromas of improved hybrid blueberries at the plump and proper stage of ripeness—right from the bush.

Just about anyone in the continental United States should be able to grow prime blueberry plants in the backyard, or as a hedge along the property border.

Blueberries and other perennial fruit plants are often regarded as difficult to grow. But attention to a few facts about blueberries, and timely observance of a few practices, should result in excellent bush growth and fruit yields.

Blueberries vary in nature. Some are vine-like with creeping branches and some spread by underground rhizomes (prostrate stems). Plants may vary in height from two to eight feet. Plant shapes range from semiupright to upright free-standing. Fruit is borne on perennial stout stems called canes.

Most cultivated blueberries are hybrids between two or more of the taller growing species and are referred to as "highbush blueberries." In the Southern United States it is also possible to grow selected hybrids of the very tall-growing blueberry species known as the "rabbiteye" or Southern highbush blueberry.

Blueberries have a mass of delicate and fibrous roots with no root hairs. The root system is confined to the

---

Gene J. Galletta and Arlen D. Draper are Small Fruit Breeders at the Fruit Laboratory, Agricultural Research Service, Beltsville, Md.

upper foot of soil, usually the upper 8 to 10 inches. The root system is not a strong competitor, consequently blueberries thrive on open porous soils with a high water table and good drainage.

Like most other plants of the Heath family—heather, azalea, cranberry, and rhododendron—blueberries grow best in acid soils (pH 3.5 to 5.0, with 4.5 about optimum).

Many acid soils are light and sandy. Nutrients and water are generally held on the organic matter fraction of such soils. Consequently, heavier garden soils such as clay loams would need a fair amount of organic matter added to the blueberry planting area to increase the acidity, porosity, drainage and tilth of the soil.

Important aspects of blueberry culture are an acid soil requirement and a need for full sunlight for best flower bud development. For these reasons, never plant blueberries in an area which has been limed in the last 2 to 5 years, or near trees which shade the bushes, rob them of moisture, or prevent free air movement around the plants.

Free Air Need

Free air movement is especially important in the spring when early morning frosts can kill the flower buds. For this reason avoid low-lying planting sites with poor air drainage and increased frost hazards.

Highbush blueberries should be spaced 4 to 5 feet apart in rows 8 to 10 feet apart. Rabbiteye blueberries should be planted 5 to 6 feet apart in rows 10 to 12 feet apart. For hedgerows, highbush can be planted as close as 2 feet apart and rabbiteye 3 feet apart, but pruning must be

279

more severe to provide the leaves and stems with adequate light.

Most highbush varieties will set fruit with their own pollen. For best fruit set, however, plant at least three varieties in any one location. Cross-pollination tends to produce larger fruits with more seeds which ripen faster and more uniformly than those which have been self-pollinated. Having several varieties in a proper ripening sequence yields fruit over a longer period.

At least two varieties of rabbiteye types must be planted together, since these varieties will not set fruit with their own pollen. The early rabbiteye varieties ripen 4 to 6 weeks after the early highbush varieties in eastern North Carolina.

Generally, highbush blueberries should be planted in areas having 160 or more frost-free days, with 800 to 1,200 hours below 45° F in the winter, and minimum winter temperatures of not less than −20°.

Rabbiteye varieties differ, but generally need 400 to 600 hours of winter chilling (below 45° F). It is not known how many frost-free days are required but the rabbiteye will grow between 30 to 37 parallels of latitude in acid soils at locations having mild winters.

Varietal choices for planting vary from region to region. The following are the better choices resulting from experience in separated test sites. New varieties are marked "trial" or "try".

Regional Varieties

Regional varietal choices for planting:

(1) North Florida, Gulf Coast, Lower Southwest, and extreme Southern California (San Diego and south)

Highbush—try the new Flordablue and Sharpblue.

Rabbiteye—Woodard, Bluegem, and possibly Tifblue; try the new Climax and Bluebelle.

(2) Coastal Plain of Georgia, South Carolina (South of Charleston), Louisiana, Mississippi, Alabama, East Texas, Southern California (Los Angeles and south)

Highbush—try Flordablue and Sharpblue.

Rabbiteye—Tifblue, Woodard, Southland, Delite, Briteblue, Climax, Bluebelle, Garden Blue.

(3) Mountain and Upper Piedmont of region 2

Highbush—Morrow, Croatan, Harrison, Murphy, Bluetta, Patriot, Bluecrop, Berkeley, and Lateblue.

Rabbiteye—same as region 2.

(4) Richmond, Va., south to Piedmont and Coastal Plain Carolinas, Tennessee, lower Ohio Valley, east and south Arkansas, lower Southwest and mid-California

Highbush—Morrow, Croatan, Harrison, Murphy. Try Bluecrop and Patriot except in coastal plain areas.

Rabbiteye—Tifblue, Woodard, Homebell, Southland, Garden Blue, Menditoo; try Climax and Bluebelle.

(5) Middle Atlantic States, Midwest, Ozark highlands, mountain areas of region 4, northern California, Oregon, Washington

Highbush (in ripening sequence)—Bluetta, Collins, Patriot, Bluecrop, Blueray, Berkeley, Darrow, Lateblue, Elliott; (also older clones like Ivanhoe, Pemberton, Burlington and Dixi are good home garden types. Additionally, Herbert and Elizabeth have high dessert quality).

(6) New England and cooler areas of Great Lakes States

Highbush—Bluetta, Collins, Patriot, Bluecrop, Blueray, Meader, Berkeley and Northland.

Many poorly drained soils, especially in the South, are infested with a root-rotting fungus called *Phytophthora cinnamomi*. Almost all highbush blueberry varieties are suscep tible to damage by this fungus. An exception is the new resistant variety from the U.S. Department of Agri culture and the Maine Agricultural Experiment Station, named Patriot.

Most rabbiteye varieties are tolerant to the root-rotting fungus. It is also possible to buy plants of highbush varieties grafted on suitable invigorating rabbiteye rootstocks like Garden Blue or Tifblue. Normally such plants are grafted to order, are more expensive, and need to be pruned differently than own-rooted plants.

Transplants

For most successful transplanting, order certified or State-inspected 2-year-old nursery plants in the 12- to 24-inch sizes. Three-year-old plants of up to 36 inches in height also transplant well. Plants older than 3 years or less than 2 years may die from water stress during the first growing season.

Before or shortly after transplanting, prune the plants to 3 or 4 strong shoots well spaced around the crown (stem base). Prune back each of the remaining shoots to remove the plump, rounded fruit buds. Cut to just above a vegetative bud (narrow and pointed bud), located preferably toward the outside of the stem.

In setting or planting, dig a hole 12 to 18 inches deep and 18 inches wide.

If your soil has a good organic matter content and the proper acidity (pH 3.5 to 5.0), thoroughly mix the soil taken from the hole and replace it along with the plant, setting the plants 1 to 2 inches deeper than they grew in the nursery. After planting, firmly press the soil around the plant with your feet, and water the area thoroughly.

Mulch the blueberry plants for at least 18 inches around the plants in all directions with 4 to 6 inches of well-rotted sawdust, peat moss, pine bark, pine straw, leaf mold, etc. Grass can be permitted to make a sod for walkways at the edge of the mulch if it is mowed during the growing season.

If you have a mineral soil low in organic matter, mix 1 cubic foot (2 to 3 shovels) of peat moss, rotted sawdust, or screened pine bark with the soil removed from the planting hole. Make sure the organic soil mixture is placed below, around and above the roots.

Sulphur can be used to make soil more acid. Add 1 pound per 100 square feet for sandy soils and 3 to 4 pounds per 100 square feet for loam soils, and work the sulphur into the soil before planting.

In areas with mild winters, you can either plant blueberries in the fall or you can plant fully dormant plants in the spring as early as the ground can be worked.

If blueberries are unmulched, cultivate them with a tined rake, fork, or hoe no deeper than 2 inches so as not to damage roots near the soil surface. Primary reasons for tilling the soil are to improve aeration and control weeds. Around mulched plants, either pull weeds by hand or gently hoe them out. Replenish the mulch at the rate of 2 inches per year.

Watering

The first year following transplanting of the blueberry plant in the field or garden is critical. The young plant is very sensitive to drying out, overwatering, fertilizer level, and weed competition.

An inch of water per week through the growing season is usually considered essential for maximum growth and fruiting of blueberries. Water is especially critical just after setting the plants, through the first two growing seasons, and at the time of flowering and fruiting.

Water can be applied from an open hose, porous hose, by sprinklers, or by surface irrigation.

Since blueberries bear their fruit

on new shoots produced during the previous growing season, it is essential to balance growth and fruit production with proper fertilization and pruning.

The purpose of fertilizing and pruning the plant during the first 3 to 4 years is to establish a number of well-spaced, stocky canes bearing many branch shoots with 6 to 12 flower buds each. During the plant's mature years, the cultural objectives are to keep the plant from getting too tall; to keep the canes branching freely; and to keep the plant producing a modest supply of new renewal canes.

Certain fertilizer salts, such as lime and chlorides, should be avoided in feeding blueberries. Don't fertilize blueberries at planting, but fertilize lightly 4 to 6 weeks later. Blueberries generally respond to nitrogen in the ammonium form, and at least once a year it is wise to apply a complete fertilizer (nitrogen-phosphorus-potassium) in a 1-1-1 or 1-2-1 ratio, such as 8-8-8 or 5-10-5. This applies especially to the East Coast. In Michigan a 2-1-1 fertilizer is preferred. For the home garden, azalea and camellia fertilizer mixes are satisfactory.

One ounce of complete fertilizer per year of plant age up to a maximum of 8 ounces per plant per year for mature plants is a good rule of thumb for fertilizing blueberries. During the first two years, the fertilizer is split and applied several times through the growing season.

From the third year on, apply the complete fertilizer just as the flower buds are breaking. Apply an ammonium nitrogen fertilizer about 6

Photos by William E. Carnahan

Top, spraying to control insects (note attire for protection against spray. Gloves also are a good idea.) Right, netting spread over frame will keep birds from eating your berries.

weeks later. Occasionally, an additional application of nitrogen fertilizer is made in early summer if the plants are quite yellow after the fruit harvest. Bear in mind that blueberries are very sensitive to excess fertilizer.

Broadcast fertilizer evenly around the plant. Extra nitrogen should be added to mulched plantings to help decompose the mulch in early years of the planting, but the plant's age and vigor in relation to the quantity of fertilizer applied must be kept in mind.

Pruning

Blueberry plants need not be pruned. However, on unpruned plants the twig growth will get thinner, the branches will shade each other out, the fruit will be very small, and the plant will die sooner than expected. Hence, it is suggested that blueberries be pruned annually during the dormant season. This will lengthen bush life, produce strong new growth, space the bearing wood evenly for best light distribution, and reduce the crop so as to increase berry size and regulate ripening time.

Prune lightly in the first two years to remove low branches, overlapping branches, and flower buds.

From the third year on, remove old canes that are weak or being shaded. Cut back very vigorous upright shoots to force branching at a lower level. Prune out overlapping canes and branches. On the remaining canes, remove the short weak shoots and tip back long shoots to about 6 to 8 buds. Reduce very heavily branched canes by a third.

Select several renewal shoots around the plant and cut them back to 12 to 18 inches if they come from the ground, and 4 to 6 inches if they arise as a branch on an older cane.

Stem borers and leaf and stem spotting fungi may prove troublesome to young blueberry plants. On bearing-age plants, problems may be caused by leaf chewing insects, bud mites, stem and leaf fungi, and fruit worms. Your county Extension office can suggest appropriate controls for these problems, and recommend varieties resistant to some pests.

Many species of birds are especially fond of blueberries. The berries can be protected with a variety of cloth barriers, nets, or cages during fruiting.

Weekly harvesting of the fully colored and plumpest berries is necessary to get the maximum flavor and fruit-keeping quality. Many rabbiteye and several highbush varieties do not attain prime flavor until they have been fully colored on the plants 5 to 10 days.

Blueberry plants will live 25 to 30 years, with at least 10 to 15 prime bearing years. Mature plants in their prime can be expected to yield 6 to 8 pounds of fruit for highbush varieties and 12 to 15 pounds for rabbiteye varieties. Yields as high as 24 pounds for highbush and 50 pounds for rabbiteye have been reported, but these are rare.

For Further Reading:

Brightwell, W. T., *Rabbiteye Blueberries*, 19 pp., University of Georgia, College of Agriculture, Experiment Station, Res. Bul. 100, Tifton, Ga. 31794. 1971. Free.

Eck, P. and N. F. Childers (eds.), *Blueberry Culture*, 378 pp., Rutgers University Press, New Brunswick, N. J. 08903. 1966. $15.

Johnston, S., J. Hull and J. Moulton, *Hints on Growing Blueberries*, 4 pp., Michigan State University, Ext. Bul. 564, East Lansing, Mich. 44824. 1967. 5¢.

Scott, D. H., A. D. Draper and G. M. Darrow, *Blueberry Growing*, 30 pp., USDA Farmer's Bul. 2254, for sale by Superintendent of Documents, U.S. Government Printing Office, Washington, D.C. 20402. 1973. 35¢.

Nut Crops—Trees for Food, Ornament, Shade, and Wood

by Richard A. Jaynes and Howard L. Malstrom

Trees are planted to give shade, to improve landscapes, and sometimes to produce food for man or wildlife. Nut trees provide all of these benefits. Wherever trees can be grown in the United States there are one or more kinds of nut trees suited for the climate.

Black walnut, pecan, and hickory can reach heights beyond 90 feet when mature. Other nut trees are low and spreading, like Chinese chestnut, butternut, Japanese walnut and heartnut. In outline they may resemble old fashioned apple trees. Almond trees are even smaller, while filberts and chinkapin chestnuts are little more than shrubs. If an evergreen is needed, large seeded nut pines are suitable.

The pecan is the most popular nut tree native to the United States. It is a species of hickory and in the same plant family as walnuts. The pecan grows throughout the southern United States from the Carolinas to Arizona and north, along river bottom land, as far as Iowa in the Midwest.

Pecan trees prefer a rich, deep, well-drained soil, and long, warm growing seasons. In the South the frost-free period extends from 190 to 220 days, but it may be as short as 150 days in the Midwest. Although trees will survive in much of the Northwest and Northeast, the cooler summers don't allow development of full kernels.

Numerous varieties have been selected for their fine nuts and other desirable characteristics. Early maturing selections are required for the Midwest. In the high rainfall area of the Southeast, disease-resistant varieties that need less spraying are preferable.

Besides pecan trees there are several other native hickories. Two, shellbark and shagbark, are notable for their sweet tasting nuts. They are hardier than pecans. Though shellbark and shagbark are not grown in commercial orchards, amateur nut growers propagate and grow several selections having large kernels that separate readily from the shell when cracked. Shellbark hickory does best on lowland and river bottom soil. Shagbark is common on thinner, more acid, upland soils.

Eastern black walnut is our most valuable native hardwood, and the nuts are harvested in huge quantities each year. The nuts' tangy flavor isn't lost even in baking.

Black walnuts, native to the eastern United States, are grown from Nebraska and Texas to southern Vermont and South Carolina. They do best on limestone-derived soils.

Persian (English) walnut is an introduced species. Like almond, filbert, and pistachio, it is adapted to the West Coast. These nut trees all originated in arid, mild climates of Europe, Asia, and the Middle East. The climate of parts of California, Oregon, and Washington is similar to that of their native habitat. Hardy members of the species are grown extensively in home and farm yards in the Midwest and East. They are often referred to as the "Carpathian" strain because many originated in the colder Carpathian mountains of Europe.

Butternut is an extremely hardy, native walnut valued for its nuts in

---

Richard A. Jaynes is Geneticist, Connecticut Agricultural Experiment Station, New Haven. Howard L. Malstrom is Associate Professor, Texas A&M University Research Center, El Paso.

Pecan Varieties

Variety	State of origin	Pollination type*	Relative production	Kernel quality	Disease resistance
		Southeast			
Chickasaw	Tex.	II	excellent	fair	good
Desirable	Fla.	I	good	good	fair
Elliott	Fla.	II	good	fair	good
Farley	—	I	fair	excellent	fair
Kernodle	Fla.	II	fair	good	fair
Mahan	Miss.	II	fair	poor	poor
Schley	Miss.	II	good	excellent	poor
Stuart	Miss.	II	good	good	good
		Southwest			
Ideal	—	II	good	good	poor
San Saba Imp.	Tex.	I	good	good	poor
Sioux	Tex.	II	good	excellent	poor
Western	Tex.	I	excellent	good	poor
Wichita	Tex.	II	excellent	excellent	poor

Variety	State of origin	Pollination type*	Kernel quality	Remarks
		Midwest		
Colby	Ill.	II	poor	Retains foliage late in fall
Fritz	Ill.	II	—	Hardy tree for extreme north
Greenriver	Ky.	II	good	Susceptible to spring frost
Major	Ky.	I	good	Good producer; susceptible to aphids
Perque	Mo.	I	good	Susceptible to aphids, squirrels and birds

* I. Pollen shed before females are receptive. II. Pollen shed after females are receptive. Interplant at least one tree from each group for best pollination.

the colder areas of our north central and northeastern states. Other species of walnut, such as the Japanese walnut and heartnut and their hybrids, are locally important.

Commercial production of filbert (hazelnut) is limited largely to Oregon's Willamette Valley where the European filbert variety Barcelona is the principal selection grown. The American filbert, hybrids with the European filbert, as well as some European varieties are grown in the eastern United States. Eastern filbert blight and cold winters restrict their wider use.

Almonds and pistachio are grown in the Sacramento and San Joaquin Valleys of California and to a limited degree in other southwestern States. Spring frosts, high humidity, and rainfall limit their success elsewhere.

American chestnut was destroyed by an introduced bark parasite in the early half of this century. Blight-resistant Oriental chestnut trees, notably the Chinese chestnut, are widely grown for their nuts. They are hardy and thrive wherever peach trees can be grown.

Chinkapin chestnuts, native to the Southeast, are small trees or shrubs

R. A. Jaynes

that bear small, tasty nuts. The Chinkapin nut size makes them especially suitable producers of wildlife food.

Macadamia is a tropical nut tree of Australia and now an important crop in Hawaii. The tree has met with some success in warmer areas of California and in Florida.

Other trees such as oak, beech, ginkgo and pine, often not thought of as nut trees, can be grown for their nut fruit. Among the nut pines, pinyon pine is native to the arid Southwest, Korean pine is hardy in the Northeast, and Italian stone pine is hardy in the Deep South.

Flowering

Pecans, hickories, walnuts, filberts, and chestnuts have both male and female parts on the same shoot, but in separate flowers. In pistachio the male and female flowers are borne on separate trees. Flowers of these species are not showy and are largely wind-pollinated, except for chestnut which is both wind- and insect-pollinated.

Walnuts and hickories, including pecans, shed pollen about a month after the buds break in spring, while filberts flower during the winter or early spring before vegetative growth begins. Chestnuts, by contrast, flower later, about two months after shoot growth begins in spring.

Many pecan, walnut, filbert, and chestnut varieties or seedlings are unfruitful unless they receive pollen from another tree of the same species. Failure of self-pollination may be due to a difference in time of pollen shedding and female receptivity on the same tree, or a pollen incompatibility. To ensure cross pollination, two or more varieties need to be present in a planting.

Almond—related to peach, plum, and other stone fruits—and macadamia have colorful flowers and are pollinated by honey bees, which are usually prevalent in residential areas. Macadamia is self-fertile. Almond flowers are self-incompatible, so two or more compatible varieties must be selected for a planting to insure pollination.

Numerous superior pecan trees have been selected from native groves. Recently, outstanding varieties have been developed from the breeding program sponsored by the U. S. Department of Agriculture at Brownwood, Tex.

Newer selections of Persian walnut, like pecan, are considerably improved over older varieties. Many of the new Persian walnut varieties bear on lateral branches or spur shoots and are capable of producing more nuts. They have also been selected for hot or for cool climates. Characteristics of several of the older and newer varieties grown on the West Coast are given in the second table.

Developing burs on a Chinese chestnut tree.

Persian Walnut Varieties for West Coast

Old varieties	Danger of spring frost damage	Relative production	Kernel quality	best adapted to
Eureka	moderate	good	excellent	cool
Franquette	none	poor	good	cool
Hartley	slight	good	good	hot
Payne	great	excellent	excellent	cool
New varieties				
Amigo*	slight	good	fair	cool
Chico*	moderate	good	good	hot
Gustine	moderate	excellent	excellent	hot
Lompoc	moderate	good	good	cool
Midland	slight	good	good	cool
Pioneer	slight	good	fair	hot
Pedro*	slight to none	good	good	cool
Serr	moderate	good	excellent	hot
Tehema	slight to none	excellent	good	hot
Viva	moderate	excellent	excellent	hot

* Good pollen producers for cross pollination with other varieties.

Many varieties of Carpathian walnut are adapted to the Midwest and East. These include Colby, Hansen, Lake and Metcalfe. Unfortunately, such selections are not readily available from nurseries.

The most widely planted almond variety is Nonpareil. Mission, Ne Plus Ultra, and Peerless are often used to pollinate it. However, the newer varieties—Davey, Karpareil, Merced, and Thompson—are also good pollinators for Nonpareil, and produce better quality kernels.

Barcelona is relatively resistant to most insects and diseases but the nuts are poorly filled. Several new hybrids yield more kernel per nut but produce smaller crops. In Washington, Noosack is commonly grown with Alpha and DeChilly as pollinators. Royal is adapted to colder areas along with Gem or Hall's Giant.

Availability of varieties of walnuts, butternuts, heartnuts, chestnut, filberts for the East, and shellbark and shagbark hickories is like that for Carpathian walnuts. Outstanding trees selected for vigor, production, nut size, and good cracking have been named and propagated, largely by amateur nut growers, but few nurseries list them.

Location, Spacing

Pecans and the large walnut trees should be spaced at least 40 to 50 feet from buildings, trees, and other obstacles if they are to remain a long time. Trees on poor soils with inadequate water will reach only half to three-quarters the size of those growing under good conditions. In areas of late spring frosts, plant on the north side of buildings to delay bud break in spring. Or plant on the upper portion of slopes to avoid frost pockets.

Smaller filbert and almond trees may be spaced about 25 feet apart. Almond is drought-hardy and tolerates poor soil, but it must not be exposed to late spring frosts. Filberts have shallow roots and can be grown on relatively shallow soils.

Chinese chestnut trees grow to about 40 feet and should be planted about 40 feet apart. They prefer an acid soil, pH 5.5, in contrast to the various walnuts which generally perform better on less acid soil, pH 6.5.

All nut trees should be planted when dormant after leaf fall and before leafing-out in spring. If the roots of a dug tree are allowed to dry, the tree will probably die. Buy young trees from reputable nurseries.

A narrow, deep hole is required to accommodate the pecan's tap root; roots of other nut trees spread out more as a rule. The hole should be large enough so tree roots are not twisted and folded back.

Once in the hole with soil filled back in, the tree should be gently lifted a little so the roots point down. Final depth of the tree in the soil should be the same as in its former location, and can be determined by the different bark color at the old soil line. Apply water to settle soil around the roots and prevent the tree from drying out.

For the first year after planting, the goal is to keep the trees alive. Ample soil moisture is the most critical factor. An inch of water per week by rain or irrigation is adequate. Excessive daily watering may waterlog the soil and kill a tree as readily as lack of water.

Maintenance of an area around the tree base free of weeds and sod will maximize tree growth. A mulch or herbicides will assist in controlling weeds and conserving moisture.

Pruning can begin the winter after the first summer of growth. Large trees like pecan and Persian walnut should be trained to a modified central leader (main trunk) rather than an open vase. A tree of this type has 5 or 6 main branches radiating from the trunk, beginning at a height of 5 feet to prevent limbs sagging to the ground. The central leader system gives greater strength and results in less limb breakage. After the first 5 years, when the tree is shaped, little pruning need be done except to thin crowded or dead branches.

As trees mature, prune out crowded branches. But don't cut back the terminal portion of pecan twigs, because these bear the fruit. The terminals can be pruned back on most walnuts because they bear on lateral twigs. If the variety does not fruit well on lateral branches, cut the main branches about a quarter of the way back.

On older pecan and walnut trees where seasonal growth is only a few inches, many small cuts may be necessary to thin fruiting wood and open up crowded areas of the tree. This type of pruning will stimulate new growth and rejuvenate the fruiting wood.

Filberts commonly form suckers at the base and grow in bush form. In the Northwest, suckers are removed to maintain single-stemmed trees. In the East, filberts grow best as multi-stemmed shrubs, but thin suckers constantly to maintain tree vigor.

Fertilizer

Shortly after planting, a handful of 10-10-10 can be broadcast around the tree, but not in the tree hole. Excessive fertilizer or a heavy narrow band can result in damage or even death of young trees. The second year and thereafter, add 1 to 4 pounds of 10-10-10 per inch of trunk diameter during winter in the South or early spring in the North. Nitrogen is the most commonly lacking nutrient; deficient trees have weak growth, pale green foliage, and small leaves. In Western areas, walnuts can suffer the effects of many nutrient deficiencies and even excessive toxic amounts of some elements.

Potassium deficiency is common

where topsoil has been removed by land leveling, construction, or where the soil is extremely sandy as in arid western areas and where excessive nitrogen has been applied in the southern United States. Boron, manganese, iron, copper, and magnesium can also be deficient.

Many States have Cooperative Extension Service laboratories, which will analyze soil and plant tissue samples. Consult local agricultural authorities for advice on taking leaf or soil samples and interpreting results as regards fertilizer needs.

Production of good nuts often depends on pest control. An itemized list of potential spoilers would be long but for each tree species there are only a few notable insects or diseases expected. Consult local agricultural authorities on identifying pests, control measures, and current pesticides registered for use.

The most common disease of pecan trees in humid areas is scab, which affects leaves and nut shucks and can ruin the crop. Crown gall, a bacterial disease which forms tumors at the base of the tree trunk, is also prevalent in some locations. Aphids, mites, spittlebug, and leaf casebearer all feed on foliage. Shuckworm and pecan weevil, as well as spittlebug, are the insects most damaging to nuts. Most pecan varieties have little resistance to scab and no resistance to the above insects. Shagbark and shellbark hickory are attacked by the same pests as pecan.

One of the most serious pests of Persian walnut in the West is the navel orange worm. The larvae eat the kernal while the nuts are still on the tree. Harvest nuts early and clean up nut husks, leaves, and dead limbs to reduce damage from this pest. The maggot of the husk fly, and codling moth larvae, can cause early season destruction of the immature kernel or stain the kernel in late season. The walnut weevil or butternut curculio may also attack the nut, but of greatest consequence is larval injury to stems and branches.

Walnut blight, caused by bacteria, is found on all walnut species, but the Persian walnut seems the most susceptible. The disease rarely kills a tree but infects trunk, limbs, shoots, and leaves, spreading to the nuts— which are destroyed. The bacteria overwinters in dead twigs infected the year before, so complete sanitation practices will help eliminate the pest.

Walnut anthracnose or leaf blotch is caused by another bacterium. Symptoms include defoliated trees and unfilled, deformed nuts. Fungicidal sprays are the usual control.

Another serious problem with all walnuts and some pecan varieties is bunch disease. Symptoms include dieback, stunting, and brooming of growth. The causal agent is a mycoplasm. Control measures are unknown.

Brown rot and shot hole fungus are two serious diseases of almond fruit. Bacterial canker is also prevalent during rainy spring weather and affects blossoms and young shoots. These diseases can be controlled with dormant sprays and fungicide applications in early spring. Insects such as brown almond mite, peach twig borer, and plant bugs can be controlled with insecticides. Red spider mite, as with other nut trees, may occur in late summer, especially under dusty, dry conditions. It can be controlled by spraying.

Insects and diseases have not been a serious problem with filberts in the Northwest. But in the East, eastern filbert blight and filbert bud mite have limited the successful growing of filbert as yard trees. Effective control measures have not yet been developed.

Chestnut blight fungus, the disease that destroyed the American chestnut, may also attack the more resistant Chinese chestnut trees. Pruning weak, shaded branches, as well as suckers, and keeping trees in vigorous condition helps control blight damage on resistant trees. Chestnut weevils (two species of curculio) infect ripening fruit, and feeding larvae cause great damage to nuts. These weevils can be controlled with sprays.

In areas where particular pests cause serious damage, yearly preventive sprays are warranted. Other pests may be controlled by careful observation, taking control measures when you see the pest or damage. Spraying for one pest often may limit other similar pests present at the same time.

Harvest, Storage

Ripening of nuts occurs from August to November, depending on the species and variety. With the exception of chestnut, most nuts have a high oil content and long shelf life. However, harvest nuts immediately after they fall from the tree, especially where there is rodent predation, rainfall, high humidity, or hot weather. Ripe nuts remaining in trees can be knocked off with poles.

Husks of pecan, shagbark and shellbark hickory, chestnut, and Persian walnut open and fall off when the nut is ripe. The husks of black and other walnuts have to be removed. All nuts should be air-dried before storage. Nuts, especially pecans, keep longer if left in the shell and refrigerated at 35° F. Shelled nuts keep well frozen.

Chestnuts, with their low oil and high carbohydrate content, have special needs and should be refrigerated at 35° to 40° under high humidity shortly after harvest. One method is to take the freshly harvested nuts, that have a high water content to begin with, mix them with slightly damp or nearly dry peat moss, and refrigerate in closed plastic bags.

Because so many selected varieties of nut trees are not readily available commercially, amateur growers may decide to do their own propagating. Trees can be grown from seed, but seedlings seldom produce as good nuts as named varieties. However, seedlings can be converted to a named selection by grafting or budding a short stick (scion) or bud of a selection to them.

Techniques used to propagate selected varieties are explained in numerous pamphlets and books. The rootstock and scion are usually of the same or closely related species.

Most temperate climate nuts germinate best if they receive a cold treatment. This may be supplied by refrigeration at 40° F, or simply by sowing the nuts in the fall and letting mother nature furnish the moist cold treatment during winter. Where temperatures well below freezing are expected, a straw or similar type mulch should cover nuts in winter. Protection from rodents, such as wire screening, may also be needed.

How often when the suggestion is made to plant a nut or fruit tree does the answer come back, "Why bother? By the time the tree bears, I'll be gone." Such a response shows an unfortunate short-sightedness and lack of concern for those who come after us. If Johnnie Appleseed or so many others had thought this way, the world would be the poorer for it. The tree planter leaves a valuable legacy behind him, not only in a harvest of wood but in valuable food sources of man and wildlife and a better place to live.

For Further Reading:

Aldrich, T. M., D. E. Ramos, and A. D. Rizzi, *Training Young Walnut Trees by the Modified Central-Leader System*, California Agri. Ext. Leaflet 2471. Copies available from University of California, Berkeley, Calif. 94720. 1975. Free.

Brison, F. R., *Pecan Culture*, Capital Printing Co., Austin, Tex. 79787. 1974. $10.50.

Subtropical Fruit Choice Wide— From Avocado to Tamarind

by Robert J. Knight, Jr., and Julian W. Sauls

Subtropical fruits number over a hundred and range from avocado and citrus to soursop and tamarind. Thus the information in this chapter has to be general rather than specific and you should seek more detailed advice from your county Extension office. Nurserymen can be most helpful too, as can experienced gardeners, garden clubs, or other specialized groups such as the rare fruit organizations active in Florida and California.

Fortunately, cultural requirements for most subtropical fruits do not differ from those of other shrubs, trees or vines grown for fruit, ornament, shade or other specialized use. In fact the outstanding ornamental value of most fruit crops fits them admirably for a dual role.

Because they come from many parts of the world with varied environmental conditions, subtropical fruit crops differ in the degree of cold they can withstand and in soil and moisture requirements. Some are adapted to warm-temperate conditions, and others are tropical plants which will tolerate brief cold spells and thus survive winters in the warmest parts of the continental United States.

When you select fruits to plant around your home, choose those known to grow well in your locality. Otherwise you may go to a lot of trouble to grow a plant that may prove disappointing despite your best efforts. Your own personal preferences, tempered by knowledge of what grows well in the area, should determine what you plant.

Most tree fruits should be planted 12 to 20 feet apart and away from the house, walks, drives, and power lines. Those tropical fruits listed as small trees or shrubs at the end of the chapter can be planted somewhat closer. Where there is significant danger of cold damage, plant subtropical fruits in the warmest part of the yard, which generally is the south side of the house.

Most failures in growing fruit trees at home can be attributed to poor transplanting or poor care. Commercial fruit growers routinely transplant fruit trees with almost no failures.

Good preparation of the planting hole is essential. Dig the hole only as deep as and about a foot wider than needed to accommodate the root system. Regardless of your soil type, it would probably benefit from the addition of liberal amounts of organic matter such as rotted manure, compost or peat.

For bare-root plants, prune off dead or damaged roots. Make a cone of soil in the center of the hole and set the plant on it, carefully spreading the roots out in the hole. For container-grown plants, remove the container and set the plant in the hole.

In either case, set the plant at the same depth it was growing in the nursery or container.

Fill the hole three-fourths full of soil. Then fill with water to settle soil around the roots and eliminate air pockets. After the water drains through, finish filling the hole with soil, then water again. A ring of soil a few inches high around the planting hole can be used to form a watering basin during the first year.

Robert J. Knight, Jr., is Research Horticulturist, Agricultual Research Service, Miami, Fla. Julian W. Sauls is Extension Horticulturist, University of Florida, Gainesville.

At planting, bare-root fruit trees should be pruned to balance the top with the reduced root system, which requires removing about a third of the top. Most people are reluctant to prune this heavily, but it's for the good of the tree. If you're unsure about doing the job right, have the nursery where you bought the tree do it for you. Container-grown plants are not usually pruned since they have an intact root system.

Training

Initial training of the fruit tree is done at planting to assure that the tree takes the desired shape. For example, the growing tips of branches are pruned off to force branching. Even so, most subtropical fruit trees are not trained appreciably, but simply allowed to develop naturally.

Mature trees are pruned to remove dead or damaged wood, or to eliminate limbs that may interfere with traffic in the yard. Such pruning can be done at any time of year.

Pruning cuts should be clean and close to the trunk to avoid leaving stubs which enable wood-rotting organisms to enter the tree. Protection with pruning paint is recommended if the cut is larger than an inch or so in diameter.

Subtropical fruit trees are not in such prominence that they require "special" fertilizers as yet. You cannot run down to the garden center and pick up a bag of "Kiwi Special" or "Atemoya and Cherimoya Food". Fortunately, subtropical fruits will grow just as well on a complete, balanced garden fertilizer such as 6-8-8, 10-10-10, or 12-12-12. However, if your area has alkaline soils or soils known to lack specific micronutrients such as iron, manganese or zinc, these may need to be supplied.

Newly planted trees should not have fertilizer until they resume active growth after transplanting. Then, fertilize sparingly and frequently until they mature and begin to produce fruit. Using 10-10-10 as an example, young trees should receive about a pound of fertilizer per year of tree age, that is, 1 pound in the first year, 2 in the second, and so on. Total fertilizer for the year should be divided into several applications so young trees receive some fertilizer every 2 to 3 months.

Mature, bearing trees can be fertilized at double that rate, or 2 pounds per year of tree age. Thus, a 10-year-old tree would receive 20 pounds per year, which would be split into 3 applications—early spring, early summer and early fall. Fertilizer can be spread on the ground under the tree and then watered in.

Lime may be needed in some cases to raise the soil pH so it is suitable for optimum tree growth. However, liming should be based on a soil test and recommendation from the county Extension agent.

Occasionally, some fruit trees may need certain micro-elements, particularly in very sandy soils or alkaline soils. Micro-elements are included in some fertilizers and are also available in nutritional sprays which are applied separately as foliar (leaf) sprays. In all cases follow recommendations of your county agent.

Mulches

Mulches around fruit trees help in weed control and water conservation. They also reduce lawn mower damage to tree trunks since you don't need to mow close to the trees.

In some cases, organic mulches can lead to fertilizer deficiencies as the micro-organisms that decompose them rob nutrients the tree could use. They also contribute to increased cold damage by inhibiting radiation of ground heat to the tree. In other cases, organic mulches increase the incidence of diseases such as foot rot and root rot. For these reasons, we recommend clean cultivation instead of mulches for citrus, avocado, lychee and some other fruits.

Mulches from your yard could include leaves and grass clippings. Or you can obtain sawdust, wood chips, pine bark, gravel and other mulches from local nurseries.

A lot of gardeners don't understand about watering plants. This is one reason why so many fruit trees die shortly after transplanting. Too little water causes the tiny root hairs to die, and the leaves then wilt for lack of water. On the other hand, too much water forces air from the soil, again causing the root hairs to die for lack of oxygen, and the leaves will wilt. For best results, water fruit trees infrequently but thoroughly.

Frequent, shallow waterings cause shallow rooting. A shallow-rooted fruit tree is subject to drought and poor growth. Consequently, when you water, water long and water well. Apply water only as fast as the soil can absorb it and keep watering until the soil is wet at least a foot down.

Newly transplanted trees need a good soaking every 2 to 4 days until they are well established. Mature trees need water every 7 to 12 days, depending on the climate and soil type. Since sandy soils don't hold much water, they require watering about once a week, while clay soils will go several days longer before drying out.

Fruit trees growing in the lawn area will compete with the lawn for fertilizer and water. In such situations pay particular attention to needs of both tree and lawn. The tree will compete much more aggressively than the grass. The grass will soon begin to thin out and may disappear completely once the tree begins to create heavy shade.

Cold protection often is required for many subtropical fruits. Young trees are more susceptible to cold than large, mature trees, but also easier to protect. Banking a mound of soil around the trunk of a young fruit tree will keep the rootstock and trunk alive even if the top should freeze. Pull the bank down in spring after cold danger is past.

Small trees can be covered with blankets, paper or plastic to prevent freezing. Lawn sprinklers have been turned on trees, but too much water can cause problems for the root system and ice can cause limb breakage. In some cases, a frame covered with clear polyethylene can be built around the tree to form a mini-greenhouse. Some slow-burning heating materials are available and work quite well; check with your county Extension agent or nurseryman.

Most subtropical fruits have enough insect and disease problems to make growing them troublesome at times. You need to learn the potential pest problems and how to control them. To do this requires a little effort on your part in order to be able to recognize the damage before it becomes serious, identify the insect or disease responsible, and take effective remedial action before the damage progresses too far to control.

Containers

Many subtropical fruits can be grown in containers in areas where freezes occur each year. The size and mobility of the containers allows the plants to be moved indoors during winter months. Thus, the plants are treated pretty much as houseplants with regard to water, fertilizer, humidity, light, and pest control.

As with houseplants, water container plants infrequently but thoroughly. Take care to acclimate the plants to the different conditions when they are moved outdoors in spring or indoors in fall. Plants going outdoors should be moved to a shady spot for a couple of weeks before being exposed to full sunlight. Reverse this process when moving them indoors in fall.

When plants are indoors, put them in areas receiving the most natural light possible. Keep them away from heaters, doors and heating ducts. Because of lower humidity indoors, you need to increase the humidity around the plants, by misting or other means.

Growing plants in containers or patio tubs will reduce plant size due to the reduced volume of soil in which they're growing. Even so, the plant may soon grow too large to bring indoors. When this happens, prune back the plant severely.

Following is information about some fruits that can be grown in many parts of the Southern and Southwestern States. The letters **Wt** (Warm-temperate), **St** (Subtropical) and **T** (Tropical) are intended to give an approximation of temperature requirements of each species. However, other factors, such as amount of rainfall and the time of year that rain comes, will also determine whether a particular fruit can be grown in your area.

Avocado (*Persea americana*). St. T. Shade tree with rough dark bark suitable for growing bromeliads and orchids. More than one variety should be planted together for cross-pollination. Plant locally adapted varieties. Will not tolerate heavy, poorly drained soils.

Banana (*Musa acuminata, Musa* hybrids). T. Rootstock may survive light freezes. Giant, treelike herb, planted for ornament where cold precludes fruiting. Many varieties have been introduced but the most widely grown are Cavendish (a commercial crop), Apple (sometimes called Ladyfinger), and Orinoco (also called Horse banana and good for cooking). The starchy cooking banana called Plantain is very tender to cold.

Carambola (*Averrhoa carambola*). T. Tree varying from small to large. Characteristic 5-angled fruit of yellow or deep orange color varies from sour to sweet and is pleasantly aromatic. Plant grafted varieties (Golden Star, Mih Tao). Cross-pollination aids fruit set.

Carob (*Ceratonia siliqua*). Wt. Small tree with attractive dark green leaves that prefers a Mediterranean climate, very dry in summer with rains during winter. Trees may be male or female, so more than one should be planted to ensure fruiting. The brown, leathery pods are rich in sugar and furnish a chocolate substitute.

Cattley guava (*Psidium cattleianum*). St. Shrub or small tree with beautiful mottled trunk and glossy dark green leaves. The small, round fruit, bright red or yellow-colored, is subacid in flavor and may be eaten fresh or made into jellies or jams. Plants grow readily from seed and are normally so propagated.

Feijoa (*Feijoa sellowiana*). Wt. Shrub. Compact, cold-resistant and most attractive, selected varieties such as Coolidge fruit well without cross-pollination, but seedlings may not do so. Flowers are edible. Fruit can be eaten fresh, and makes a firm jelly.

Fig (*Ficus carica*). Wt. Small tree. Adapted to a wide range of climates, fig will not tolerate nematodes. Where these are a problem, heavy mulching and occasional application of an approved nematicide, according to prescribed rules, will help. Lemon, Brown Turkey, and Celeste varieties are recommended.

Guava (*Psidium quajava*). T. Small tree. Somewhat weedy unless pruned to shape it, the guava can be attractive, particularly when in bloom. Fruit of some seedlings and selected varieties is excellent for jelly, while that of varieties such as Ruby x Supreme and Indian Red is good to eat out-of-hand. Fruit flies are a problem where abundant.

Jaboticaba (*Myrciaria cauliflora*). St. Shrubbery tree. Grows slowly but where well established produces abundant crops of black, grapelike fruit excellent to eat fresh or use in jellies or wines.

Kiwi, Yangtao (*Actinidia chinensis*). Wt. Vine. Not successful in warmer parts of Florida, this deciduous species is sensitive to nematode damage. Flowers of named varieties (females, for example Hayward) must be pollinated in order to fruit, so a pollinator should grow nearby. Because of its excellent quality this fruit should be planted wherever it can be grown well.

Longan (*Dimocarpus longan*) is a lychee relative that bears clusters of attractive, smooth, golden brown, sweet-flavored fruit that is less tart than lychee fruit. The tree is less demanding as to soil and moisture than lychee, and makes a shade tree of stately proportions. Kohala, from Hawaii, bears large fruit of good quality.

Loquat (*Eriobotrya japonica*). Wt. Small tree. The dark green, deeply ribbed leaves of this tree combined with its tendency to produce fragrant creamy-white flowers over a period of months make the loquat a universally valued ornamental. The excellent fruit quality of grafted varieties such as Wolfe, Gold Nugget (Thales), and Champagne make these worth the effort needed to find them. Fruit is excellent eaten fresh, but may also be made into pie, jam, and jelly.

Lychee (*Litchi chinensis*). T. Tree. Somewhat finicky, demanding slightly acid, well-drained soil, with abundant moisture and no salts in soil or water, this tree covered with its bright red fruit is a sight to remember where it grows well. Long popular in Southeast Asia, the fruit has many American devotees. It may be eaten fresh or dried like raisins. The most dependably productive varieties are Sweetcliff and Mauritius.

Mango (*Mangifera indica*). T. Tree. Of the many existing varieties, take the time to select one that appeals to you: Carrie, Irwin, Glenn, Keitt, and Tommy Atkins are outstanding. Blooming trees can cause allergic reactions; do not plant near bedroom windows or air conditioner intake. The mango is one of the world's most popular fruits.

Passion fruits (*Passiflora edulis*, purple, and *P. edulis* f. *flavicarpa*, yellow). St. T. Vines are ornamental. The purple-fruited form is sensitive to nematodes and soil-borne fungus disease, but withstands more cold than the yellow-fruited form, which is disease-resistant. Self-pollinating types should be planted where possible, otherwise fruit production may be sparse.

Pineapple (*Ananas comosus*). T. Perennial herb. This bromeliad makes an attractive house plant where outdoor temperatures are too low for it. The plant can be moved to a porch or patio during warm weather. Given enough light it will eventually flower, then produce a fruit of fine quality provided conditions are warm enough.

Pomegranate (*Punica granatum*). Wt. Small tree that tolerates extremes of heat and alkaline soils, but thrives under a wide range of conditions. Needs full sun for best performance. Wonderful and Sweet are the varieties best known for their fruit quality. Other varieties are grown primarily as ornamentals.

Tamarind (*Tamarindus indica*). T. Large tree related to the carob, with very acid fruit in pods. Pulp of the tamarind, an essential ingredient of many chutney recipes, also is used to make a refreshing ade-like drink. Where the climate is warm enough for it, this tree is easy to grow.

Annonaceous Fruits

Atemoya (*Annona cherimola* x *A. squamosa*). T. Moderate-sized tree, a hybrid of the cherimoya and the sugar-apple, that combines the excellent fruit quality of the cherimoya with the fitness for low elevations of the sugar-apple. Flowers abundantly in warm weather, but may need to be hand-pollinated to assure fruit set. Desirable varieties are Kaller (African Pride) and Bradley. Others are under test.

Cherimoya (*A. cherimola*). St. Small tree adapted to high elevations in tropical South America, producing a large green fruit with a sweet, delicately aromatic pulp that surrounds many smooth dark seeds. Does not grow well in southern Florida but is more successful in California where it withstands temperatures as low as 25° F.

Soursop or guanabana (*A. muricata*). T. Small tree, very sensitive to sudden cold spells, that bears a large, rough fruit with a refreshing acid flavor that is excellent in drinks and sherbets. Should be planted in a sheltered location.

Sugar-apple (*A. squamosa*). T. Small tree that bears a soft-pulped, many-seeded fruit similar to the cherimoya but without that fruit's fine aroma. Grows well at sea level in southern Florida and other areas of similar climate.

Cactus Fruits

Indian fig (*Opuntia ficus-indica*). Wt.

295

Large treelike cactus with smooth flat joints and few spines. Yellow flowers in spring are followed by large red or yellow fruit. Bristles can be irritating; handle fruit with care. Prefers a dry climate and does not thrive in humid situations.

Citrus

Calamodin (*Citrus blancoi*). Wt. St. Small tree of great ornamental value that grows and fruits well in small containers. The fruits resemble small oranges but are acid and not good to eat out-of-hand. Flavor is excellent for drinks and marmalades.

Grapefruit (*Citrus paradisi*). St. Medium to large-size tree, excellent for shade and for growing orchids and hanging plants, providing up to 300 pounds of excellent breakfast or juice fruit per year. Varieties include Duncan (white, seedy pulp, excellent flavor), Marsh (white, seedless) and Ruby (pink pulp, seedless).

Kumquat (*Fortunella japonica*). Wt. Shrub or small tree, very cold-tolerant, extremely attractive when in fruit. Nagami is the most common variety, with oblong fruit, deep orange in color having a thick edible skin and an acid pulp. Adapted to candy making or use in marmalades.

Lemon (*Citrus limon*). T. Small tree that remains in active growth all year and thus is less cold-resistant than the tangerine or even the orange. Of irregular growth habit, the lemon must be pruned from time to time to promote an attractive shape. Eureka, Lisbon and Villa Franca all bear similar fruit, of acceptable commercial quality; Eureka makes a smaller tree than the others. Novelties are Meyer with a less acid fruit, and Ponderosa, which bears very large, mild-flavored lemons.

Lime (*C. latifolia*). T. Small tree that bears large, juicy green fruit useful in drinks, pies, and as a condiment. The most disease-resistant and dependably productive variety is the seedless-fruited cultivar known as Tahiti, Persian, or Bearss. Less resistant to disease and cold, and bearing smaller seedy fruit of a delectable flavor, is the Key or Mexican lime, *C. aurantifolia*. (A hybrid between the Key lime and the Kumquat, the Limequat produces a valuable acid fruit in areas too cold for the lime itself. Eustis fruits well in the open as well as in containers.)

Orange (*C. sinense*). St. Tree of moderate size, probably the most popular of all citrus fruits, available in a number of varieties that ripen at various seasons. Hamlin is one of the earliest, ripening in November, followed by Pineapple and Washington Navel, which ripen from December to February, and then by Valencia, which ripens in April or later and can be "stored on the tree" into the summer months.

Tangelo (*C. reticulata* x *C. paradisi*). St. Tree, hybrid between tangerine and grapefruit, bearing fruit which combines characters from both parents. Vigorous and cold-resistant. Several varieties are available. Minneola and Orlando need to be planted near other citrus trees for cross-pollination. The Temple tangor (*C. sinensis* x *C. paradisi*) bears a sweet, juicy fruit similar to tangelos.

Tangerine (*C. reticulata*). St. Tree of attractive growth habit, fairly resistant to cold, whose beauty is enhanced by the waxy, deep orange-colored fruit in season. Dancy ripens before Christmas, as does Clementine, which can be "stored on the tree" in good condition for months. Closely related are the cold-hardy and early dwarf Owari Satsuma, which ripens from October to Christmas, and the Kara, Honey and Kimow mandarins.

Persimmons

Black-sapote (*Diospyros digyna*). T. A tropical Mexican and Central American persimmon that grows well in southern Florida. The dark brown pulp is rich in vitamin C, and also a source of calcium and protein. It was important in the diet of Central America before Columbus.

Japanese persimmon (*Diospyros kaki*). Wt. Small tree, attractive even when out of fruit with its large, hairy leaves; highly ornamental when the bright orange-colored fruit is ripening. Trees grafted on *D. lotus* or the native American *D. virginiana* are available. Fuyu bears fruit that is non-astringent even before fully ripe. Fruit of Hachiya and Tane Nashi is astringent until fully ripe, but then delectable. In dry climates, fruit may be sun-dried to make a fine-flavored product.

PART 4

Home Food Preservation

The Whys of Food Preservation
by Edmund A. Zottola and Isabel D. Wolf

The telephone rings in a county Extension office. A harried voice says, "My garden is growing more than my family can eat, what shall I do with it? How can I keep it from spoiling?"

How many times during the gardening season is this scenario repeated in an Extension office? Too often to count! The answers to these questions are readily available in the many bulletins, folders, and leaflets on food preservation available from county, State, and Federal Extension agencies. The publications tell how to preserve food safely and wholesomely, but do little else to explain why directions must be followed precisely. Let's take a look at the whys.

To understand food preservation, first consider the sources. Home garden food comes from plants: sources of raw food are living, biological entities, continuing to metabolize after they are harvested. Plants also provide a source of food for microorganisms which can grow on or in them, spoiling food before it can be eaten. The primary objective of food preservation is to prevent food spoilage by preservng food until it can be used by people.

Historically, food preservation and processing assured a food supply and prevented starvation. This is probably the major reason why food is processed today in many developing countries. In the United States, however, affluence and a plentiful food supply now influence the reasons for food preservation. Today, Americans live many miles from rural areas where food is produced. Consequently, food must be preserved to assure the nonfarm population an adequate supply.

Our people want a food supply that is safe, high in quality and appearance, adequate nutritionally, and reasonably priced. Many consumers try to obtain these food attributes by returning to the "old ways" of growing and preserving food themselves.

To understand food preservation, let's look at five causes of food spoilage or deterioration (four are biological, the fifth physical or mechanical):

(1) The primary cause of food spoilage in the United States is microbiological. Micro-organisms are small living organisms such as yeast, molds, or bacteria. They are the chief causes of microbial spoilage.

Related to microbiological spoilage of food and also a concern in food preservation is microbiological foodborne disease. There are two types. Salmonellosis is an example of a food infection where food may not support growth of the micro-organisms but merely serves to transfer it from the source to the human host. In the second type, the micro-organism grows in the food and produces a poison or toxin which when eaten, causes illness symptoms. Staphylococcal food poisoning is the most common of the second type in the U.S.

Severity of the major types of foodborne disease in the United States varies from the finality of botulism to the mild discomforts of *Clostridium perfringens* food poisoning. Food preservation techniques, followed precisely, prevent food-borne disease.

(2) The second cause of food spoilage is vermin such as rodents, rats, mice and insects that attack the food and eat or contaminate it before humans can use it. These vermin ruin millions of pounds of food each year.

Edmund A. Zottola is Extension Food Microbiologist and Isabel D. Wolf is Extension Specialist, Foods and Nutrition, at the University of Minnesota, St. Paul.

WHAT ELSE CAN WE DO WITH TOMATOES ?!!

(3) Have you noticed how an apple, left at room temperature, eventually gets soft, wrinkles, and dries out? This spoilage is called senescence: an aging process caused by continued respiration of the apple, eventually making it useless as food. Other foods also spoil this way.

(4) Related to senescence is chemical deterioration of food. The development of rancid flavor in high fat-containing foods is a chemical reaction which brings about an undesirable change. Loss of color or bleaching and loss of vitamins, while food is stored, are chemical deteriorations that can be controlled with proper preservation methods.

Both senescence and chemical deterioration are conveyed by organic compounds called enzymes. These enzymes are produced by all living organisms and their function is to speed up or cause the metabolic reaction necessary for the organism's continued existence. The enzymes will continue to act after the plant is harvested, and bring about deterioration of the food unless controlled or destroyed. Preservation methods have been developed to control or destroy these organic catalysts.

(5) The last cause of food spoilage concerns food handling. Physical or mechanical damage to the food causes bruising, crushing, cutting, and wilting or water loss. These mechanical defects, besides detracting from the food's appearance, allow easier entry of micro-organisms, insects, and other vermin to cause spoilage and aging.

Food preservation processes have been developed to slow down, prevent, or stop completely these processes of food spoilage.

An inherent part of food preservation is the package containing the food before or after processing.

Packaging provides a convenient method of handling food, prevents contamination during and after processing, bars vermin infestation, supplies a container for storage, and is a necessary part of preservation. An example would be a mason jar with proper seal for pickling.

What are the major methods available for home preservation of food? How are they carried out? Why do they prevent spoilage, food-borne disease, and give desired attributes of safety, quality, appearance, nutrition, and economy? The economics of food preservation will be developed in a subsequent chapter. Let's explore the following available methods for home preservation of food:

• Control of temperature of the food, heat or cool
• Control of the food's acid content
• Control of moisture content of the food

The major method used for home preservation of food is temperature control. This includes canning with a pressure canner or a boiling water bath, blanching food before freezing, refrigerating food, and freezing it. Micro-organisms which cause disease and food spoilage are sensitive to environment temperature variations. By increasing the food's temperature, micro-organisms are destroyed. When the temperature is decreased, their growth is inhibited.

Let's look at the temperature scale

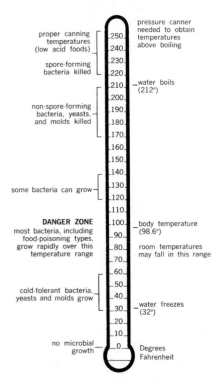

EFFECT OF TEMPERATURE ON MICRO-ORGANISMS

- proper canning temperatures (low acid foods): 240–250 — pressure canner needed to obtain temperatures above boiling
- spore-forming bacteria killed: 220–230
- 212° — water boils
- non-spore-forming bacteria, yeasts, and molds killed: 180–200
- some bacteria can grow: 120–170
- DANGER ZONE — most bacteria, including food-poisoning types, grow rapidly over this temperature range: 60–120
- 98.6° — body temperature
- room temperatures may fall in this range: 70–80
- cold-tolerant bacteria, yeasts and molds grow: 30–50
- 32° — water freezes
- no microbial growth: 0–20

Degrees Fahrenheit

illustration and see how temperature affects micro-organisms. To prevent growth of micro-organisms in food and subsequent microbial spoilage, food must be kept out of the temperature range that allows growth. This is most commonly achieved by refrigerating or freezing. Refrigeration slows down or stops microbial spoilage. Freezing stops it completely.

Enzymatic activity, while slowed down by freezing, is not stopped in many vegetables and these enzymes must be destroyed by blanching the vegetables before freezing to prevent enzymatic deterioration. Successful preservation by freezing must deactivate any enzymes that might be in the food as well as rapidly lower the food temperature to below freezing to stop microbial activity.

Freezer burn, a common problem with frozen foods, comes from improper packaging. Food moisture is lost in freezer burn, which results in undesirable flavor and texture changes. Freezer burn can be controlled by proper packaging, proper storing temperature, and avoiding long-term storage.

Refrigeration or storage above freezing, but below room temperature, preserves food for days and sometimes weeks. Refrigerated storage slows down activities of enzymes in the food and reduces metabolism of the contaminating micro-organisms.

Preservation of food by reduced temperature, refrigeration, or freezing is achieved because enzyme activity and microbial deterioration are slowed down or stopped.

Increasing the temperature of food to achieve preservation also results in destruction of the micro-organisms that produce spoilage and disease. Time and temperature regulate this preservation. Theoretically, since food will be stored at temperatures which will allow most microbes to grow, the ideal heat treatment needed to preserve the food would be one that completely sterilizes the food, that is, kills all attendant micro-organisms. To achieve complete sterilization, for example, every particle of food in a jar must reach the required temperature and be held there long enough to destroy all micro-organisms.

Heat Transfer

The time required for heat to penetrate to the center of the food in a container (the slowest heating point) is extremely important. Heat is transferred through food in containers by two mechanisms: conduction and convection. The mechanism involved depends on the consistency and amount of liquid in the food. The heat penetration rate is also influenced by size of the container, type of heating medium (wet steam vs. dry air), ratio of solid to liquid, kind and size of solid material in container, amount of fat, and amount of salt and sugar.

For example, pumpkin or squash can be home canned in two forms: strained or cubed. University of Minnesota research has shown that the time required for the center of a pint jar of strained squash (which heats by conduction) to reach sterilization temperature is three to four times as long as for a pint jar of cubed squash (which heats by convection). The same is true of creamed corn (heats by conduction) and whole kernel corn (heats by convection).

Methods and recipes recommended by Extension agencies take into account all of these factors and must be followed precisely to assure a safe and wholesome product.

Why is it necessary to heat-process pint jars of string beans at 240° F for 20 minutes in a pressure canner when tomatoes can be successfully heat-processed in a boiling water bath? This brings up the second method of preserving food at home, controlling the food's acid content. This method is most commonly used in combination with heat processing.

Most foods contain naturally occurring organic acids. Some foods contain more of these acids and are called acid or high acid foods. These organic acids have the ability to limit, inhibit, or prevent the growth of many of the micro-organisms producing spoilage and disease. The degree of inhibition is related to the amount of acid present.

A method used for measuring acid content is called pH. A measure of pH is a determination of the hydrogen ion concentration which reflects the amount of acid or alkali present in the system. A scale from 0 to 14 is used. A pH of 7 is considered neutral, above 7 alkaline, below 7 acidic. Very few foods have a pH above 7.

The classification of foods in the acid range below 7 is extremely important. Above pH 4.6 most of the spoilage type micro-organisms can grow, as well as the dreaded *Clostridium botulinum* (see discussion following on botulism). In foods with a pH greater than 4.6, it is necessary to heat-process the food at temperatures above boiling to obtain the desired level of sterility.

There are some types of bacteria that produce entities called endospores or spores which are extremely resistant to environmental stresses. They are a means of assuring survival in bacteria, although not themselves a reproductive mechanism. One growing or vegetative cell will produce one spore, which under proper growth conditions will germinate and produce one cell. This one cell continues to grow and can produce millions of bacterial cells. Destruction of the resistant spore necessitates the use of temperatures above that of boiling water (212° F).

BOTULISM. One of the most notorious of the spore-forming bacteria is *Clostridium botulinum*. When growing in food this bacteria can produce a deadly poison which causes *botulism*, a deadly illness. The mortality rate is 56 percent. This bacteria and its spore are present in soil throughout the world and as a result contaminates most of the food we eat. But the spore only germinates and grows where there is suitable food, no air, and a pH above 4.6. These conditions exist in canned low acid foods.

To assure botulism-free home-canned foods, it is absolutely essential that low-acid foods be canned in a pressure canner at temperatures above 212° F. The poison produced by this bacteria is one of the most potent poisons known to humans. It has been estimated that 1 cup (8 ounces) is sufficient to kill all the humans on earth. It is not something to take chances about. All home canning procedures recommended by Federal and State Extension agencies

pH VALUE OF VARIOUS FOODS

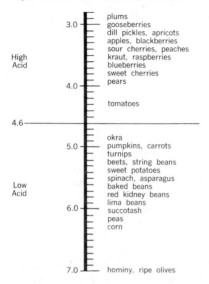

are adequate to destroy this dreaded spore-forming bacteria.

Commercial canneries, which are regulated by the Food and Drug Administration, use similar processes to assure safe canned foods. In the past 50 years, 75 percent of the reported cases of botulism in the United States have been caused by home-canned food while less than 10 percent have been caused by commercially canned food. If proper home-canning procedures are followed, botulism from this source could virtually be eliminated as a cause of death in the U.S.

On the lower side of pH 4.6, acid content of the food will prevent growth of *Clostridium botulinum* and most of the other spore-forming bacteria. Thus, these foods can be preserved by using a lower heat treatment. The most common types of spoilage micro-organisms associated with acid foods are yeasts and molds. These organisms are acid-tolerant and can grow in an acid enviroment. They are killed at a lower temperature than spore-forming bacteria. Acid foods only need a heat treatment in a boiling water bath for a specified time to destroy the microbes present.

Fermentation

Preservation of food by controlling the acid content can be achieved in two ways. One is to naturally ferment the food — turning cabbage into sauerkraut. The other is to add an organic acid to the food to reduce the pH — adding vinegar to cucumbers to make pickles. Some foods such as berries and fruits naturally contain enough organic acids so their pH is below 4.6, and preservation of these foods requires only a boiling water bath heat treatment or freezing.

In a natural fermentation, lactic acid bacteria convert fermentable carbohydrates in the food to lactic acid. In this way the pH is reduced and most bacterial growth inhibited. When cabbage is fermented to sauerkraut, the cabbage's pH is reduced from pH 6.8 during the fermentation to less than 3.5. Cucumbers can also be fermented to pickles by a similar process; however, most pickles are made by direct acidification of the cucumbers.

Direct acidification, that is, adding vinegar which contains 4 to 5 percent acetic acid, is the most common method of making cucumber pickles in the United States. It is easier, quicker, and foolproof. Often the natural fermentation will go astray. Other undesirable microbes may grow, bringing unwanted changes in the food: spoiling rather than preserving. Other foods made by fermentation include wine.

Regardless of the method used to control the pH of food, to successfully preserve food by this method it is absolutely essential to heat-process or freeze to prevent spoilage by yeasts and molds. An example of spoilage in cucumber pickles not heat-treated after acidification is the development of cloudiness and bubbling. This common spoilage is caused by microbes that would be destroyed by heat processing.

Drying

Control of the moisture content of food is one of the oldest preservation methods. Removal of water from the food prevents growth of most microbes and slows down enzymatic deterioration.

Water removal from food can have several forms. The oldest and most primitive method for removing water is sun drying, which requires long hot days with low humidity to dry food evenly and quickly. Over-drying and uneven drying will result in nutrient destruction, microbial growth, and other undesirable changes. Drying of food in the home can be done and a later chapter tells how.

Water activity or availability of water in foods can also be controlled by adding compounds to the food which tie up the water chemically, making it unavailable for use in an enzymatic reaction or for use by the micro-organisms. The two most common home ingredients used in this way are sugar and salt. In making fruit preserves, the high sugar content ties up the water and helps prevent growth of many micro-organisms.

Methods used for preserving food in the home are combinations of the basic techniques discussed here. Make certain the recipe you follow is from a reliable source. Follow it precisely and be assured of a safe and wholesome food supply.

How to Minimize Quality Losses
by Gerald D. Kuhn and Louise W. Hamilton

All methods of preserving food will alter, if not lower, the quality of fresh fruits and vegetables, but these changes can be minimized. For practical understanding of how to minimize quality changes you need to know something about: (1) natural differences in raw food, (2) natural causes of quality deterioration, and (3) how various food handling, preparation, packaging and storage techniques and conditions affect quality retention.

Because of insufficient research information on home preserved foods, some research findings of commercially processed food have been used to foster an increased awareness of factors affecting the quality of home preserved food.

Natural differences in quality and nutrient content in raw food often exceed losses caused by preserving food. Extreme variations (tenfold or more) in vitamin A and C content have been found in some fruits and vegetables. These natural differences are known to be caused singly or collectively by differences between varieties, climate changes between seasons in the same region, and between regions in the same year. They are also influenced by some cultural practices, and maturity of crops at harvest.

Because of color, texture and especially flavor qualities, some varieties are more suited for canning; others are more suitable for freezing. Few varieties possess the all-purpose criteria needed to preserve them as either high-quality canned or frozen products.

Gerald D. Kuhn is Professor, Food Science Extension, and Louise W. Hamilton is Professor, Foods & Nutrition Extension, The Pennsylvania State University, University Park.

There is no single, ideal maturity for harvesting or preserving fruits and vegetables. Obviously, the quality of canned or frozen products made from any maturity other than ideal for that specific commodity lacks the ultimate potential of satisfaction for eating.

Three natural causes account for most quality changes in freshly harvested fruits and vegetables — respiration, enzymes and oxygen. Their effects generally increase with the time between harvest and preservation, and at higher temperatures.

Freshly harvested fruits and vegetables are living organisms. They are sustained by cellular reactions known as "respiration," in which native enzymes use oxygen and components in food and give off heat, water and carbon dioxide. The energy needed to sustain respiration is obtained from a storehouse of natural components in food. As a result, natural flavors and vitamins are diminished, sugars and sometimes acids decrease, and texture may soften. The speed of respiration and the onset of its effects differs among various fruits and vegetables, and is generally lowered by refrigerator temperatures.

Besides enzymes relating to respiration, fruits and vegetables sometimes contain other cellular enzymes associated with reducing food quality. These enzymes act with oxygen to cause rapid discoloration of bruised, peeled and sliced tissue; oxidation of flavors, and excessive softening.

The roles of oxygen in respiration and other enzyme reactions have been noted. Oxygen may react still another way to lower food quality: autoxidation, where sensitive flavor, color and vitamin components are altered by oxygen. This causes fading of colors, off-flavors and, reduced vitamin retention. These reactions occur quickly during canning and drying of foods, and also account for slow changes during storage of preserved food.

Handling Raw Foods

Because of the effects of respiration, other enzymes, autoxidation, and desiccation (loss of water), the raw product temperature and the time between harvest and preservation are of utmost importance to preserving high-quality fruits and vegetables.

Ideal handling conditions vary with the product. Small berry and stone fruits, asparagus, green beans, beets, broccoli, corn and leafy greens, to name a few, should be preserved the same day of harvest, for highest quality. Apples, peaches, pears, plums and tomatoes, if harvested at firm maturity, should be ripened a few days before preserving them. Other fruits and vegetables may be stored from a week to months before preservation without significantly lowering their quality.

From a nutritional standpoint, if after harvest the handling and preservation of these crops are carefully planned and implemented, quality of the preserved products can be equal to food prepared from fresh market fruits and vegetables.

Fruits and vegetables are recognized as significant sources of vitamins A, the B family, C, minerals, and fiber.

Vitamin A, pro-carotene, is heat-stable and insoluble in water but is subject to a minor loss caused by autoxidation. Therefore, losses in home preserved foods are insignificant.

Of the water-soluble vitamins, ascorbic acid (C) and thiamin (B_1) are subject to serious loss upon heating foods. Riboflavin (B_2) is subject to loss when raw or preserved foods are exposed to light.

Significant losses of water-soluble vitamins and minerals occur when food contacts water, such as in washing, blanching, cooking or canning foods. As would be expected, losses due to leaching of water-soluble vitamins and minerals rise with increased cut or exposed food surface areas, repeated exposures to water, and more time in water, especially when heated.

The effects on fiber in preserved food is much the same as in fresh cooked foods.

Canning

Home canned foods should have a bright color, characteristic texture, pleasant flavor and contain nutrients naturally present. When all steps of scientifically based canning directions are carefully followed, color, texture and flavor will be optimum. These observable signs of quality are also an index to nutrient retention of that canned food.

Proper pretreatment of the fruit or vegetable being canned is essential for top quality. This includes using only good, wholesome food at peak eating quality. Any trace of moldy or otherwise spoiled food should be completely removed. All food must be washed thoroughly.

Removing the skins from peaches before canning.

Removal of skins from thin-skinned foods such as tomatoes and peaches is important. A short blanching time in boiling water loosens the skins, and a short cooling time in cold or ice water stops the cooking of the food. Because water leaches out vitamins, dilutes color and flavor, and results in a mushy product, do not allow foods to soak unnecessarily in either hot or cold water.

Air is an enemy of light-colored foods, and should be kept from the peeled or cut food. This can be done with a commercial antioxidant (ascorbic acid), lemon juice, or a simple solution of two tablespoons each of vinegar and salt per gallon of water. Peel or cut food directly into such a solution for maximum protection from air. When enough food for a canner load is prepared, drain and rinse the food for best flavor.

The packing method used is one of the most important factors in quality canned products. Foods that are heated before being put in jars have better quality than foods packed raw in jars. Heating destroys enzymes and removes much of the air from food tissues.

Enzymes must be destroyed quickly for top-quality food. Removal of air not only shrinks the food, but leads to better color, flavor and nutrient retention, and too, fruits are less likely to float. Hot packed jars also have higher vacuums, thus less oxidation can take place.

Tomatoes and tomato juice are less likely to separate if tomatoes are cut and heated at once. Cutting activates enzymes in tomatoes which then change the pectins in the tomatoes and causes separation. Heating right after cutting inactivates the enzymes and minimizes separation.

Sugar, salt and vinegar play important roles in quality canned foods. While sugar is not needed to safely can fruits, it contributes to better color, flavor and texture. Likewise, vegetables may be safely canned without salt, but when salt is used, better flavor results. Salt is essential for slow process pickles; vinegar is needed for quick process pickles. Canning or pickling salt should always be used. Iodized salt will inhibit proper fermentation and react with the starch in some foods, causing them to turn pink or blue.

Sugar concentration in sirups is important to appearance of the fruit. With a light sirup, fruits are less likely to float than with a heavier sirup. And with a light sirup, fewer calories are added. Then too, the cost is less than with a medium or heavy sirup.

Headspace

Using the proper amount of headspace pays off in quality products. Adequate headspace is needed so food has enough room for expansion during heat sterilization.

Generally, more headspace is needed for foods heat-sterilized in a pressure canner than those in a boiling water canner, because of increased expansion of foods at the higher temperature.

With too little headspace, liquid is more likely to cook out of the jars during heat sterilization. This increases the danger of seal failure due to food particles that may be trapped at the interface of the sealing surface. With too much headspace, air remaining in the jar can cause darkening of the food, oxidized flavor, and lower vitamin retention, especially in food at the top of the jar.

Pressure canning low-acid foods is receiving increased emphasis today, mainly based on the greater safety of this technique. The method also offers an additional advantage that is little known to home canners—improved quality and nutrient retention.

Generally an 18° F (10° C) rise in canner temperatures increases the destruction of bacteria tenfold. At the same time, chemical changes that affect color, flavor, texture and nutrients of the food are only doubled.

This means that for every 18° F (10° C) increase in canning temperature, bacteria are destroyed five times faster than are vitamins, flavor, color and texture. Accordingly, in contrast to boiling water canning, heat sterilizing in a pressure canner at 240° F (116° C) (10 p.s.i.) destroys bacteria about 17 times faster than chemical changes are effected, accounting for better nutrient and quality retention of pressure canned low-acid foods.

Be cautious about interpreting this as an advantage in pressure sterilizing acid foods. This technique can increase quality and nutrient losses because of the total time the jars of food will be near or above boiling water temperature.

For example, with tomatoes it takes about 5 minutes until steam begins to escape, 10 minutes to exhaust the air, about 5 minutes to build up pressure, 5 minutes for heat sterilizing at 240° F (116° C) and 30 to 45 minutes cooling before opening the canner. That adds up to about 55 to 80 minutes at temperatures near or above 212° F (100° C), as compared to the recommended 35 minutes for heat sterilizing in a boiling water canner. Assuming the average product temperature is elevated above 212° F (100° C) about half of this total time, the loss in quality could be at least doubled.

Properly managed heat sterilization has a great deal to do with the quality of canned food. If the temperature is too high or the time of sterilization too long, the result is a product with poor color, soft texture, and less flavor. Tomatoes canned in the pressure canner are an example.

After heat sterilization, a quality canned food will have liquid covering the solid food in the jar. The lid must allow air, but not liquid, to escape during heat sterilization. A low liquid level in the jar may indicate the lid did not function properly, pressure was allowed to fluctuate in the pressure canner, or the canner wasn't cooled completely before opening, or raw pack was used. Foods not covered with liquid will have poorer quality.

Jars must be air-cooled naturally after heat sterilization to maintain quality as well as for safety. Delayed cooling in tightly enclosed areas, such as cardboard cartons, will decrease quality of the product.

Storage conditions are vital to quality of canned foods. If stored in a light, rather than a dark place, light-sensitive colors will darken gradually. Some colors, such as carotenoids in tomatoes and carrots will gradually fade with prolonged storage.

Storage temperatures between 40 to 50° F (4.5° to 10° C), if the spot is dry, are best for quality retention. Quality losses are increased when canned foods are stored at higher temperatures. For example, about a third of the vitamin C is lost if foods are stored a year at temperatures of 80° F (26° C) or higher.

Storage temperatures between 50° to 70° F (10° to 21° C) are acceptable.

Properly canned and stored foods should be safe to eat for more than two years. However, for top quality, it is best to can only the amount to be used in one year.

Top quality home canned foods have rich, jewel-like colors, characteristic of the food canned. Colors are those of well-prepared foods ready to be served, rather than fresh uncooked foods or over-cooked foods. Light-colored foods should retain their color with no signs of darkening at tops of jars. Fruits should look neither under-ripe and hard, nor over-ripe and mushy. Vegetables ought to look young and tender, rather than old and starchy. Vegetables and fruits should be free of stems, cores, seeds, or pieces of skin, and be of uniform size, shape and color.

Freezing

Freezing, like canning, does not improve food quality. Top quality fresh fruits and vegetables are essential for premium frozen products. Quality factors include a suitable variety for freezing, optimum maturity, and freshness of the product. Even with high quality, fresh produce, it is imperative to freeze foods on the day when they are at their peak of maturity or ripeness for eating fresh.

Selection, sorting and trimming of produce and the quality control steps as related to color, flavor, and nutrient retention described for canning high quality fruits and vegetables apply equally to preservation by freezing. However, most vegetables must be blanched before packaging and freezing to prevent slow but accumulating effects of enzyme activity and autoxidation. These effects, if not prevented, will cause discoloration, oxidized flavors (sometimes described as tasting grassy or hay-like), and increased loss of vitamins, especially A and C. Proper blanching recommendations are contained in the USDA Home and Garden Bulletin No. 10, *Home Freezing of Fruits and Vegetables*.

The quality of packaging materials used is reflected in overall quality of the frozen product. Packaging materials must be moisture-vapor-proof. This means there is no transfer of liquids or vapors from the inside to the outside, or from the outside to the inside of the frozen packages.

In addition, fruits and vegetables must be packaged in containers with as little air inside as possible. Air left inside will oxidize the food, causing deterioration of color, flavor and nutrients.

Proper sealing is essential for packaging quality frozen foods. If the seal itself is not moisture-vapor-proof, it becomes the weakest part of the package and poorer quality results.

Adequate labeling helps assure that each package of food is used while at top quality. Packages kept too long will be of poor quality. Even with the finest produce and use of the best preparation and packaging procedures, retention of quality in frozen food is affected by how quickly food is frozen, the temperature of food stored in the freezer, and how long it is frozen before eating.

For top quality, avoid freezing per day, more than two pounds of fresh packaged food per cubic foot capacity of your home freezer. Food packages to be frozen should be spread one package deep over the bottom or other areas designated by your freezer manufacturer. The freezer should be regulated to a uniform temperature between 0° F (—18° C) and minus 5° F (—21° C). A freezer temperature fluctuation of more than 5° F should be avoided if possible. Freezing too slowly, temperatures above 0° F, and temperature fluctuations in freezers increase the ice crystal size in frozen food, lower the quality and shorten the shelf life of food. Frozen foods should be used according to the guidelines in the 1974 Yearbook of Agriculture, Shoppers Guide.

Quality frozen fruits and vegetables should have the natural color, texture and flavor of the individual food. There should be a minimum of ice crystals inside the package and no sign of freezer burn.

Frozen foods should be higher in ascorbic acid and thiamin than canned foods. Vitamin C, including that leached into the juices, approaches 90 percent of the value of raw fruits.

Drying

Quality fruits and vegetables can

be sun-dried or dried inside using an oven or a food dehydrator. As with other methods of preservation, quality of the final product will depend greatly on quality of the fresh food being dried. Drying does not improve the quality of any food.

A top quality dried product reflects suitable pre-treatment before drying, and adequate drying under proper conditions. Also, appropriate storage after drying is vital to keep moisture from re-entering the dried product.

Fruits may be dried more easily than vegetables, because of their high sugar content and since not as much moisture must be removed to get a quality product. When properly dried, fruits should be leathery and pliable, and have a color characteristic of the fruit. Excessive darkening indicates a less desirable, poorer quality product.

Vegetables must be dried until they are brittle. The color should be characteristic of the vegetable and not excessively dark.

To sum up, nutrient content of foods depends on natural differences, control of deterioration, and handling techniques of food preparation as well as preservation. Therefore, overall nutrient content of a specific fruit or vegetable, whether fresh or preserved, may be about the same.

Top quality, garden fresh foods, served in season, provide the greatest satisfaction when served fresh as table-ready food. Individual preferences for market fresh, frozen canned, or dried food will differ. Since nutrient content is nearly the same, the choice is up to the consumer. Even more important is the choice of vegetables and fruits for a good diet.

For Further Reading:

Food editors of Farm Journal, *How to Dry Fruits and Vegetables at Home*, Countryside Press and Doubleday & Company (Dolphin Books), 1975. $2.95.

Hamilton, Louise W., Kuhn, Gerald D., Rugh, Karen A., with the food editors of Farm Journal, *Home Canning—The Last Word*, Countryside Press and Doubleday & Company (Dolphin Books), 1976. $2.95.

U.S. Department of Agriculture, *Home Canning of Fruits and Vegetables*, H&G Bul. No. 8, on sale by Superintendent of Documents, U.S. Government Printing Office, Washington, D.C. 20402. 45¢.

U.S. Department of Agriculture, *Home Freezing of Fruits and Vegetables*, H&G Bul. No. 10, on sale by Superintendent of Documents, U.S. Government Printing Office, Washington, D.C. 20402. 75¢.

Economics of Home Food Preservation, or Is Do-It-Yourself Back to Stay?
by Ruth N. Klippstein

Back-to-basics may turn out to be the theme song of the 1970's as some of the simple activities of the past are rediscovered and practiced. Do-it-yourself is in. Nowhere is the trend more apparent than in the area of home production and preservation of the family food supply. Ten years ago no one would have dreamed the lowly canning jar lid would be the subject of Federal hearings. No one would have believed that an estimated one in four U.S. families would be raising and preserving a portion of their food supply.

What motivates people to return to home food production and preservation? Are their expectations realistic? How extensive is their gardening? Will they continue a second year? Do they preserve any of their crop? These were among the questions posed by Stuhlmiller, How and Stone of Cornell University in 1975 to a group of gardeners in five upstate New York counties.

When asked whether they gardened to save money, to have better quality food or just for a hobby or recreation, three-fourths of the 2,800 who replied hoped to save money, 54 percent considered gardening a hobby, while only 46 percent gardened for fresher food. Most said they preserved at least some of the food they grew.

If this study is indicative of the country as a whole, it is important to realistically assess whether home food production and preservation can save substantial amounts of money and whether the satisfaction gained warrants the cost of time and energy expended.

Ruth N. Klippstein is Professor, Division of Nutritional Sciences, Cornell University, Ithaca, N.Y.

The actual costs of home food preservation, for example, should be considered. The cost of home grown food should be compared to the cost of similar food purchased for preservation in quantities at local farms or markets. The quality of the home preserved items should be realistically analyzed against readily available commercially preserved food.

There is no such thing as free food. Someone, somewhere, has to pay for it in time, energy, know-how, and at least some outlay of dollars. Home production in amounts needed for food preservation requires a longtime commitment of family resources. Beginners should realize that realistic goals and reasonable skills in the field and kitchen are essential to make home preservation pay off.

There are no general statistics citing the average dollar-cost needed to grow a given amount of fresh produce in a home garden. Conditions between individual gardens, weather, soil type, skill of operator, and geographic areas vary too much for valid comparisons.

Extension specialists at Michigan State University, however, have computed the actual cost of raising tomatoes under home gardening conditions in East Lansing, Mich. They found it costs 12¢ to grow the amount of tomatoes (2½ to 3 pounds) needed for one quart, canned. A similar cost analysis for green beans showed that beans cost 30¢ for the amount needed for a quart. Only the expendable cost —seed, fertilizer, pesticides and water —was considered.

Adding the expense for needed tools, hoses and other capital items raised the cost another 33¢ a quart if the cost were absorbed in one season or 2¢ if amortized over a 20-year

period. Unfortunately, the first-year gardener will find that the outlay for tools must be spent the first year so that return for the investment requires a commitment to gardening over many years.

Additional expenses are necessary if the bounty is to be preserved at home. Equipment for preserving tomatoes is minimal but equipment for canning vegetables and for freezing may be costly. Homemakers needing to invest in canners, a pressure canner, and home freezer will find that the dollar cost per package of food preserved during the first years of preservation may be higher than the cost of comparable food at the corner supermarket.

The costs of canning peaches, tomatoes and green beans in upstate New York were calculated by the author in 1975 and updated for price changes in 1976 using a number of different cost variables. She found that those who canned tomatoes could realize substantial savings, while the cost of purchasing peaches and preserving them at home approximated the cost of the commercially canned peach.

Determinations of the true cost of frozen food must consider the initial cost of the freezer plus the cost of operation and repair. Containers, plastic bags and boxes, or foil are additional costs.

Evelyn Johnson in her Outlook Talk of 1975 quoted staff at Virginia Polytechnic Institute and Cornell University as reporting a cost of 20¢ to 24¢ per pound of food frozen just for the convenience of freezing and storing food at home. Add to this the price of the food being frozen for the correct cost of home-preserved frozen food.

Freezing is probably the most satisfactory method of home food preservation, the most versatile and the easiest to do. But for all except the very best managers who use the freezer intensively, the home freezer is more a convenience than a money saver.

Time, Energy Costs

Raising a garden takes time over a significant number of months. As a hobby for table use, gardening can be a real pleasure. Skillful persons with the right tools and knowhow can handle a garden of the size needed for home food preservation with a few hours of work a week, once the plants are well established. Novices can expect to spend a significantly greater amount of time per week during the four or five month growing season in northern areas and even more in areas with longer growing seasons.

The author, an experienced gardener, kept records of the hours spent cultivating and harvesting a 20 by 40-foot garden, planted primarily for fresh consumption. Only three foods —tomatoes, green beans and cucumbers—were raised in amounts sufficient for a limited amount of home preservation. Over 40 person-hours were required. The actual grocery store value of the garden food consumed by the family of three was $45. Food given as gifts and preserved raised the dollar value to $75.

Gardening often helps stretch cash income, but the dollar return is low for hours of effort. And poor weather may cause crop failures and small

Needed equipment for home canning. Left, water bath canner, and right, pressure canner.

Cost of Home Food Preservation

Method	Time	Energy		Dollar cost from kitchen to table	Quality satisfaction
		Fuel	Human effort		
Freezing	Minimal low	High	Low	Very high	Very high
Canning	Moderate	Moderate	High	Moderate	Moderate to high
Drying	High	Moderate to high	Moderate	Moderate to high	High (specialty items)
					Low, if only method available
Pickling	High	Low	Moderate	*Depends upon type chosen	High
Storage (Unprocessed)	Low to moderate	Low	Moderate (Checking/ culling)	Low	Moderate to high

*Some (such as quick dill pickles) are quick to make, take little effort, and use inexpensive ingredients. Others require prolonged brining over several days' time plus expensive sugar and other ingredients.

yields, regardless of effort. Food preservation also is time-consuming.

Satisfactions. Why garden or preserve food at home? Most gardeners will cite a number of reasons:
—The best of good fresh food with no unknown additives or ingredients
—Healthy exercise
—Family pleasure working together toward a goal
—The joy of giving
—The challenge of growing a seed into edible food
—Prestige

There is no one right answer to the question, "Does it pay to raise and preserve my own food?" It depends upon your personal goals. You may not save a significant number of dollars. You will work hard. And you probably will experience one of the most exciting activities possible—raising at least some of the food your family uses.

"We grew it" are heady words which bring people back to home food production and preservation year after year.

Beginner's Guide to Home Canning
by Frances Reasonover

Canning is probably the most economical and practical method of preserving food at home. Among other things it is a way to save food that otherwise might be wasted.

Cost of home canning depends on the kinds and sources of food canned as well as the processing methods, containers, and equipment used. Other cost factors—labor, energy, water and added ingredients—make exact cost figures impossible to apply generally, but studies are reporting averages that show canning to be economical.

The wise homemaker will can only the amount to be used within a year. Food held longer will be safe to eat if it has a good seal and no signs of spoilage, but there may be nutrient or quality loss, especially if stored at temperatures above 70° F.

As a beginner canner you need to know something about micro-organisms, including yeasts, molds and bacteria, on the food, in water, air and soil, as causes of spoilage in foods. Knowing about these minute forms of life, which are so abundant everywhere, will help make the work safer as well as more interesting.

In addition to the action of these minute organisms, the spoiling of fruits and vegetables is hastened by natural changes in color, flavor and texture of the food. These changes result from the action of enzymes or micro-organisms found in nature which break down and decompose foodstuffs.

Bacteria are the most serious foes to combat in canning because they are more difficult to kill by heat than either molds or yeasts.

Acid in canned food is expressed

Frances Reasonover is Extension Food and Nutrition Specialist, Texas A & M University, College Station.

as pH value. Foods having a pH of 4.5 or lower are called high-acid foods and those with a value of 4.6 or higher are termed low-acid foods.

Since few bacteria thrive in acids, their destruction is less difficult in fruits than in vegetables (with the exception of tomatoes).

Botulism is a deadly poison caused by a toxin from the growth of spores (seeds) of the bacteria, *Clostridium botulinum*. These spores will produce a deadly toxin in low-acid foods in the absence of air (oxygen) inside a sealed jar. Therefore, the spores must be destroyed by processing under pressure at 240° F. The length of time has been determined by scientists for each individual food.

Clostridum botulinum will not grow in foods with a pH of 4.5 or lower, so high-acid foods may be processed safely in boiling water at 212° F.

Low and High Acid Foods

Low-acid vegetables	High-acid fruits and vegetables
Asparagus	Apples
Beans—snap or shelled	Apricots
Beets	Berries
Carrots	Cherries
Corn	Grapefruit
Potatoes	Peaches
Pumpkin	Pineapple
Squash	Rhubarb
Sweet potatoes	Tomatoes

Yeasts, mold and non-spore forming bacteria are readily controlled by processing at 212° F.

Most canning equipment and supplies may be purchased at hardware stores, housewares departments, and from mail order companies. Jars and lids are available in many retail stores.

Canning Jars

Select standard canning jars made of tempered glass that can withstand high temperatures. The manufacturer's name or symbol in glass will identify the product. With careful handling, jars last an average of about 10 years. Avoid using antique jars because there can be hair-line cracks not visible to the eye, causing jars to break.

Use canning jars in sizes suitable for the product canned and your family's needs. Canning jars generally are sold in half-pint, pint and quart sizes with wide and narrow mouths. Large-mouth jars are convenient for packing such foods as whole tomatoes and peach halves. Quart jars are convenient for vegetables and fruits where your family has four or more members.

Examine the sealing edge of jars for nicks, cracks, or sharp edges that would prevent a seal. Discard any with these imperfections.

One-trip jars from purchased canned foods should not be used because they generally are not tempered to withstand the high heat required for home canning, and may break when subjected to the heat. Tops of these jars may not fit standard canning lids, thus preventing a good seal.

Closures—jar lids and rings come with new canning jars. The sealing compound of lids recommended for one use only will not hold a seal effectively after the first use.

Select lids appropriate for the jars being used. You may find the two-piece units (flat lid with sealing composition and ring), one-piece lids, or flats with separate gaskets made of metal or plastic. Always follow the instructions for pretreatment as indicated on the box or container by the manufacturer. If no name is indicated on the lid, use a black wax marking pencil or crayon and mark the identity on each lid. If there are problems, contact the manufacturer whose name and address is on the box or container.

Screw ring bands may be reused if kept clean and dry in a protective container with a tight-fitting lid. Never use bands with rust, or pried up or bent edges.

If you have extra lids, store them protected in a dry, cool place.

One-piece zinc caps lined with white porcelain, with rubber rings, may be used. The caps may be reused if they have not cracked, spread or bent at the edges and are clean, like new. The rubber rings are effective only once because they tend to dry and deteriorate with age, often become porous, and sometimes crack.

If you have jars with bail wire clamps, sometimes called "lightning"-type jars, be sure they are not in the "antique" class. Lids for these jars are all glass, and rubber rings are used between the jar and lid for sealing. A wire clamp holds the lid in place during processing; after processing, the short spring wire of the clamp is snapped down to complete the seal.

A *boiling water bath canner* is needed for processing high-acid foods, such as fruits, tomatoes, tomato and fruit juice, and pickles.

Water bath canners in several sizes are available on the market. The container must be deep enough for a rack

Home canning equipment.

to hold the jars off the bottom of the canner. The depth allows water to be over the jars of food by at least 1 to 2 inches. Keep 1 to 2 inches of space above the water to allow for boiling; this prevents water from boiling over.

The canner must have a tight-fitting lid. Or you can use a large kettle with a tight-fitting lid, and a wooden or wire rack to hold jars off the bottom. There should be free circulation of water to every part of the surface of the jar and lid.

If you are going to buy a water bath canner, check the height, and the lid to be sure it is tight-fitting. The rack preferably should have dividers so jars will not touch each other or fall against the sides of the canner or each other during processing.

A steam pressure canner is absolutely essential in canning low-acid foods, such as vegetables, and insures the destruction of spoilage microorganisms.

Ten pounds pressure is used for processing food in standard canning jars at sea level. This pressure corresponds to 240° F.

The steam pressure canner is made of heavy metal that withstands high pressure developed by steam. It consists of a kettle with a tight-fitting lid equipped with an accurate weight or dial gage to register the pounds of pressure in the canner. The lid must lock or seal to prevent escape of steam.

The canner must have a safety valve, petcock or steam vent that can be opened or closed to permit exhausting (venting), and a pressure gage. It must have a rack to hold jars at least ½ inch from the bottom of the canner.

A dial gage indicates pressure on a numbered instrument.

A weighted gage has no dial, but automatically limits pressure with weights preset for 5, 10, and 15 pounds pressure.

The pressure is adjusted for high

Scalding tomatoes in a blancher.

altitude. For information on canning at altitudes above sea level, see the later chapter by Carole Davis.

To insure the canner's proper working condition, check the dial gage for accuracy each year—or if a canner or lid has been roughly handled or dropped, the dial gage glass broken, or any parts are rusty. The manufacturer or your county Extension office can give information on testing availability. Study and follow the manufacturer's directions for using your pressure canner.

Run through the process of operating the pressure canner on your range in a trial run before you get into the canning season, to be sure everything is working properly. Make a note of the dial setting of the range if you use an electric range for holding pressure steady.

Trying to use a pressure canner obtained from garage, rummage, or auction sales or handed down to you from someone's attic may prove dangerous. You may not have any idea as to the care, handling, or storage of the canner. A manufacturer manual on care, use and replaceable parts usually is not available. Old-old canners did not have complete information—manufacturer's name, address or model number—on the appliance.

General kitchen equipment is helpful in any needed washing, peeling, coring and slicing in the preparation of fruits and vegetables. Examples are, a vegetable brush for cleaning vegetables, a blancher or wire basket for scalding fruits and vegetables such as tomatoes and peaches to loosen skins for peeling, and a colander for washing delicate fruits such as berries.

A food mill is handy for making purees and straining fruits for making juices, and a strainer for straining juice. A long handled fork or plastic spatula aids in fitting and packing food and removing air bubbles. A wide-mouth funnel is very convenient for filling jars, and a jar lifter helps you avoid burns in handling hot jars. Use an automatic timer to time processing accurately.

The number of pints of preserved food you will get from a given quantity of fresh food depends on the quality, variety, and maturity of the fruit or vegetable; on the size of the pieces, and on the packing method used.

Selection of good sound fruits and vegetables is of paramount importance. The quality of canned fruits and vegetables will be no better than quality of the raw food used. For best flavor retention, preserve only those vegetables that are young, tender, and freshly gathered.

Work Fast

All steps, from beginning to end, of any lot of canning should be carried through as rapidly as possible. A good slogan is "two hours from harvest to container".

Work fast with small amounts of food at a time, especially vegetables with high starch content such as corn and peas which lose quality rapidly. Any delay will result in loss of flavor and nutritive value.

Sorting and grading should be done very carefully, according to size and degree of maturity and ripeness.

Use only uniformly well-ripened products. Discard all defective products and use together those of the same size.

Dirt in seeds, bits of food, or sirup contains bacteria that is hardest to kill, and encourages yeasts and molds to grow on the outer surfaces. Wash fruits and vegetables thoroughly before canning.

Scalding, peeling and coring—some fruits, such as peaches and tomatoes, are scalded in order to peel them smoothly.

Follow up-to-date recommendations, available in U.S. Department of Agriculture or Extension publications, for detailed procedures in preparing fruits and vegetables for canning.

Packing Methods

You can pack food hot or raw in jars. Hot-packed food is heated thoroughly before it is packed into jars. Raw-packed food is placed raw in jars. Watery and soft foods such as tomatoes are pressed gently to make their own juice.

Air, a poor conductor of heat, should be removed from the jar. Remove air bubbles by gently moving the blade of a plastic spatula or plastic knife around the jar—being careful that the food is not broken. Add more boiling liquid if necessary to get a proper fill.

When filling jars, you will find the jar-filling funnel easy to manage. This makes it possible to avoid spills of seeds, bits of food, or sirup that could prevent sealing. But even when using a funnel you still need to wipe the jar rim.

Prepare the lids and sealing of jar according to the manufacturer's directions. When using a flat metal lid, place the composition side on the rim of the jar. Add the ring band and screw it down until firm, but not hard enough to cut through the sealing compound. The lid will have enough "give" to let air escape during processing. This is called venting and means heating to remove air from jars.

When using porcelain-lined zinc caps, fit the wet rubber ring on the jar shoulder, but do not stretch it more than necessary. Screw the cap firmly and turn it back ¼ inch.

Use a jar lifter or tongs and place the filled jars on the rack in the canner. Fill and place jars in the canner one at a time to keep jars as hot as possible while filling the canner.

Water Bath

Before you begin preparing the food, fill the water bath canner half full of hot water. This permits water to heat while you prepare the food. Put a large kettle or teakettle of water on to boil. The water should be boiling when hot-pack food is put into the canner.

Place raw-pack jars in water that is hot (180° to 190° F), just below boiling. Then bring it to a boil after adding jars.

As the rack of jars is lowered into the water, the water level will rise. If more water is needed to have the jars completely covered by 2 inches of water, add boiling water.

Prepare only enough jars of food at one time to fill the canner. Work rapidly, allowing as little time as possible between filling and closing the jars and getting them into the canner.

Start counting processing time as soon as the water in the canner reaches a gently rolling boil. Put the lid on the canner. Set your timer or clock and make a written note of starting time and final time. Keep the water boiling all during the processing period. If water boils down, add boiling water sufficient to keep it at the required height. When pouring water, avoid letting it hit tops of the jars.

Process for the recommended length of time. Do not cut processing time.

Pressure Canner

Follow the manufacturer's directions for operation of your pressure canner before, during and following processing. Supplement these directions with information in U.S. Department of Agriculture or Extension publications.

Count processing time as soon as the pressure reaches 10 pounds or the proper pressure adjusted for altitude. Be sure to hold pressure steady.

At end of the processing time, remove the canner from the heat. Allow the canner to cool until the gage registers zero to avoid breakage of jars and loss of liquid from jars. After a minute or two, open the petcock

Heating beans thoroughly before placing them in jars. Here, cut beans are covered with water to be followed by five minutes of boiling.

gradually and remove the cover. If a weighted gage is used, nudge the weight slightly. If no steam escapes, pressure is down. Tilt the far side of the lid upward so steam escapes away from your hands and face. Because food in the jars may be boiling vigorously, leave jars in the canner about five minutes and then remove them.

After Processing

When you remove hot jars from the canner, use a jar lifter, or protect your hands with cooking mits, pot holders or canvas gloves. Set the jars upright to cool on a rack, such as a cake rack, or a bread or cutting board, with double layers of dry cloth or newspapers beneath the jars. If jars are placed on a cold surface or wet cloth, the difference in temperatures may cause the glass to crack.

Avoid placing jars in a draft, but leave two or three inches between them so air can circulate freely. Avoid further tightening of lids that have sealing compound, since this usually breaks the seal—unless the lid manufacturer states it is safe to tighten.

If your processing temperature was not held steady and liquid boiled out in processing, do not open the jar to add more. Leave the sealed jar just as it is.

Do not cover jars because this slows down cooling and food continues to cook. If you have an air conditioning vent that will direct cold air on jars, cover the vent during this canning session.

After 12 hours, check the seals. The vacuum may cause a loud snap of the two-piece vacuum seal while it cools, which is an indicator of an airtight seal. If the center of the lid holds down when pressed and the lid does not move, it is sealed.

Tap the center of the lid with a spoon—a clear, ringing sound indicates a good seal; a thudding sound indicates the possibility of an imperfect seal.

If there is a sealing failure, you will need to reprocess the jars. Remove the lid, heat the food and liquid, fill a clean jar and use a new lid. Process the full length of time. If only a few jars did not seal, you may elect to refrigerate and use the food within a day or two or freeze it.

Once the jar is sealed, allow it to set until cold. Then remove the screw ring band, wash and store in a dry place for reuse. For safety make a routine check of canned foods each month.

Label and Inventory

Write name of product and date canned on a gummed label or the lid of each jar with a felt tip pen. Keep a record of food canned, date, number of quarts or pints, and a place for you to check them off as you use them. This can be your guide for next year's preservation plan. Use food preserved for the current year, readying a storage place for next season's garden produce.

Canned foods stored in a dry, dark, cool temperature (70° F or below) will retain good eating quality for a year. Home canned foods stored in a warm place near direct sunlight, hot pipes, above a range or refrigerator, or in kitchen cabinets may lose some eating quality within a few weeks. Dampness may corrode lids and cause leakage so that the food spoils.

The main cause of spoilage in canned foods is improper processing. Bulging jar lids, or a leak, may mean gas is present and the food spoiled.

Before opening home canned foods, wash jars and lids and carefully inspect the jars. Bacteria, yeasts and molds should have been destroyed if the food was properly processed.

When you open the container, look for such danger signs as spurting, cloudy or frothy liquid, an "off" odor, deterioration, or slimy texture. A foamy or murky appearance and patches of mold are visible signs of

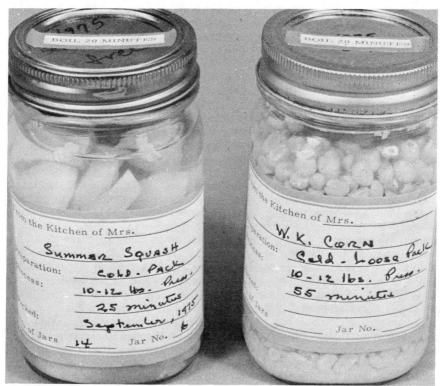

ALWAYS label home-canned foods.

spoilage. That ordinary looking mold on home-canned food may indicate the presence of a much more deadly problem: botulism.

The odor in good jars of food should be pleasant and characteristic of the product. Do not use food which looks or smells bad, or if there is any doubt as to its safety.

Destroy food if any of these signs are obvious; discard out of reach of humans and animals.

All low-acid, home-canned food should be boiled 10 to 20 minutes to insure destruction of botulism-causing toxin for added safety. Heating denatures the toxin so that it does not react with the body. Never taste home canned food before cooking it.

Successful results largely depend upon the accuracy with which up-to-date directions are followed.

Safety is best assured when you exercise special care as you prepare and pack food into canning jars, fitting jars with properly pretreated lids, and heating jars of food to a high enough temperature for a sufficient length of time to kill micro-organisms that cause spoilage.

A Primer on Home Freezing for the Beginner
by Charlotte M. Dunn

Every homemaker knows that meals must be planned to get the most out of the food dollar and to provide the family with a well-balanced diet. The freezer, more than any other household appliance, can help secure these results. The more you learn how to use it in relation to your own family, the greater the returns.

Freezing is a quick, convenient and easy way to preserve foods in the home. Plan ahead to manage your time and energy for preserving food directly from harvest. Freeze limited amounts at one time so the work is spread over several days of picking, rather than squeezed into one long tiring period of time. Be practical about what you attempt.

Your own observation has taught you that some foods "spoil" more quickly than others, so the rate of speed at which they must be frozen varies with their individual temperaments. A good rule for home freezing is: two hours from garden or orchard to container, and the faster the better!

Most food that is highly perishable at normal temperatures can be quick frozen. Even delicate fruits and vegetables can be frozen, with only a few exceptions such as tomatoes (stewed tomatoes can be frozen) and those vegetables that lose crispness such as radishes, celery, cucumbers and salad greens.

Decide what you will freeze on the basis of availability of foods, family needs and taste, freezer space, cost of freezer storage, and availability of alternate methods of storage.

It is essential to start with high quality raw material. As garden foods mature, process without delay. Quality of the frozen food will be only as good as the quality of the food before freezing. Freeze foods at their peak of eating quality to preserve flavor, texture, and appearance as near those of the fresh product as possible.

Do not ignore details of the recommended procedures for preparing foods for freezing. Seemingly unimportant steps can make the difference between a low quality and a superior frozen product.

Before you begin freezing foods at home it's important to know exactly which process to use and what the process is doing to the food.

Micro-organisms grow on food, causing it to spoil. The common growths are simple yeasts, molds, and bacteria. Because these micro-organisms are everywhere—in the air, water, soil and on all surfaces they contact—they naturally occur on all foods. Storing and preserving foods properly controls or inhibits the growth of micro-organisms, thus maintaining both quality and safety of the food.

Cleanliness and sanitary methods are as important in handling foods for freezing as in preparing them for immediate use.

All foods contain chemical substances called enzymes. They are essential to life, and continue their chemical activity after the fruits and vegetables mature or are harvested.

If allowed to work after a food reaches its peak of maturity, enzymes destroy the food's physical properties, thus changing its color, flavor and texture.

When perishable food is not preserved by one of the recommended ways, enzymes within the cells of the food continue to live and cause spoilage.

Charlotte M. Dunn is Food and Nutrition Specialist, University of Wisconsin-Extension, Madison.

What Freezing Does

Freezing and storage even at very low temperatures will not inactivate any of the common enzymes. At 0° F, the recommended temperature for storing frozen foods, enzymes are not inactivated but only slowed down. In two to three months they will produce off-odors and bad taste. This temperature only checks the growth and reproduction of destructive bacteria. The faster a food is properly prepared frozen, the sooner both enzymes and bacteria are rendered harmless.

Just about every kind of food you or I will freeze contains moisture or water, and the process of freezing food involves the freezing point of water. As temperature of the surrounding air goes below the freezing point of water, the water progressively crystallizes out in the form of pure ice. Size of the crystals which form is determined by the span of time during which freezing takes place. When the temperature is lowered slowly, the crystals expand considerably. If the freezing is sharp and sudden, the crystals retain approximately the same size as the original water molecules.

In case you have doubts about how well a food will freeze, test it before freezing large quantities. To test, freeze three or four packages and sample the food a couple of weeks later. This will show the effect of freezing but not the effect of storage. Some varieties of the same kind of food freeze well, others do not.

Much of the success you have with your home freezer will depend on how you prepare, package, wrap and seal foods. Protecting frozen food is as important as freezing food of high quality.

You will need general kitchen utensils plus a steel, aluminum or enamel kettle large enough to hold at least one gallon of boiling water, with a tight fitting cover. Use a mesh basket, a strainer, or large squares of cheesecloth to hold one pound of vegetables in the boiling water.

Steaming of cut, sliced or green leafy vegetables is recommended and will preserve more nutrients than water does.

You will need a container to hold ice water for quick chilling of vegetables to stop cooking action. Drain thoroughly in a colander and turn out on absorbent towels.

It is false economy to skimp on wrappings and containers. They should protect the food from cold air, which is dry, so as to retain the moisture in foods and prevent freeze burn and dehydration. Select them according to the use they will be put to.

Most freezer containers on the market today are easy to seal, waterproof, and give satisfactory results. Rigid plastic containers, bags, and jars with wide tops are favorites.

Moisture- and vapor-resistant wraps, which are exceptionally effective at excluding oxygen, include heavyweight aluminum foil, coated and laminated papers, polyethlene films, saran, and polyester films. They should be strong and pliable so the wrap will adhere readily to irregularly shaped objects, and eliminate as much air as possible to avoid frost accumulation inside. Careful wrapping is of no avail if the package breaks. It should be easily sealed, either by heatsealing or freezer tape.

Freezer bags are available, and freeze-and-cook bags that withstand temperatures from below 0° F. to above the boiling point. The freeze-and-cook bags are suitable for freezing and reheating food. Points to consider include the size convenient for your use and the cost.

Materials not moisture-proof and vapor-proof, and thus not suitable for packaging foods to be frozen, are ordinary waxed papers, cartons from ice cream or milk, and plastic cartons from cottage cheese or gelatin products because they crack easily.

Compare price, durability, shape and reusability in selecting containers, keeping in mind their convenience and the economical use of freezer space.

Retaining the vitamins and other nutrients depends on how fruits and vegetables are handled before freezing, on storage temperature in the freezer, and on how you cook them. Always follow up-to-date recommendations available from the U.S. Department of Agriculture or county Extension office.

Select foods of top quality. A freezer is not magic—it does not improve food. Its function is to preserve quality and food values and to prevent spoilage.

Choose vegetables and fruits suitable for freezing, and the best varieties for freezing. Because growing conditions and varieties vary greatly across the country, check with your county Extension office to find out which varieties are best for freezing.

Freeze fruits and vegetables when they are at their best for table use. If possible, freeze those that are ripened on the tree, vine or bush. Fruits should be ripe but firm.

Enzymatic changes continue after harvest, lowering quality and nutritive value. If stored at too warm temperatures, fruits can lose vitamin C, turn brown, lose flavor and color, and toughen.

Don't delay in harvesting vegetables since asparagus, corn, peas, snap beans, and lima beans all deteriorate rapidly in the garden after reaching their peak.

Observe cleanliness while you work, to avoid contaminating foods.

Prepare vegetables for freezing by blanching them in boiling water for recommended times. County Extension offices will have information on specific times for various foods.

Blanching vegetables is absolutely necessary to inactivate enzymes that cause undesirable changes in flavor and texture. This brief heat treatment reduces the number of micro-organisms on the food, enhances the green color in vegetables such as peas, broccoli and spinach, and displaces air trapped in the tissues.

Pack food in containers as solidly as possible to avoid air pockets, leaving the necessary head space for expansion. Press out as much air as possible, with your hands or by using a freezer pump. Then seal the plastic bags by twisting the open end, folding it over. Freezer rubber bands, twist-seals, or freezer tape are satisfactory for sealing bags.

Label packages clearly and carefully with name of product, date when frozen, number of servings or poundage, and any information that will help you. Special pens are made for marking frozen food products. Or you can use a wax pencil or crayon.

Speed is important in preparing food and getting it into the freezer, so as to maintain quality. Put only the amount of unfrozen foods into the freezer at one time that will sharp freeze within 24 hours.

Allow at least one inch between packages of unfrozen food in the freezer for circulation of cold air. Leave packages in freezing position for 24 hours before stacking them close together.

Uniform freezing temperature and keeping frozen products at 0° F or lower will maintain quality. Different foods have varying storage periods, so keep your frozen food inventory changing.

Use a freezer thermometer in your freezer. Check your freezer door and wall plug daily to avoid any catastrophe.

A freezer can pay wonderful dividends with considerable thought and planning by the homemaker.

Pressure Canners, Vital for Low-Acid Foods
by Nadine Fortna Tope

Use of a pressure canner for preserving low-acid foods is not new. Pressure canners for home canning were first marketed in the early 1900's. In 1917, the U. S. Department of Agriculture announced that use of a pressure canner at 10 pounds pressure (240° F) was the only safe method for canning vegetables. Today's recommendations are essentially the same.

A temperature of 240° F or 10 pounds pressure at sea level is needed to kill spoilage organisms in a reasonable time, especially the spores of *Clostridium botulinum*. These spores, if not killed, can produce the most deadly toxin known to man.

The commercial canning industry stringently follows safe canning practices. Their safety record is excellent. *A Complete Course in Canning* by Lopez (1975) says that since 1925 four deaths have been reported from the consumption of more than 800 billion cans of commercially processed foods. The record for home canning is much worse—450 deaths in a fraction of the number of cans. Unsafe practices were probably used in preserving the deadly home-canned food.

Methods like open kettle, oven, and boiling water bath canning for low-acid foods are not sufficient to kill *Clostridium botulinum* spores. Educators talk to many people even today who still use unsafe practices because "that's the way my mother always did it", or because they are unfamiliar with pressure canners.

In this chapter, the need for safe pressure canning procedures will be discussed along with the rationale for using care in preparing low-acid foods for canning.

Acidity of a particular food is the most important factor in determining which canning method should be used —pressure or boiling water bath.

Acidity is measured and stated much the same way we express length or weight. Acidity (pH) refers to acid strength, not the *amount* of acid present.

For example, citric acid, an acid found in oranges, grapefruit, and other citrus fruits, is a weak acid compared to hydrochloric acid, a very strong acid. The measure used to express acid strength is pH. The pH scale runs from 1 to 14 with 7 as the neutral point. Substances with pH below 7 are called acidic, while those above 7 are called basic or alkaline foods. The lower the pH, the more acid the food.

Acidity or pH of a food affects the length of time it must be processed at a particular temperature to make it safe. The more acid the food (the lower the pH), the shorter the time required for processing.

Almost all foods are acid in nature. Hominy is an example of a food that is neutral or slightly alkaline.

Foods are further categorized as high acid or low acid because the *C. botulinum* spore will not grow at pH levels of 4.6 or below. High acid foods (pH 4.6 or below) include tomatoes and all fruits except figs. (See chart).

Those with a pH above 4.6 are the low-acid foods. All vegetables except tomatoes and those that have been pickled or fermented are low-acid.

Safe processing times have been established at 240° F for low-acid home canned products, since at this temperature the processing times are reasonably short and texture of the

---

Nadine Fortna Tope is Extension Specialist in Food Conservation and Preparation, North Carolina State University, Raleigh.

Fruit and Vegetable Acidity

High acid foods	Low acid foods
Lemon Juice	Figs, Pimentos
Cranberry Sauce	Pumpkins
Gooseberries	Cucumbers
Rhubarb, Dill Pickles	Turnips, Cabbage, Squash
Blackberries	Parsnips, Beets, Green Beans
Applesauce, Strawberries	Sweet Potatoes
Peaches	Spinach
Raspberries, Sauerkraut	Asparagus, Cauliflower
Blueberries	Carrots
Sweet Cherries, Apricots	Potatoes Peas
Tomatoes	Corn

resulting product remains good. The heat-resistant C. *botulinum* spore has been known to survive many hours of boiling at 212°. Once food reaches 240°, the spore is killed when held for the recommended number of minutes.

Holding of produce to be canned for long periods in warm summer temperatures gives bacteria ample time to multiply into vast numbers, thus increasing the chances of spoilage.

For example, one cell can multiply into a billion cells in just 15 hours of holding under favorable conditions.

Salt and spices added to low-acid canned products in amounts recommended do not appreciably alter processing time. Salt may slightly lower the heat resistance of some micro-organisms but not enough to present a problem if omitted for dietary reasons.

Fats and oils, if added, may reduce the rate of destruction of bacterial spores. Spores of C. *botulinum* have been known to survive beyond all reasonable expectation when heated in oil suspensions. Thus, adding oil or fat to a product being canned could be dangerous and is not recommended.

The type, consistency, and piece size of food and how it is packed in the jars are important factors which affect processing time. In preparing jars of food for the pressure canner, follow directions carefully. Do not use jars larger than the directions specify.

Determining the safe processing time for a food product involves two important steps.

First, the rate of heat penetration is measured by finding the spot in the jar that takes the longest time to heat. This is referred to as the "cold spot". Times will depend on the type of food (squash vs tomato juice) as well as how it is prepared (whole kernel vs cream style corn).

The second step is done in a laboratory. A known amount of some live bacterial spores is put into the "cold spot" of the jar of food. These jars are then heated and the amount of time needed to kill the spores is determined.

How Food Is Heated

Heat is a form of energy which flows from hot to cold substances. This flow occurs by convection, conduction, and radiation. In a pressure canner, convection and conduction are the primary methods of heat transfer.

Convection heating occurs in thin liquids and in gases like air and steam. As molecules are heated, they become lighter and rise to the top of the jar, displacing cooler ones toward the bottom. This movement is visible in water that is being heated in a clear glass container. Convection heating occurs best in liquids like fruit and vegetable juices or broths. These heat rapidly and thus have shorter processing times.

Small quantities of starch either added or leached from vegetables slows down the convection and increases processing time. For example, jars of liquid containing pieces of green beans or peas would heat more slowly than apple juice because the pieces would interfere with convection.

A tightly packed jar takes longer to heat than a loosely packed one which allows some convection heating. Thus it is important not to over-pack jars as this will decrease the rate of heat penetration.

Some types of food heat by a combination of conduction and convection. One example is a peach half in thin sirup. The sirup heats by convection while the peach heats by conduction. Another example is cream style corn. Initially, the liquid is thin and heats by convection. As the liquid thickens, it heats by conduction.

Conduction occurs when heat is transferred from one particle or substance to another right next to it. This is the slowest type of heat transfer. Foods that mat together, like spinach, or viscous material like mashed pumpkin, heat by conduction. The larger the pieces of food or the thicker the puree, the slower the heat penetration.

Pressure Canners

A pressure canner is a kettle made from a material, usually aluminum, that is strong enough to safely withstand the pressures used in home canning. The lid is built so it can be locked to the base of the canner. On one type, metal in the sealing edge is ground smooth so little or no leakage occurs between the lid and the base. Care should be taken to avoid damage to the sealing surfaces which could ruin the canner.

Other canners have a gasket made of a rubber-like substance that prevents leakage of steam. The gasket should be washed in hot suds, rinsed, and dried thoroughly after use.

All pressure canners include a safety plug or fuse. One type has a metal fuse that melts when the temperature is too high. If the canner is used properly the fuse should never need replacing. Another type of canner has a rubber-like safety plug. Care should be taken to replace the plug when the rubber gets hard. As the rubber hardens, it takes a greater pressure for it to blow out. Some older type canners have a petcock that serves as a vent and safety valve.

In canners that have a pressure gage, vents serve to exhaust air from the canner. The air is exhausted by venting for 10 minutes after steam starts escaping. All the air must be exhausted before the canner is sealed because the steam has much more heat energy. For example, air in a $212°$ F oven feels just warm while $212°$ steam from a teakettle will burn you.

Be sure to read and follow the instructions with your canner.

All canners should have some type of rack in the bottom. A rack keeps the jars from touching the bottom of the canner and breaking. It also aids in transferring the heat more evenly within the canner by permitting water and steam circulation.

There must be enough water in the canner to provide steam throughout processing. Two quarts of water is usually recommended, although this may vary depending on size of the canner and the quantity of jars.

Pressure canners have either a dial gage, a pressure control or a combination of these. The dial pressure gage indicates the pressure and corresponding sea level temperature. The control type canner has a precision weight that sits on the vent pipe and jiggles to regulate pressure. A third type is a combination gage and control.

The dial pressure gage measures steam pressure. The tube in the pressure canner gage operates like a New Year's Eve noisemaker, which is a flat paper tube rolled up. Blowing into the tube causes it to become more round and unroll.

The pressure gage works the same way except not as dramatically. The gage is made of a partially flattened metal tube. When pressure is applied, the tube becomes more round and straightens slightly. The needle (pointer) moves as the tube straightens. The gage is calibrated to indicate pressure. Pressure is controlled by adjusting burner heat to maintain the desired pressure. This type of gage should be checked yearly or after suspected damage, such as dropping, to be sure it functions properly.

A pressure control consists of a precision weight that rests on a specially designed vent pipe. It automatically maintains an even pressure and temperature inside the canner. Pressure builds inside the canner until the upward force (steam pressure times seat area) is greater than the downward force of the weight on the seat area. At this point, the control weight is lifted, releasing steam and reducing pressure, until the upward force equals the weight. The pressure inside again increases slightly, lifts the weight, and releases pressure.

The repeated lifting and reseating or jiggling of the control weight indicates that the pressure is being controlled.

The burner is adjusted so the control jiggles at least several times a minute. Excessive jiggling will deplete the supply of water in the canner.

There are two types of pressure controls. One type is a single weight with 3 holes which fit on the vent pipe. The diameter at the base of the hole (seat area) is different for each of the 3 pressures—largest for 5 pounds pressure, and smallest for 15 pounds pressure.

The second type has 1 seat area and a 3-piece weight. For 5 pounds pressure, the small center weight is used. One additional ring or weight is added for 10 pounds pressure, and a second ring or weight is added for 15 pounds pressure.

With care, the pressure control remains accurate throughout the canner's lifetime. Be sure that seat areas where the weight and the vent pipe make contact are not damaged or excessively worn; this could affect the canner's performance.

The combination gage is not as common as the other two types. It has a sliding piston which pushes up on a spring. As pressure inside the canner increases, the piston is pushed up. Rings on the piston indicate pressure. If the heat is not regulated correctly, pressure builds up to beyond 15 pounds, at which point the weight is lifted to release the excess pressure. It will jiggle audibly, indicating to the user that the pressure is too high. This system serves as a gage as well as a safety device.

The combination gage should be kept clean and dry when not in use, to prevent corrosion. It also must be checked yearly to be sure the piston slides easily and indicates the correct pressure.

Effect of Altitude

Atmospheric pressure is like the thickness of frosting on a cake. Where it is thickest it weighs more per square inch than where it is thin. At sea level, where the atmosphere is the thickest, it is heavier than atop a mountain.

As altitude increases, atmospheric pressure or its weight per square inch decreases. Altitude affects the boiling point of water. Where altitude is least, at sea level, water boils at 212° F. As altitude increases the boiling point of water decreases.

The same is true in a pressure canner. Under 10 pounds pressure at sea level, water boils at 240° F. As altitude increases, the temperature in a pressure canner at 10 pounds of pressure is less than 240°. This difference is enough to affect the safety of canned products at altitudes above 2,000 feet.

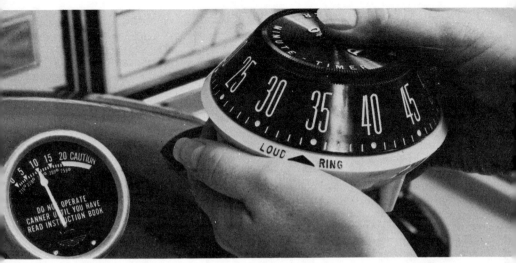
Fred Farout

Processing time for a particular vegetable is the time it takes to heat the coldest part of the jar to a temperature of 240° F, and mantain it long enough to kill any *C. botulinum* spores present. At an altitude of 2,000 feet, it takes 11 pounds of gage pressure for water to boil at 240°. For each additional 2,000 feet increase in altitude, 1 pound of pressure should be added.

For pressure canners that have the pressure control, the 15 pounds pressure weight should be used at altitudes above 2,000 feet for canning low-acid food.

At the end of processing, pressure inside the jars as well as inside the canner is 10 pounds. The pressure inside the canner should be allowed to drop slowly. If pressure inside the canner is released too rapidly, pressure inside the jars will be great enough to force the contents, especially liquid, out of the jars. This may prevent a jar from sealing if a piece of food lodges on the top of the sealing rim. It may even break the jar.

Remove the lid from the canner as soon as the pressure drops. Jars should then be taken out and allowed to cool to room temperature quickly. A type of non-toxic spoilage called flat sour can occur if the jars are allowed to stand in the canner for long periods.

When using the steam-pressure canner, the pressure given is for altitudes less than 2,000 feet above sea level. If you live in an area with a higher altitude, it is necessary to make an adjustment in pressure. See the next chapter for details.

Processing times and pressures should be adjusted for altitude change.

Home Canning of Fruits and Vegetables
by Carole Davis

Canning in the home is increasing in popularity as a method of food preservation. Economic considerations are causing consumers to look for ways to stretch their food dollars. By having their own gardens and canning the harvested produce, they often can save money. Eating quality of home-canned products encourages some individuals to can. The activity of growing or obtaining produce and preserving it in the home sometimes gives people a sense of personal achievement and satisfaction.

In canning, food is preserved by applying heat to prepared food in containers so that micro-organisms that cause spoilage or food poisoning are destroyed and enzymes that cause undesirable quality changes in the flavor, color, and texture of food are inactivated. Preservation of food by canning also depends on sealing the food in sterile, airtight containers to prevent it from coming in contact with micro-organisms in the environment.

Canning is not a difficult technique, but it must be done properly to avoid spoilage and food poisoning, such as the often fatal botulism. It is extremely important that only tested reliable instructions are used, such as those found in U.S. Department of Agriculture publications. Based on scientific research, specific instructions have been developed for preparing, packing, and processing each food. Instructions should be followed *exactly* from beginning to end—without taking any shortcuts or altering any recipes.

Acidity of the food is the chief factor in influencing the time and temperatures necessary for processing. The more acid the products, the easier spoilage organisms are destroyed by heat. Acid foods—such as tomatoes, fruits, and pickled vegetables—can be safely processed at the temperature of boiling water in a boiling-water-bath canner. If spoilage organisms are not killed by adequate processing, they will continue to grow and could reduce the acid in the canned product, thus encouraging the growth of more dangerous organisms, such as *Clostridium botulinum*.

Low-acid foods—all vegetables, except for tomatoes—require a more severe heat treatment than acid foods to kill the organisms. The only safe way to can these foods is with a steam-pressure canner, one with a weighted or dial gage, to obtain temperatures above boiling. *Clostridium botulinum* is extremely dangerous in these low-acid foods because if it is present, and the heat treatment has been insufficient, it can grow and produce a deadly toxin in the sealed containers.

Processing times are based on sea level conditions where water boils at 212° F or, when under 10 pounds of pressure, at 240°. As the altitude increases, the temperature at which water boils decreases. Therefore, at altitudes above sea level you need to make adjustments in canning instructions to insure that foods are adequately processed.

When using a boiling-water-bath canner you must add more time in processing, as given in the table.

At altitudes above sea level it takes more than 10 pounds of pressure for the temperature of boiling water to reach 240° F.

When using the steam-pressure canner, the pressure must be in-

---

Carole Davis is a Supervisory Food Technologist, Consumer and Food Economics Institute, Agricultural Research Service.

Altitude Corrections for Boiling Water Bath

Altitude (feet)	Increase processing time if the time recommended is:	
	20 minutes or less	More than 20 minutes
1,000	1 minute	2 minutes
2,000	2 minutes	4 minutes
3,000	3 minutes	6 minutes
4,000	4 minutes	8 minutes
5,000	5 minutes	10 minutes
6,000	6 minutes	12 minutes
7,000	7 minutes	14 minutes
8,000	8 minutes	16 minutes
9,000	9 minutes	18 minutes
10,000	10 minutes	20 minutes

Screw ring band and disk.

creased although the time remains the same as that recommended for sea level. At altitudes above 2,000 feet, process as follows:

Altitude	Pounds of pressure
2,000 feet	11
4,000 feet	12
6,000 feet	13
8,000 feet	14
10,000 feet	15

Equipment

Use jars made especially for home canning so the jars will be the right size for the processing time and temperature used, properly heat tempered, and resistant to mechanical shock. Always check jars before using to be sure they do not have nicks or cracks. Wash jars in hot, soapy water and rinse well.

It is important to use standard jar closures. They are designed to fit home canning jars correctly, and are made from suitable materials to provide a proper seal. One of the most popular types is the two-piece lid with a metal ring or band and a flat metal disk with a sealing compound. Flat metal disks can be used only once since they may not seal properly if reused. Metal bands may be used repeatedly if they are not rusted or dented.

Wash and rinse lids and bands. Metal lids with sealing compound may need boiling or holding in boiling water for a few minutes; follow the manufacturer's directions. Porcelain-lined zinc caps with rubber shoulder rings can also be used to seal jars. Rubber rings should be used only once.

Water-bath canners are readily available on the market. However, any large metal container may be used for a water-bath canner if it is deep enough so the water is well over the tops of the jars and has space to boil freely. Allow 2 to 4 inches above jar tops for brisk boiling. The container or canner must have a tight fitting cover and a rack to allow water to circulate under the jars.

The steam-pressure canner is made of heavy metal and has a cover which fastens to make the pan steam-tight. The cover is fitted with a safety valve, a petcock or vent, and a gage —either weighted or dial. All parts of the canner must be clean and in good working order. Check the gage before the canning season, and also during the season if canner is used often. The weighted ones need only be checked to determine if they are thoroughly clean. A dial gage can be tested for accuracy by a county Extension agent or an equipment manufacturer.

A pressure saucepan may be used for canning pint jars of food. However, 20 minutes must be added to the processing time recommended for a particular food canned with the pressure canner. This is because pressure saucepans heat and cool more rapidly than pressure canners do. Thus additional time is needed to compensate for the otherwise reduced exposure of the food to heat.

Selecting and Preparing

Home-canned foods will be no better than the raw products with which you begin. Fruits and vegetables should be of good quality with no bruises or soft spots. Be sure to choose fresh, firm, ripe fruits and young tender vegetables. Use them before they lose their freshness. Do not use overripe produce because some foods lose acidity as they mature, and the recommended processing time may not be adequate.

Wash all fruits and vegetables thoroughly, but gently, to remove dirt which contains bacteria. Wash small quantities at a time under running water or through several changes of water. Lift fruits and vegetables out of the water so the dirt will not resettle on the food. Do not let fruits or vegetables soak, as they may lose flavor and food value. Peel and cut or slice produce as indicated in instructions for each specific fruit or vegetable.

Fruits and vegetables may be packed raw into jars, or preheated and packed hot. Raw or cold pack means that raw, unheated food is placed in jars and covered with boiling hot sirup, juice, or water. When foods are hot packed they are heated in sirup, water or steam, or juice for a specified length of time and then packed hot into jars.

Most raw fruits and vegetables can be packed fairly tightly into containers because they cook down during processing. However, raw corn, lima beans, and peas should be packed loosely because they expand during processing.

Hot food should be packed fairly loosely. It should be at or near the boiling temperature when packed. There should be enough sirup, water, or juice to fill in around the solid food in the container and to cover the food. Food at the top of the container may darken if not covered with liquid.

Do not overpack containers as this may result in underprocessing. It is necessary to leave headspace between the lid and the top of the food or liquid in the jar because there will be some expansion of food during processing. The amount of headspace varies with the product, style of pack, and method of heat sterilization, so follow directions for each fruit or vegetable.

When using the flat metal lid with sealing compound, put the lid on a clean jar rim, with sealing compound next to the glass. Then screw the metal band down firmly. The lid will still have enough "give" to let steam escape during processing. Do not tighten the band further after removing the jar from the canner.

When using the porcelain-lined zinc cap, fit the wet rubber ring down on the jar shoulder. Fill the jar and wipe clean the rubber ring and jar rim. Screw the cap down firmly and turn back ¼ inch. When the jar is removed from the canner, tighten the cap to complete the seal.

Processing Fruits

Sugar or sugar-water sirup is often added to fruits to help them hold their shape, color, and flavor. Sugar can be added in the dry form to very juicy fruits.

To make sugar sirup—mix sugar with water or juice extracted from the fruit. Proportions for 3 types of sirup are as follows:

Type of Sirup	Sugar (cups)	Liquid (cups)
Thin	2	4
Medium	3	4
Heavy	4¾	4

Heat sugar and water or juice together until sugar is dissolved.

Fruit may be canned without sweetening—in its own juice or in water—for special diets. Processing time is the same for unsweetened fruit as for sweetened because sugar is not needed to prevent spoilage.

Process fruits by the boiling-water-bath method. Work only with the quantity of food needed for one canner load at one time.

As each jar is filled, adjust the lid, and place the jar on the rack in the water-bath canner about one-half full of hot or boiling water for raw or hot pack, respectively. Be sure the water is 1 to 2 inches over the tops of the jars, and there is an additional 1- to 2-inch space to allow the water to boil freely.

Cover the canner and when the water comes to a rolling boil, start to count the processing time. Boil gently and steadily for the recommended time for the fruit you are canning. A definite length of time is recommended for processing each kind of fruit.

When the processing time is completed, immediately remove the jars from the canner with a pair of jar tongs. Adjust the jar lids if necessary. Cool the jars on a rack or folded towel away from drafts.

Processing Vegetables

A steam-pressure canner must be used for processing all vegetables except tomatoes and pickled vegetables. Work only with the quantity of vegetable needed for one canner load at a time. As each jar is filled, adjust the lid, and place the jar in the pressure canner containing 2 to 3 inches of hot or boiling water for raw or hot pack, respectively, to keep food hot.

Top, hot packing peaches. Bottom, covering peaches with a boiling sirup solution.

The manufacturer's directions for general operation of the canner you are using should be followed. A few pointers on the use of any canner follow:

—Use 2 to 3 inches of boiling water in the bottom of the canner.
—Set filled containers on rack in canner.
—Fasten canner cover securely.
—Allow steam to escape from open petcock or weighted gage opening for at least 10 minutes to drive all air from canner. Then close petcock or put on weighted gage.
—When pressure reaches 10 pounds (240° F), start counting processing time. Keep pressure constant by regulating heat under the canner.
—When processing time is completed, remove the canner from heat immediately. Cool undisturbed at room temperature until the pressure registers zero. After a minute or two, slowly open the petcock or remove the weighted gage. Unfasten the cover and tilt the far side up so steam escapes away from you.
—Remove containers from canner with jar tongs.
—Adjust lids if necessary.
—Cool jars on a rack or folded towel away from drafts.

Day-After Check

Jars should be examined after they have cooled, but within 24 hours after processing, to be sure a seal has been obtained. To test a jar that has a flat metal lid, press the center of the lid; if the lid is down and will not move, it is sealed. Turn jars with porcelain-lined zinc caps partly over in your hands; if they do not leak, they are sealed.

When jars are thoroughly cooled, metal screw bands should be carefully removed. Wipe outside of jars clean, and label jars to show date and contents. Store in cool dry place. If you find a jar that did not seal, use food right away or re-can the food immediately; empty the jar, pack and process the food as if it were fresh.

Look for Spoilage

Check dates on jar labels to be sure you first use food that has the earliest processing date. Before opening any jar for use, look at it carefully for spoilage signs. If it leaks, has a bulging lid, spurts liquid when opened, or has an off-odor or mold, then do not use it. *Do not even taste it.* Destroy it out of the reach of children and pets.

Canned vegetables may contain the toxin that causes botulism without showing any visible signs of spoilage. Therefore, boil all home-canned vegetables covered for at least 10 minutes before tasting or serving. Heating generally makes any odor of spoilage more evident.

If the food appears to be spoiled, foams, or has an off-odor during heating, destroy it.

Cooling snap beans on a rack. Leave space between jars so air can circulate.

How to Can Cut Green Beans*

1. Select green beans:
   Choose young, tender beans
   Allow 1½ to 2½ pounds of fresh beans for each quart to be canned
2. Prepare green beans:
   Wash beans
   Trim ends
   Cut into 1-inch pieces
3. Pack into jars:

   To pack raw—
   Pack raw beans *tightly* into jar
   Leave ½-inch space at top of jar
   Add ½ teaspoon salt to pints; 1 teaspoon to quarts
   Cover with boiling water to ½ inch from top of jar
   Wipe jar rim clean
   Adjust jar lid
   Process in pressure canner at 10 pounds pressure (240° F)
   Pints—20 minutes
   Quarts — 25 minutes
   (At altitudes above sea level, increase pressure according to instructions in early part of chapter)

   To pack hot—
   Cover cut beans with boiling water; boil 5 minutes
   Pack hot beans *loosely* into jar to ½ inch of top
   Leave ½-inch space at top of jar
   Add ½ teaspoon salt to pints; 1 teaspoon to quarts
   Cover with boiling water to ½ inch from top of jar
   Wipe jar rim clean
   Adjust jar lid
   Process in pressure canner at 10 pounds pressure (240° F)
   Pints — 20 minutes
   Quarts—25 minutes
   (At altitudes above sea level, increase pressure according to instructions given earlier)
4. Allow pressure to return to 0
5. Remove jars from canner
6. Complete seals, if necessary

* These instructions are for green beans only. Procedures and processing times are specific for each vegetable. See USDA Home and Garden Bulletin No. 8, *Home Canning of Fruits and Vegtbles*, for directions for canning other vegetables.

How to Can Peaches*

1. Select peaches:
   Choose fresh, firm, ripe fruit, with no soft spots or bruises
   Allow 2 to 3 pounds of fresh peaches for each quart to be canned
2. Prepare peaches:
   Wash peaches
   Dip in boiling water; then in cold water
   Slip off skins
   Cut in halves and remove pits. Slice if desired
3. Prevent darkening:
   Drop peeled peaches into solution of 2 tablespoons each of salt and vinegar per gallon of water
   Drain just before heating or packing raw
4. Pack into jars:

   To pack raw—
   Pack raw peaches in jar to ½ inch of top
   Cover with boiling sirup (See sirup table)
   Leave ½-inch space at top of jar
   Wipe jar rim clean
   Adjust jar lid
   Process in boiling-water bath—
   Pints—25 minutes
   Quarts—30 minutes
   (At altitudes above sea level, increase processing time according to table near start of chapter)

   To pack hot—
   Heat peaches through in hot sirup (See sirup table). If fruit is very juicy, heat it with sugar, adding no liquid
   Pack hot fruit in jar to ½ inch of top
   Cover with boiling sirup
   Leave ½-inch space at top of jar
   Wipe jar rim clean
   Adjust jar lid
   Process in boiling-water bath—
   Pints—20 minutes
   Quarts—25 minutes
   (At altitudes above sea level, increase processing time according to table near start of chapter)
5. Remove jars from canner
6. Complete seals, if necessary

* These instructions are for peaches *only*. Procedures and processing times are specific for each fruit. See USDA Home and Garden Bulletin No. 8, *Home Canning of Fruits and Vegetables*, for directions for canning other fruits.

Freezing Your Garden's Harvest
by Annetta Cook

The growing season brings an abundance of fruits and vegetables freshly harvested from your garden. The unmatchable sweetness of peas cooked fresh from the pods, the tender-crisp texture of fresh broccoli, the delectable flavor of sweet juicy strawberries are irresistible. It is always a disappointment when the growing season is over. You may have more produce than you were able to use within a short time, so why not savor its just-picked freshness during the autumn and winter months—freeze it!

Of all the methods of home food preservation, freezing is one of the simplest and least time-consuming. The natural colors, fresh flavors, and nutritive value of most fruits and vegetables are maintained well by freezing. However, to freeze foods successfully—that is, to preserve their quality—produce must be carefully selected, prepared and packaged, and properly frozen. Be sure to use reliable home-freezing directions such as those found in U.S. Department of Agriculture publications. Unless recommended practices and procedures are observed, the food's eating quality will be a disappointment.

The first consideration before deciding whether to freeze the garden's harvest is whether your freezer can maintain temperatures low enough to preserve quality of the food during freezer storage. Storage temperatures must be 0° F (–18° C) or below to help prevent unfavorable changes in the food, including growth of bacteria. The temperature control of your freezer should be adjusted so the warmest spot in the freezer will always be at 0° F or lower. Freezers and most two-door refrigerator-freezer combinations are best suited for long storage of home-frozen fruits and vegetables since they can be set to maintain this temperature.

Proper preparation of produce is also important to insure high eating quality of frozen vegetables and fruits. Vegetables, except green peppers and mature onions, maintain better quality during freezer storage if blanched, or heated briefly, before freezing.

Blanching is necessary to prevent development of off-flavors, discoloration, and toughness in frozen vegetables. Besides stopping or slowing down the action of enzymes responsible for these undesirable changes, blanching also softens the vegetable, making it easier to pack into containers for freezing.

Fruit does not need to be blanched before freezing. However, most fruits require packing in sugar or sirup to prevent undesirable flavor and texture changes in the frozen product. Sugar, either alone or as part of the sirup, plus the acidity of fruit retards enzyme activity in fruit stored at 0° F or below.

Packaging Material

Material selected for packaging fruits and vegetables for freezing must be moisture-vapor-proof or moisture-vapor-resistant to keep the food from drying out and from absorbing odors from other foods in the freezer. Loss of moisture from the food causes small white areas called "freezer burn" to develop. These areas are not harmful, but if extensive they can cause the food to become tough and lose flavor.

Suitable packaging materials include rigid plastic food containers,

---

Annetta Cook is a Food Technologist in the Consumer and Food Economics Institute, Agricultural Research Service.

plastic freezer bags, heavy aluminum foil, freezer paper or plastic film, glass freezer jars, and waxed freezer cartons. Collapsible, cardboard freezer boxes are frequently used as an outer covering for plastic bags to protect them against tearing.

Select packaging materials suiting the shape, size, and consistency of the food. Rigid containers are suited for freezing all foods, but are especially good for fruit packed in liquid. Non-rigid containers are best for fruits and vegetables packed without liquid. Paper, plastic, or foil wraps are ideal for freezing bulky vegetables such as broccoli, corn on the cob, and asparagus.

Rigid containers with straight sides and flat bottoms and tops stack well in the freezer. They take up less freezer space than rounded containers, containers with flared sides, and bulky, wrapped packages or plastic bags without protective outer cartons. Containers with straight sides or those that are flared, having wider tops than bottoms, are preferred for easy removal of the food before thawing. If the opening is narrower than the body of the container, the food will have to be partially thawed so you can get it out of the container.

Freezer containers and bags are available in a variety of sizes. Do not use those with more than ½-gallon capacity for freezing fruits and vegetables since the food will freeze too slowly, causing poor quality food.

Choose a container that will hold enough food for one meal for your family. You may wish to put up a few smaller packages for use when some family members are not home or to go with your family-size packages when guests are present for meals.

Pack foods tightly into containers. Since most foods expand during freezing, leave headspace between the packed food and closure.

For fruits that are in liquid, pureed, or crushed and packed in containers with wide openings, leave ½-inch headspace for pints, 1-inch headspace for quarts. If containers with narrow openings are used, leave ¾-inch headspace for pints, 1½-inch headspace for quarts.

For fruits and vegetables packed without liquid, leave ½-inch headspace for all types of containers. Vegetables that pack loosely, such as asparagus and broccoli, require no headspace.

Any container for freezer use must be capable of a tight seal. Rigid containers should have an airtight-fitting lid.

Press out all air from the unfilled parts of plastic bags. Immediately twist the top of each bag and securely tie it with a paper- or plastic-covered wire twist strip, rubber band, or string to prevent return of air to the bag.

Some bags may be heat-sealed with special equipment available on the market. Follow the manufacturer's directions.

Edges and ends of paper, foil, or plastic wraps should be folded over several times so the wrap lies directly on top of the food and all air has been pressed out of the package. Seal the ends with freezer tape to hold them securely in place.

Selecting and Preparing

Grow varieties of fruits and vegetables that freeze well. Your county Extension office can provide information on suitable varieties that grow well in your locality.

Produce selected for freezing should be of optimum eating quality. Freezing only preserves the quality of produce as it is at the time of freezing. It never improves quality.

Fruits to be frozen should be firm and ripe. Underripe fruit may have a bitter or off-flavor after freezing. Pick berries when ripe and freeze them as soon after picking as you can. Some fruits—apples, peaches, pears—may need to ripen further after harvesting. But take care they don't get too ripe. Frozen fruit prepared from overripe fruit will lack flavor and have a mushy texture.

Choose young, tender vegetables for freezing. Since vegetables lose quality quickly after harvest, freeze them as soon as possible for maximum quality. The sugar in corn, peas, and lima beans is rapidly lost when held too long before freezing. If you must hold vegetables and ripe fruits for a short while, refrigeration will help retain the just-picked freshness better than leaving produce at room temperature.

Wash small quantities of fruit gently in cold water. Do not permit fruit to stand in water for any length of time since it will become water-soaked and lose flavor and food value. Drain fruit thoroughly.

Peel fruit and remove pits or seeds. Halve, slice, chop, crush, or puree fruit as indicated in the instructions for each specific fruit. Some fruit, especially berries, may be left whole, but remove stems or hulls. Work with small quantities of fruit at a time, particularly if it is fruit that darkens rapidly. Two to three quarts is an adequate amount to handle at one time.

Pack fruit by sirup pack, sugar pack, or unsweetened pack. Most fruit has better texture and flavor with a sweetened pack. Apples, avocados, berries, grapes, peaches, persimmons, and plums can all be frozen satisfactorily without sweetening, but the quality is not quite as good as freezing in sirup or sugar. An unsweetened pack will give as good a quality product for gooseberries, currants, cranberries, rhubarb, and figs as a sweetened pack.

Sirup pack. Make a sugar sirup by dissolving sugar in water. A 40% sirup (3 cups of sugar to 4 cups of water) is recommended for freezing

most fruits. Sirups containing less sugar are sometimes used for mild-flavored fruits; those with more sugar for very sour fruits. The type of sirup to use is specified in the directions for freezing each fruit. Allow ½ to ⅔ cup of sirup for each pint of fruit. Cut fruit directly into the freezer container, leaving the recommended headspace. Add sirup to cover fruit.

Sugar pack. Cut fruit into a large bowl. Sprinkle with sugar. The amount of sugar to use is specified in freezing directions for each fruit. Mix gently until juice is drawn from the fruit and all the sugar is dissolved. Pack fruit and juice into freezer containers.

Unsweetened pack. Some fruit may be packed dry, without added liquid or sugar. Other fruit, particularly if it darkens rapidly, can be covered with water to which ascorbic acid has been added. Crushed fruit or sliced fruit that is very juicy can be packed in its own juice without added liquid.

For all packs except the dry, unsweetened pack, liquid—either sirup, juice, or water—should completely cover the fruit. This prevents the top pieces from changing color or losing flavor due to exposure to air in the headspace.

A small crumpled piece of waxed or parchment paper placed on top of the fruit helps keep it pressed down

Slicing strawberries before freezing in a sugar pack.

in the liquid once the container has been sealed. The paper should loosely fill the headspace area. Do not use aluminum foil since acid in the fruit can cause the foil to pit (form holes), and tiny pieces of foil may drop into the food.

Anti-darkening. Many fruits darken during freezing, particularly if not kept under liquid. Darkening occurs when the fruit is exposed to air. Since a small amount of air is in the liquid as well as the tissues of fruit, some darkening can occur even when the fruit is submerged in liquid. To help retard darkening during freezer storage, add ascorbic acid (vitamin C) to the fruit during preparation.

Ascorbic acid is available in several forms from drug stores, some freezer locker plants, and some grocery stores that sell freezing supplies. Crystalline ascorbic acid is easier to dissolve in liquid than powder or tablet forms. The amount of ascorbic acid to use is given in the directions for those fruits where use of ascorbic acid is beneficial. Ascorbic acid mixtures containing sugar, and sometimes citric acid, also are available. Follow the manufacturer's directions for use of these products.

In preparing vegetables, wash a small quantity of the vegetable gently in several changes of cold water. Lift the vegetable out of the water each time so all dirt will settle to the bottom of the sink or pan.

Shell, husk, or peel and trim. Some vegetables such as lima beans, corn on the cob, and asparagus require sorting for size, since blanching times depend on size of the pieces.

Blanch the vegetable (this is not necessary for green peppers and mature onions). Most vegetables are blanched by heating them in boiling water. A blancher consisting of a tall kettle, basket, and cover is convenient to use and can be purchased at most department or farm supply stores. However, any large pan which can be fitted with a wire or perforated metal basket and covered is suitable.

To insure adequate blanching, immerse a basket containing a small amount of the vegetable (1 pound) into a large amount of boiling water (at least 1 gallon). Start timing once the vegetable has been immersed and the kettle is covered. Blanching time will vary with the vegetable and the size of the pieces, so follow the recommended blanching times for each vegetable.

Cool the vegetable by immersion in a large quantity of cold or iced water. Rapid cooling is necessary to stop the food from cooking. Cool the vegetable for about the same length of time as it was heated. Once cooled, do not leave the vegetable standing in water, as loss of flavor and food value can occur. Drain the cooled vegetable thoroughly before packaging.

Other methods of blanching and cooling are recommended for some vegetables. For example, mushrooms are heated by sauteing, tomatoes by simmering in their own juice. These foods are cooled by setting the pan of food in cold or iced water to speed cooling.

Freezing and Storing

After packing and sealing containers, label them with the name of the food, type of pack (for fruits), and date of freezing. Freeze food soon after packing, placing a few packages at a time in the freezer as you have them ready.

Freeze food at 0° F or below. Do not load the freezer with more food than can be frozen in 24 hours. Usually 2 to 3 pounds of food per cubic foot of freezer capacity can be frozen at a time. Place packages on freezing coils or plates or in fast-freeze section of freezer, leaving a space between each package. Loading the freezer in this manner enables the food to be frozen quickly. Freezing foods too slowly can result in loss of quality.

Once food has frozen, stack containers. Keep freezer surfaces relatively free from frost to insure maximum operating efficiency of your freezer.

Fruits and vegetables stored at 0° F or below will maintain high quality for 8 to 12 months. Unsweetened fruit loses quality more rapidly than sweetened fruit.

Keeping food longer than the recommended time will not make it unsafe to eat, but some quality loss can occur.

Thawing

Home-frozen fruits and vegetables are convenient and easy to use since most of their preparation is done before freezing. Thaw frozen fruit in the refrigerator, or at room temperature in a pan of cool water. Leave fruit in the unopened freezer container.

A pint package of fruit frozen in sirup will take about 6 to 8 hours to thaw in the refrigerator, or ½ to 1 hour in a pan of cool water. Fruit in sugar packs takes less time. Unsweetened packs need more time than sirup packs. For best eating quality, serve fruit with a few ice crystals remaining.

Cook most frozen vegetables without thawing first. (Corn on the cob and leafy vegetables require partial thawing to insure even cooking.) Add the vegetable to boiling salted water. Use 1 cup of water and 1 teaspoon of salt for each quart of vegetable with these exceptions: Use 2 cups of water for lima beans; water-to-cover for corn on the cob. Cover the saucepan during cooking. Cook the vegetable only until tender. Avoid overcooking.

Consult timetable in freezing directions for recommended times for cooking home-frozen vegetables.

How to Freeze Strawberries*

1. Select strawberries:
   Choose firm, ripe red berries with a slightly tart flavor
   Allow about 1½ quarts fresh strawberries for each quart to be frozen
2. Prepare strawberries:
   Wash berries in cold water; drain well
   Remove hulls
3. Pack into rigid freezer containers:

   To pack in sirup—
   Prepare ahead of time a 50 percent sirup by dissolving 4¾ cups sugar in 4 cups of water; this will make 6½ cups sirup
   Add about ½ cup sirup to each container
   Put berries into prepared containers

   To pack in sugar—
   Add ¾ cup sugar to each quart berries
   Mix gently until sugar is dissolved and juice is drawn from berries
   Pack strawberries with juice in containers

   To pack unsweetened—
   Put berries into containers
   For better color, cover with cold water containing 1 teaspoon ascorbic acid per quart of water

   For all packs—
   Press fruit gently down in each container; add liquid (sirup, juice, or water) to cover fruit, unless fruit is packed dry, unsweetened
   Leave recommended amount of headspace (See earlier reference)
   Put a small piece of crumpled waxed paper on top of berries to keep them down in liquid
   Wipe all liquid from top and sides of containers
   Seal tightly with lid
   Label with name of fruit, type of pack, and date of freezing
4. Freeze strawberries:
   Immediately after packaging, place berries in freezer set at 0° F or below; leave space around each container for faster freezing
   Do not freeze more than 1 quart of berries per cubic foot of freezer capacity at a time
   Stack containers of berries once frozen; store at 0° or below

* These instructions are for strawberries only.

How to Freeze Green Peas*

1. Select green peas:
   Choose bright-green, plump, firm pods with sweet, tender peas (do not use immature or tough peas)
   Allow 4 to 5 pounds fresh peas for each quart to be frozen
2. Prepare green peas:
   Shell peas
   Wash shelled peas in cold water; drain
3. Blanch green peas:
   Bring 1 gallon water to a boil in a large kettle
   Put peas (1 pound) in blanching basket
   Lower basket into boiling water
   Cover kettle and heat peas 1½ minutes
   Chill peas promptly in cold or iced water 1½ minutes
   Drain cooled peas
4. Pack green peas:
   Pack drained, blanched peas in freezer containers (See reference on containers in early part of chapter)
   Leave ½-inch headspace between peas and closure
   Seal containers tightly
   Label each package with name of vegetable and date
5. Freeze green peas:
   Immediately after packaging, place peas in freezer set at 0° F or below; leave space around each container for faster freezing
   Do not freeze more than 2 to 3 quarts of peas per cubic foot of freezer capacity at a time
   Stack packages of peas once frozen; store at 0° F or below

* These instructions are for green peas only. Preparation procedures and blanching times are specific for each vegetable. See USDA Home and Garden Bulletin 10, *Home Freezing of Fruits and Vegetables*, for directions for freezing other vegetables.

For Further Reading:

Home Freezing of Fruits and Vegetables, U.S. Department of Agriculture H&G Bul. No. 10, on sale by Superintendent of Documents, U.S. Government Printing Office, Washington, D.C. 20402. 75¢.

Jellies, Jams, Marmalades, Preserves
by Catharine C. Sigman and Kirby Hayes

Changing fruit into a variety of products such as jellies, jams, marmalades, and preserves can be most rewarding. These products serve as a good way to use fruit that is not completely suitable for canning or freezing, while adding variety and economy to the home food preservation plan.

Jams, jellies, and preserves are similar in that they are preserved using sugar, and all are jellied or partially jellied. Each differs from the other due to the fruit used, ratio of ingredients, and methods of preparation.

Jelly is made using fruit juice. It is clear and firm enough to hold its shape when removed from the jar.

Jam is made from crushed or macerated fruit. Less firm than jelly, it spreads more easily.

Conserves are jams made from a mixture of fruits including citrus. Sometimes nuts and raisins are added.

Preserves are whole fruits or large pieces of fruit in a sirup that varies in thickness.

Marmalades are usually made from pulpy fruits, with skin and pulp suspended in a clear, jellied liquid. For citrus marmalades, the peel is sliced very thin.

Butters are made by cooking fruit pulp with sugar to a thick consistency which spreads easily.

Jellied fruit products need a balanced ratio of fruit, acid, pectin, and sugar for best results.

Fruit provides the characteristic color and flavor, and furnishes at least part of the pectin and acid that combines with added sugar to give the desired gel. Full flavored fruits are needed to offset the dilution of flavor by the large proportion of sugar used.

Pectin is the actual jellifying substance and is found in many fruits in adequate quantity. If pectin is lacking, apple juice extract or commercial pectin may be used. All fruits have more pectin when underripe.

Commercial pectin is available in both liquid and powder forms. It is essential to follow the manufacturer's instructions or tested recipes as in U.S. Department of Agriculture publications. These preparations generally bring higher yields plus the advantages of being able to use fully ripe fruit, with a shorter cook time and more uniform results.

Acid content varies among fruits and is higher in underripe fruits. Acid is needed both for gel formation and for flavor. When fruits are low in acid, lemon juice or citric acid may be used. Commercial pectins also have added acid.

Either beet or cane sugar in fruit products acts as a preserving agent, helps in forming the gel, and enhances the finished product's flavor. In preserves, sugar aids in firming the fruit or fruit pieces.

Sweeteners such as brown sugar, sorghum and molasses are not recommended since their flavor overpowers the fruit flavor and their sweetness varies.

Other than artificial sweeteners, suitable sugar replacements are light corn sirup and light, mild honey. Neither can substitute fully for sugar on a one-to-one basis. For best results use a tested recipe, but if one is not available replace about ¼ to ½ of the sugar with corn sirup or honey. Longer boiling (for recipes without

---

Catherine C. Sigman is Extension Home Economist—Foods, University of Georgia, Athens. Kirby Hayes is Professor, Department of Food Science and Nutrition, University of Massachusetts, Amherst.

pectin) may be required since additional moisture is being added.

Fruits for jellied products without added pectin must be hard ripe and full flavored, or in a proportion of ¾ fully ripe and ¼ underripe, in order to provide the needed pectin. If liquid or powdered pectin is used, fully ripe fruit is best.

After sorting to remove overripe or undesirable fruit, wash in cold running water or several changes of cold water. Prepare fruit according to the specific recipe, discarding any spoiled or bruised portions. Only the amount needed should be prepared to prevent quality loss.

Jam and Jelly Equipment

Water bath canner
Jelly thermometer
Timer
Widemouth funnel
Large, flat-bottomed kettle (8-10 qt.)
Measures
Measuring cup and spoons
Food chopper or masher
Long-handled spoon
Colander
Ladle
Jar lifter
Jelly bag and cheesecloth
Jelmeter
Canning jars and fittings

Extraction

In jelly making, juice is extracted either by crushing, by limited heating using small amounts of water, or by longer cooking with measured amounts of water. Heating aids in pectin extraction for those recipes not using added pectin.

The prepared fruit is put in a damp jelly bag or several thicknesses of damp cheesecloth, tied, and hung to drip. The clearest juice will be free run, but yields increase if the bag is pressed or twisted. Re-straining this juice is recommended. Do not squeeze or press.

Jams, jellies, and preserves can be made with added pectin or without it, depending on the fruit. Fruits such as raspberries, strawberries, and peaches generally need added pectin. Apples, crabapples, currants, plums, grapes, and quinces—if not overripe —contain enough pectin and acid for good gel strength.

Pectin content can be checked visually by mixing 1 tablespoon of cool cooked fruit juice and 1 tablespoon of denatured alcohol and mixing. Fruit high in pectin will form a jellylike mass while fruit low in pectin will show little clumping. **Caution:** Do not taste; the mixture is poisonous.

Pectin may also be tested using a jelmeter. This graduated glass tube measures the rate of fruit juice flow through the tube, giving a rough estimate of the amount of pectin present.

Jellied fruit products made without added pectin require less sugar per fruit unit and need longer boiling to reach the end point. The yield of finished product is less than that with added pectin.

Pectin added to fruit, either in powder or liquid form, must be used in recipes designating the type. Powdered pectin is mixed with the unheated fruit juice or unheated crushed fruit.

Liquid pectin is added to the boiling fruit juice or fruit and sugar mixture. The boiling time of 1 minute for both types is used and must be accurately timed. Regardless of type, or whether pectin is used, you must follow directions closely, taking accurate measurements.

When It's Done

One of the largest concerns when making jelly without added pectin is to know when it is done, or judging the end point. Two of the most frequently used methods for testing doneness of jelly without added pectin are the temperature test and the spoon or sheet test.

The temperature test is the most scientific method and probably the most dependable. Before cooking jelly, take the temperature of boiling water with a jelly or candy thermometer. Cook the jelly mixture to a temperature 8° F higher than the boiling point of water. If cooked to this point, the jelly mixture should form a satisfactory gel. Cook other jellied mixtures to a temperature 9° higher than the boiling point of water.

To get an accurate reading, place the thermometer in a vertical position with the bulb completely covered by the jelly mixture but not touching the bottom of the kettle. Stir jam, preserve, conserve, and marmalade mixtures before taking the temperature. Read the thermometer at eye level.

To test the jellying point by the spoon or sheet test, dip a cool metal spoon into the boiling jelly mixture and lift the spoon so the sirup runs off the side. When the sirup no longer runs off the spoon in a steady stream, but two drops form together and sheet off the spoon, the jelly should be done.

Pouring hot jelly mixture into canning jars.

Once the jellying point is reached, quickly pour jelly into sterilized containers. When sealing jelly with lids, use only standard canning jars and new lids. Pour the boiling hot jelly mixture into sterilized hot jars, leaving ⅛ inch head space. Wipe the jar rims clean, place hot metal lids on jars with the sealing compound next to the glass, screw the metal bands down firmly, and stand the jars upright to cool.

The paraffin seal is recommended only for jelly. Pour the boiling hot jelly mixture into sterilized hot containers, leaving ½ inch head space. Cover hot jelly with hot paraffin to make a single thin layer ⅛ inch thick. Paraffin should touch all sides of the container. Prick air bubbles in the paraffin.

Heat processing of jams, preserves, conserves, and marmalades is recommended, especially in warm or humid climates. Place filled jars on a rack in a water bath canner or other large container filled with hot water. The water should be an inch or two over the tops of the jars. Cover canner. Bring the water to a rolling boil and boil gently for five minutes.

Remove the products from the canner immediately when the processing time is up. Place the containers on a rack or folded cloth away from drafts to cool.

Let jellied products stand overnight to avoid breaking the gel. Remove screw bands, and label the containers with the name of the product and the date. Store in a cool, dry place. Jellied products have a much better flavor and color if stored only for a short time.

If It Doesn't Gel

What if the jelly doesn't gel? Try using it as a topping for pancakes or ice cream, or try recooking the mix-

How to Prevent Problems With Jellied Products

Problem	Cause	Prevention
Formation of crystals	Excess sugar	Test fruit juice with jelmeter for proper proportions of sugar
	Undissolved sugar sticking to sides of kettle	Wipe side of pan free of crystals with damp cloth before filling jars
	Tartrate crystals in grape juice	Make grape jelly stock, and let tartrate crystals settle out before making jelly. Then strain through two thicknesses of cheesecloth to remove crystals
	Mixture cooked too slowly or too long	Cook at a rapid boil. Remove from heat immediately when jellying point is reached
Syneresis or "weeping"	Excess acid in juice makes pectin unstable	Maintain proper acidity of juice
	Storage place too warm or storage temperature fluctuated	Store in a cool, dark and dry place
	Paraffin seal too thick	Seal jelly with a single thin layer of paraffin ⅛ inch thick. Prick air bubbles in paraffin
Too soft	Overcooking fruit to extract juice	Avoid overcooking as this lowers the jellying capacity of pectin
	Incorrect proportions of sugar and juice	Follow recommended instructions
	Undercooking causing insufficient concentration	Cook rapidly to jellying point
	Insufficient acid	Avoid using fruit that is overripe. Lemon juice is sometimes added if fruit is acid deficient
	Making too large a batch at one time	Use only 4 to 6 cups of juice in each batch of jelly
Too stiff or tough	Overcooking	Cook jelly mixture to a temperature 8° F higher than the boiling point of water or until it "sheets" from a spoon
	Too much pectin in fruit	Use ripe fruit

How to Prevent Problems With Jellied Products—*Continued*

Problem	Cause	Prevention
Cloudy	Green fruit (starch)	Use firm, ripe fruits or slightly underripe
	Imperfect straining	Do not squeeze juice but let it drip through jelly bag
	Jelly allowed to stand before it was poured into jars or poured too slowly	Pour into jars immediately upon reaching jellying point. Work quickly
Bubbles	Kettle was not held close to top of jar as jelly was poured, or jelly was poured slowly and air became trapped in hot jelly	Hold kettle close to top of jar and pour jelly quickly into jar
	May denote spoilage. If bubbles are moving, do not use	Follow recommended methods to get airtight seal
Mold (denotes spoilage; do not use)	Imperfect seal	Use recommended methods to get airtight seal
	Lack of proper sanitation	Sterilize jelly glasses and all equipment used

ture. To remake jelly without added pectin, heat the jelly to boiling and boil for a few minutes until the jellying point is reached. Remove the jelly from the heat, skim, pour into hot, sterilized containers and seal.

To remake with powdered pectin, measure ¼ cup sugar, ¼ cup water, and 4 teaspoons powdered pectin for each quart of jelly. Mix the pectin and water and bring to a boil, stirring constantly. Add the jelly and the sugar, stir thoroughly, and bring to a full rolling boil over high heat, stirring constantly. Boil the mixture hard for 30 seconds, remove from the heat, pour into hot containers and seal.

To remake with liquid pectin, measure ¾ cup sugar, 2 tablespoons lemon juice, and 2 tablespoons liquid pectin for each quart of jelly. Bring the jelly to a boil over high heat. Add the sugar, lemon juice and liquid pectin and bring to a rolling boil, stirring constantly. Boil the mixture hard for 1 minute. Remove the jelly from the heat, skim, pour into hot, sterilized containers and seal.

High quality jellied products depend on many factors so there may be several possible solutions to problems in making these products. Some common problems and their prevention are given in the table.

For Further Reading:

How To Make Jellies, Jams and Preserve at Home, H&G Bul No. 56, on sale by Superintendent of Documents, U.S. Government Printing Office, Washington, D.C. 20402. 55¢.

Pickles, Relishes Add Zip and Zest
by Isabelle Downey

Pickles or relishes can add zip and zest to your meals, snacks and party refreshments. They contain small amounts of nutrients, depending on ingredients used in making them. But they have little or no fat and are low in calories, except for the sweet varieties.

Sun-drying, salting, smoking and pickling were methods used in ancient times for preserving food. Pickling is still popular today.

Pickling is preserving foods in vinegar or brine or a combination of the two. Other ingredients are sometimes added to make pickles crisp and spicy.

Relishes and some pickles can be made in a few hours. Other pickles may take three to six weeks.

There are four basic classifications of pickle products, depending on ingredients used and method of preparation.

Brined pickles are sometimes called fermented pickles and take three weeks or longer to cure. Dilled cucumbers, sauerkraut and some vegetables are often prepared this way. Cucumbers change from a bright green to an olive or yellow green while the interior becomes uniformly translucent. Sauerkraut is tart and tangy in flavor, creamy-white in color, and crisp and firm.

Fresh pack pickles are also called the quick process. This method is very popular for the family with limited time. Ingredients are combined and put directly in the jar to be heat processed, or combined and heated a short time before being placed in the jar for heat processing.

Fruit pickles are usually made of whole fruits simmered in a spicy, sweet-sour sirup. Some of the favorites are peach, pear and watermelon rind.

Relishes are made from chopped fruits or vegetables (or a combination), with seasonings added and cooked to a desired consistency. They can be hot, spicy, sweet or sour, depending on the recipe used. Corn relish, chili sauce, catsup, chow-chow and chutney are popular examples.

Always use a tested recipe; one that is current and reliable. Too little of one ingredient and too much of another could cause the pickles to be unsafe to eat. Read the complete recipe before starting the preparation, and be sure you understand exactly what you are to do. Check to see you have all the ingredients. Accurate measurements and weights are most important in making pickles and relishes if a quality and safe product is to be the result.

Use only good quality fruits and vegetables. Select tender vegetables and firm ripe fruit. Pears and peaches may be slightly underripe for pickling. The pickling type cucumber is the variety you will want to use. The salad (slicing) variety does not make a crisp pickle. Contact your county Extension office for the variety grown in your area.

Wax-coated cucumbers bought from the vegetable counter are not suitable for pickling because brine cannot penetrate the wax. Besides, cucumbers for pickling should be used within 24 hours after gathering. If they are kept—even refrigerated—longer than 24 hours before the pickling process begins, you may have a poor quality product.

Always remove the blossom. This may contain fungi or yeasts which could cause enzymatic softening of the cucumber. If whole cucumbers are

---

Isabelle Downey is Home Economist-Food Preservation, Cooperative Extension Service, Auburn University, Auburn, Ala.

to be brined, you may want to leave a ¼-inch stem.

Do not use vegetables or fruits that have even a slight evidence of mold or decay.

In preparing fruits and vegetables to be pickled, wash them thoroughly in cold water whether they are to be peeled or not. Lift out of the water each time, so soil that has been washed off will not drain back over them. Rinse the pan thoroughly between each washing. This is a good time to check again to see if you have fruits or vegetables that should not be used. Too, you can sort as to size, shape and color. This makes for a uniform pack and attractive product.

Ingredients

SALT—Pure granulated salt with no noncaking material or iodine added is best. This is sold as pickling salt, "barrel" salt, and "kosher" salt. Pickling salt is sold at the grocery store and "barrel" salt from many farm supply stores.

Table salt contains noncaking materials that may interfere with fermentation during brining. It also may make the brine cloudy. Iodized salt may darken pickles. Never use ice cream salt or rock salt—they are not food-pure.

VINEGAR—Use a 4 to 6 percent acidity (40 to 60 grain) cider or white vinegar. Read the label, for if it does not have the amount of acidity listed, it should not be used for making pickles or relishes. Some vinegar has 19 percent acidity—this must be diluted. Directions are on the label. Don't use homemade vinegar since the acidity is not known.

Cider vinegar, used in most recipes, has a good flavor and aroma but may discolor pears, cauliflower, onions; therefore white distilled vinegar is used for these. If a less sour product is preferred, choose a recipe that has more sugar. Do not use less vinegar than the recipe specifies.

SUGAR—Granulated, white sugar is used in most pickles. However, some recipes have brown sugar as an ingredient and say so.

SPICES AND HERBS—Always use fresh spices and herbs for best flavor. They deteriorate and lose their pungency in heat and humidity. If they are not to be used immediately, store them in an airtight container in a dark, dry, cool place.

Whole spices, if left in the jar with the pickles, will darken them; therefore they can be tied in a thin cloth bag and removed just before pickles are packed into the jar. Ground spices tend to darken pickles and relishes.

WATER—It is best not to use hard water in brining. If you have hard water, boil it in a stainless steel or uncracked enamel container for 15 minutes. Remove from heat, cover, and let sit for 24 hours. Remove any scum which might have formed. Slowly pour water from the container so that sediment will not be disturbed. The water is now ready to use.

Equipment

Having the right kind, size and amount of equipment and tools can save you time and energy. Check these the day before you plan to make your pickles. Otherwise you may not have what you need.

For fermenting or brining use a crock or stone jar that has never had fat or milk in it. An unchipped enamel-lined pan, glass or stainless steel are also O.K. Do not use plastic.

To cover vegetables while they are in a brine, you will need a heavy plate or large glass lid that fits inside the container. Use a filled jar of water to hold the cover down, so that vegetables are kept below the surface of the brine. Be sure the jar has a tight fitting lid.

For heating pickling liquids, use utensils of unchipped enamelware, stainless steel, aluminum or glass. Do not use copper, brass, galvanized or iron utensils; these metals may react with acids or salts and cause undesirable color changes in pickles or form undesirable compounds.

Among small utensils that will help you do the job are measuring spoons, stainless steel spoons, measuring cups, household scales, sharp knives, vegetable peelers, large trays, canning tongs, ladle with a lip for pouring, slotted metal spoons, footed colander or wire basket, canning funnel, food chopper or grinder, and non-porous cutting board.

All pickles and relishes should be processed in a boiling-water bath canner. Any large metal or enamel container may be used if it:

- Is deep enough to allow 2 inches or more of water above the tops of the jars, plus extra space for boiling
- has a close-fitting cover
- Is equipped with a wire or wooden rack

A steam-pressure canner can be used if it is deep enough. For this purpose, set the cover in place without fastening it. Be sure the petcock is wide open so that steam escapes and pressure is not built up.

Standard home canning jars are used for pickles and relishes. Do not use jars and lids from commercially canned foods. They are designed for use on special packing machines and are not suitable for home canning.

Select jars free from nicks, chips or any defects. As you wash the jars in warm soapy water and rinse them, run your finger around the jar opening to see if there is a defect. If there is, the jar will not seal.

Look at each new metal lid to be sure the sealing compound is even and smooth. Check the metal screw band to see that it is not bent or rusty. Bands can be used over and over again. As for pretreatment of lids and bands, follow the manufacturer's directions. Read these even if you have used that brand before; the directions may have changed.

When using rubber rings get clean, new ones that are the right size for the jars. Do not test these by stretching. Follow the manufacturer's directions as to pretreatment needed.

It is always best to follow current, reliable procedures as in U.S. Department of Agriculture or Extension publications. This insures a quality product and one that is safe to eat. Time, energy and money may be wasted if you use outdated or careless canning procedures.

Fill the jars firmly and uniformly with the pickle product. Avoid overpacking so tightly that the brine or sirup is prevented from filling around and over the product. Slide a plastic spatula down each side of the jar to remove any air spaces. Add enough liquid to cover the pickles. Be sure to allow head space at the top of the jar as recommended in the recipe. This means there is no food or liquid in that space.

Wipe the rim, inside and top, and threads of the jar with a clean, damp cloth to remove any particles of food, spices, seeds or liquid. A small particle may prevent an airtight seal.

The two-piece metal cap (flat metal lid with sealing compound and metal screw band) is the most commonly used closure. Read the manufacturer's directions on treatment needed to close the lid. These vary from one manufacturer to another.

When using a porcelain-lined zinc cap with shoulder rubber ring, screw the cap down firmly against the wet rubber ring, then turn it back one-fourth inch. Immediately after processing and removal of the jar from the canner, screw the cap down tight to complete the seal.

If liquid has boiled out of a jar during processing, do not open it to add more liquid, because spoilage organisms may enter. This applies to 2-piece lids also. Seal the jar as it is.

347

Heat Treatment

All pickle products require heat treatment to destroy organisms that cause spoilage and to inactivate enzymes that may affect flavor, color and texture. Adequate heating is best achieved by processing the filled jars in a boiling-water bath.

Spoilage organisms are in the air and there is danger of them contaminating the food as it is transferred from boiler to jar. This can happen when even the utmost care is taken. Therefore, boiling-water bath processing is needed.

After adjusting the lid, put the jar on the rack into the actively boiling water.

Now that the jar is in the water bath canner, fill the next jar. Continue until all jars are in the canner. Be sure to leave a small space around each jar. This allows the water to circulate. Water should come 2 or more inches above jar tops; add boiling water if necessary.

Cover the canner with a close-fitting lid and bring the water back to boiling as quickly as possible. Start to count the processing time when the water returns to boiling, and continue to boil gently and steadily for the recommended time according to the recipe.

When time is up, slide the canner from the hot range unit. Close windows and doors so that a draft will not be blowing on jars as they are removed. As you remove the lid, be sure to do this away from you so that you will not be burned by steam. Remove one jar at a time, using your canning tongs. Complete the seals if the manufacturer so directs. Set jars upright, away from a draft, and several inches apart, on a dry cloth or wire rack to cool. Do not cover with a cloth.

For fermented (brined) cucumbers and fresh-pack dills, start to count the processing time as soon as all the filled jars are in the actively boiling water. This prevents development of a cooked flavor and loss of crispness.

Most pickle and relish recipe processing times are given for altitudes less than 1,000 feet above sea level. If you are 1,000 feet or above, you need to increase the recommended processing time. See table in canning chapter by Carole Davis.

After 12 to 24 hours, check to make sure the jars have an airtight seal. Read the manufacturer's directions but if these are not given, here are some general ways to tell if the seal is airtight. For the metal lid with a sealing compound and the metal screw band, if the center of the lid has a slight dip or stays down when pressed, the jar is sealed. Another test is to tap the center of the lid with a spoon. A clear, ringing sound means a good seal. A dull note, however, does not always mean a poor seal. Another way to check for an airtight seal is by turning the jar partly over. If there is no leakage, the jar may be stored.

If the porcelain-lined zinc cap with rubber ring has been used, check for airtight seal by turning the jar partly over. If there is no leakage, the seal is tight.

If the jar is not sealed, use the product right away or recan it. To recan, empty the jar, repack in another clean jar, use a new lid, and reprocess the product as before.

If metal bands are used, these can be removed from jars after 24 hours if you want to.

Wipe jars with a clean, damp cloth. Make a label for the jar. Put the name of the product and date on the label.

Store canned pickles and relishes in a dark, dry, cool place where there is no danger of freezing. Freezing may crack the jars or break the seals, and let in bacteria.

Before using, always examine each jar for signs of spoilage. A bulging lid or leakage may mean that the contents are spoiled.

When a jar is opened look for other signs of spoilage such as:
- Spurting liquid
- Mold
- Disagreeable odor
- Change in color
- Unusual softness, mushiness or slipperiness

If there is ever the slightest indication of spoilage, do not eat or even taste the contents. Dispose of the contents so they cannot be eaten by humans or animals. Also dispose of the lid.

After emptying the jar of spoiled food, wash the jar in hot, soapy water and rinse. Boil in clean water for 15 minutes.

Pickle Problems

Why are pickles soft or slippery?
This generally results from microbial action which causes spoilage. Once a pickle becomes soft it cannot be made firm again. Microbial activity may be caused by
—Too little salt or acid
—Cucumbers not covered with brine during fermentation
—Scum not removed from brine during fermentation
—Insufficient heat treatment
—Seal is not airtight
—Moldy garlic or spices

Blossoms, if not removed from the cucumbers before fermentation, may contain fungi or yeasts responsible for enzymatic softening.

Why are pickles shriveled?
—Using too strong a vinegar, sugar or salt solution at the start of the pickling process. In making the very sweet or very sour pickles, it is best to start with a dilute solution and increase gradually to the desired strength
—Overcooking
—Overprocessing

Why are pickles dark?
—Use of ground spices

—Too much spice
—Whole spices left in jar
—Iodized salt
—Minerals in water, especially iron
—Overcooking

What causes garlic to turn purple or blue?
—Garlic contains anthocyanins, a water soluble pigment also found in beets. This changes color very easily and with the acid condition in pickles, turns blue or purple in color

How can you tell if sauerkraut is spoiled?
—Undesirable color
—Off odors
—Soft texture

Why does kraut get soft?
—Insufficient salt
—Too high temperatures during fermentation
—Uneven distribution of salt
—Air pockets caused by improper packing

Why does kraut get pink?
This is caused by growth of certain types of yeast on the surface of the kraut due to:
—Too much salt
—Uneven distribution of salt
—Kraut improperly covered or weighted during fermenation

Why does kraut turn dark?
—Unwashed and improperly trimmed cabbage
—Insufficient juice to cover fermenting cabbage
—Uneven distribution of salt
—Exposure to air
—Long storage period
—High temperature during fermentation, processing and storage

For Further Reading:

Making Pickles and Relishes at Home, H&G Leaflet #92, on sale by Superintendent of Documents, U.S. Government Printing Office, Washington, D.C. 20402. 35¢.

Wine Making (with a note on vinegar)
by Philip Wagner and J. R. McGrew

Grapes are the world's leading fruit crop and the eighth most important food crop in the world, exceeded only by the principal cereals and starchy tubers. Though substantial quantities are used for fresh fruit, raisins, juice and preserves, most of the world's annual production of about 60 million metric tons is used for dry (nonsweet) wine.

Wine is of great antiquity, as every Bible reader knows, and a traditional and important element in the daily fare of millions. Used in moderation, it is wholesome and nourishing, and gives zest to the simplest diet. It is a source of a broad range of essential minerals, some vitamins, and easily assimilated calories provided by its moderate alcoholic content.

In its beginnings, winemaking was as much a domestic art as breadmaking and cheesemaking. It still is, wherever grapes are grown in substantial quantity. Though much wine is now produced industrially, many of the world's most famous wines are still made on what amounts to a family scale, the grapegrower being the winemaker as well.

Production of good dry table wine for family use is not difficult, provided certain essential rules are observed.

The right grapes. Quality of a wine depends first of all on the grapes it is made from. As is true of other fruits, there are hundreds of grape varieties. They fall in three main groups.

First, there are the classic *vinifera* wine grapes of Europe. These also dominate the vineyards of California, with its essentially Mediterranean climate. But several centuries of trial have shown that they are not at home in most other parts of the United States.

Second, there are the traditional American sorts such as Concord, Catawba, Delaware, and Niagara, which are descendants of our wild grapes and much grown where the *vinifera* fail. They have pronounced aromas and flavors, often called foxy, which, though relished in the fresh state by many, reduce their value for wine.

Third, there are the French or French-American hybrids, introduced in recent years and now superseding the traditional American sorts for winemaking. The object in breeding these was to combine fruit resembling the European wine grapes with vines having the winter hardiness and disease resistance of the American parent. They may be grown for winemaking where the pure European wine grapes will not succeed.

What wine is. Simply described, wine is the product of the fermentation of sound, ripe grapes. If a quantity of grapes is crushed into an open half-barrel or other suitable vessel, and covered, the phenomenon of fermentation will be noticeable within a day or two, depending on the ambient temperature. It is initiated by the yeasts naturally present on the grapes, which begin to multiply prodigiously once the grapes are crushed.

Fermentation continues for three to ten days, throwing off gas and a vinous odor. In the process, the sugar of the grapes is reduced to approximately half alcohol and half carbon dioxide gas, which escapes. Fermentation subsides when all the sugar has been used up. The murky liquid

---

Philip Wagner is Proprietor, Boordy Vineyards, Riderwood, Md., and author of *Grapes Into Wine.* J. R. McGrew is with the Agricultural Research Service, Beltsville, Md.

is then drained and pressed from the solid matter and allowed to settle and clear in a closed container.

The resulting liquid is wine—not very good wine if the constituents of the grapes were not in balance, and readily spoiled, but wine nevertheless.

Beneath the apparent simplicity, the evolution of grapes into wine is a series of complex biochemical reactions. Thus winemaking can be as simple or as complex as you wish to make it. The more you understand and control the process, the better the wine. The following instructions cover only the essentials of sound home winemaking.

Under Federal law the head of a household may make up to 200 gallons of wine a year for family use, but is first required to notify the Treasury Department's Bureau of Alcohol, Tobacco and Firearms on Form 1541.

Making Red Wine

The grape constituents which matter most to the winemaker are (a) sugar content of the juice, and (b) tartness or "total acidity" of the juice.

Sugar content is important because the amount of sugar determines alcoholic content of the finished wine. A sound table wine contains between 10% and 12½% alcohol. The working rule is that 2% sugar yields 1% of alcohol. Example: a sugar content of 22% yields a wine of approximately 11% alcohol.

California grapes normally contain sufficient sugar. Grapes grown elsewhere are often somewhat deficient, and the difference must be made up by adding the appropriate amount of ordinary granulated sugar which promptly converts to grape sugar on contact with the juice.

Saccharometer and hydrometer jar. Instrument floats at zero in plain water. It floats higher according to sugar content of grape juice.

Sugar Correction Table

What the saccharometer shows	For wine of 10% by volume, add	For wine of 12% by volume, add
	Ounces of sugar per gallon	
10	11.8	16.2
11	10.1	14.8
12	8.9	13.3
13	7.4	11.9
14	5.9	10.4
15	4.6	8.9
16	3.0	7.5
17	1.5	6.0
18		4.3
19		2.9
20		1.4

Note: The result is not precise, yield of alcohol varying under the conditions of fermentation.
—Adapted from *Grapes Into Wine* by Philip M. Wagner.

In using non-California grapes, you need to test the sugar content in advance. That is done by a simple little instrument called a saccharometer, obtainable at any winemakers' shop. This is floated in a sample of the juice, and a direct reading of sugar content is taken from the scale. The correct amount of sugar to add, in ounces per gallon of juice, is then determined by reference to the sugar table.

If *total acidity*, or tartness, is too high and not corrected, the resulting

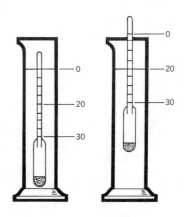

wine will be too tart to be agreeable. Again, California grapes are usually within a satisfactory range of total acidity. Grapes grown elsewhere are often too tart, and acidity of the juice should be reduced.

In commercal winemaking this is done with precision.

The home winemaker rarely makes the chemical test for total acidity but uses a rule of thumb. He corrects the assumed excess of acidity with a sugar solution consisting of 2 pounds of sugar to 1 gallon of water—adding 1 gallon of the sugar solution for every estimated 4 gallons of juice. This sugar solution is in addition to the sugar required to adjust sugar content of the juice itself.

In estimating the quantity of juice, another practical rule is that 1 full bushel of grapes will yield approximately 4 gallons. The winemaker therefore corrects with 1 gallon of sugar solution for each full bushel of crushed grapes.

A hand-crank grape crusher.

The pigment of grapes is lodged almost entirely in the skins. It is during fermentation "on the skins" that the pigment is extracted and gives red wine its color.

How to proceed. Crush the grapes directly into your fermenter (a clean open barrel, plastic tub or large crock, never metal). Small hand crushers are available, but the grapes may be crushed as effectively by foot—wearing a clean rubber boot. Then remove a portion of the stems, which may otherwise give too much astringency to the wine.

Low-acid California grapes are quite vulnerable to bacterial spoilage during fermentation. To prevent spoilage and assure clean fermentation, dissolve a bit of potassium metabisulfite (known as "meta" and available at all winemakers' shops) and mix it into the crushed mass. Use ¼ ounce (⅓ of a teaspoonful) per 100 pounds of grapes.

Also use a yeast "starter". This comes as a 5 gram envelope of dehydrated wine yeast, also obtainable at winemakers' shops. To prepare the starter, empty the granules of yeast into a shallow cup and add a few ounces of warm water. When all the water is taken up, bring it to the consistency of cream by adding a bit more water. Let stand for an hour, then mix it into the crushed grapes.

After the meta and yeast are added, cover the fermenter with cloth or plastic sheeting to keep out dust and fruit flies, and wait for fermentation.

If non-California grapes are used, test and make the proper correction for sugar content. Then correct the total acidity by adding sugar solution as described earlier. In using non-California grapes, it is desirable, but not necessary at this point, to add a dose of meta. A yeast starter is advisable.

As fermentation begins, the solid matter of the grapes will rise to form a "cap". Push this down and mix with the juice twice a day during fermentation, always replacing the cover.

When fermentation begins to subside and the juice has lost most of its sweetness, it is time to separate the turbid, yeasty and rough-tasting new wine from the solid matter. For this purpose a press is necessary, preferably a small basket press though substitutes can be devised.

Be ready with clean storage containers for the new wine, several plastic buckets, and a plastic funnel. The best storage containers for home winemaking are 5-gallon glass bottles or small fiberglass tanks.

Beware of small casks and barrels for several reasons. They are usually leaky. They are sources of infection and off-odors that spoil more homemade wine than any other one thing. And there is frequently not enough new wine to fill and keep them full. Wine containers must be kept full; otherwise the wine quickly spoils. Using glass containers, you can see what you are doing.

With the equipment assembled, simply bail the mixture of juice and solid matter into the press basket. The press basket serves as a drain, most of the new wine gushing into the waiting buckets and being poured from them into the containers. When the mass has yielded all its "free run", press the remainder for what it still contains.

Fill the containers *full*, right into the neck. Since fermentation will continue for awhile longer, use a stopper with a fermentation "bubbler" which lets the gas out but does not let air in. When the bubbler stops bubbling and there are no further signs of fermentation, replace it with a rubber stopper or a cork wrapped in waxed paper.

Store the wine for several weeks at a temperature of around 60° F. Suspended matter in the wine will begin to settle, and at this temperature certain desirable reactions continue to take place in the wine itself.

At the end of this period, siphon the wine from its sediment, with a plastic or rubber tube, into clean containers. At the same time dissolve and add a bit of the meta already referred to at the rate of ¼ level teaspoon per 5 gallons of wine. This will protect against off odors and spoilage but does not otherwise affect the wine.

Clarifying

Next, transfer the containers to a place where the wine will be thoroughly chilled, even down to freezing. This precipitates more suspended matter and unwanted ingredients, and encourages clarification.

Assuming that the wine was made in early fall, hold it in cool storage until after the first of the year. By then it should have "fallen bright" and be stable. To test its clarity, hold a lighted match behind the bottle.

The wine is then siphoned once again from its sediment, and another

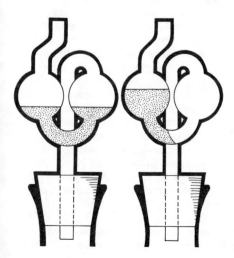

Fermentation bubbler fitted to jar. Left, water is poured in to level shown. Right, position of water immediately before a gas bubble passes through.

dose of meta added at the same rate of ¼ teaspoon per 5 gallons.

If the wine is brilliantly clear, one container of it may then be siphoned into wine bottles, corked or capped, and is ready for immediate use. Despite the common impression, most wine does not gain greatly by aging once it is stable. It continues to evolve, but not necessarily for the better.

The rest of the wine is held until after the return of warm weather to make sure there will be no resumption of fermentation, which would blow corks if the wine was bottled. By mid-May that hazard will have passed, and the wine is ready for its final siphoning, its final dose of the same quantity of meta, and bottling.

Fining. If in January the wine is not brilliantly clear, it should be "fined". This consists of dissolving in a small amount of hot water and mixing in, at the time of siphoning, ordinary household gelatin at the rate of ¼ ounce (2 teaspoonsful) per 5 gallons. This will turn the wine milky when mixed in and will slowly settle, dragging all impurities and suspended matter with it. In two weeks to a month the process of "fining" will be complete. The wine is then ready to be siphoned from the fining sediment and treated as above.

Making White Wine

As we have seen, red wine is fermented "on the skins" in order to extract the coloring matter and other ingredients lodged in the skins. In making white wine, the grapes are crushed and the fresh juice immediately separated by pressing so that it may ferment apart from the skins.

This fresh juice is checked for its sugar content and acidity, as in preparing to ferment red wine, and the proper corrections are made immediately after pressing. Likewise, a yeast "starter" is added.

The fermentation takes place in the same 5-gallon glass containers that are later used for storage. But as fermenters they are filld only two-thirds full as a precaution against any overflow or unmanageable formation of bubbles.

When the primary fermentation has run its course, the several partly-filled bottles are simply consolidated —filled full and equipped with bubblers. Subsequent siphoning from sediment, chilling, and dosing with meta are carried out as with red wine.

If fining is necessary, it differs in one respect: before mixing in the gelatin, mix in an equal amount of dissolved tannic acid to remove the impurities. Tannic acid is obtainable at drug stores or winemakers' shops. as a powder. This provides better settling out of suspended matter.

Dry table wine is a food beverage, to be used with meals. **Sweet wines** are more like cordials.

The making of sweet wines takes advantage of a characteristic of the yeast organism, namely, that its activity dies down and it usually ceases to ferment sugar into alcohol after a fermenting liquid reaches an alcoholic content of around 13%. The secret, then, is to add an *excess* of sugar when correcting the juice of crushed grapes before fermentation. When fermentation ceases, there is still some residual sugar in the juice. From then on the still-sweet new wine is treated much as other wine.

The three important differences are: (1) the wine is siphoned from its sediment immediately after fermentation, without the waiting period at 60° F; (2) the chilling begins as soon as possible; and (3) the dose of meta added then and at each subsequent siphoning is doubled (½ teaspoon per 5 gallons instead of ¼ teaspoon) to guard against spoilage and against any accidental resumption of fermentation.

Dry table wines made from other fruits are rarely successful, but agreeable sweet wines may be made from

Making Sweet Wine

Fruit	Average sugar level	Sugar needed per gallon to make a sweet wine (ounces)	Average acid	Gallons of sugar water [1] to add per gallon
Grapes (eastern)	12–20	1¼–2	med. to high	0–1
Grapes (Calif.)	16–20	1–1½	low [2] to med.	0
Apples	13	2–2¼	low [2] to high	0–½
Apricots	12	2–2½	med. to high	0–¼
Blackberries	6	2–3	high to very high	1 or more
Blueberries	8	2¼–3	low to med.	0
Cherries (sour)	14	2–2¼	high to very high	1 or more
Cherries (sweet)	18	1½–2	medium	0
Pear	12	2¼–2½	med. to high	0–¼
Plum (Damson)	14	2–2¼	med. to high	0–¼
Plum (Prune)	17	1½–2	med. to high	0–¼
Peach	10	2–2½	med. to high	0–¼
Raspberries	8	2½–3	high to very high	1 or more
Strawberries	5	2–3¼	med. to high	0–½

[1] To maintain proper sugar level when the acidity is reduced by adding water, it is easier to make up a sugar solution by dissolving 3 pounds of sugar in enough water to fill a 1-gallon jug.
[2] Addition of some acid (citric or tartaric) may help. This can be done "to taste" after the active fermentation is over.

them. The point to remember is that most fruits are lower in sugar than grapes and higher in acid. Corrections for both are almost always necessary, plus sufficient excess sugar to leave residual sweetness after fermentation.

These fruits, with the exception of apple juice, are fermented in a crushed mass in order to obtain a maximum extraction of characteristic odors and flavors. Once fermentation is concluded, they are treated like sweet grape wine. The table will serve as a rough guide to their relative sugar content and total acidity.

Vinegar

If a cork happens to pop out unnoticed and air reaches the wine for several weeks, there is a good chance that bacterial action will begin to convert the alcohol in the wine into acetic acid. Once the presence of acetic acid can be detected (a vinegar-like odor) the wine will lose its appeal as wine. A usable vinegar can be retrieved by encouraging the process to go to completion.

Vinegar produced from an undiluted wine will be overly strong, so an equal volume of water should be added. The container should be less than three-quarters full and closed with a loose cotton plug or covered with a piece of light cloth to keep out fruit flies.

If wine vinegar is your desired goal and no wine has started to sour, use a vinegar starter. A selected strain of vinegar starter can be purchased from some winemakers' shops, or a wild starter may be used. Frequently the water in an air-bubbler will have a vinegar-like smell. This can be used to start a batch of vinegar. The wine is diluted with an equal volume of water and the container partly filled and covered as above.

A warm, but not hot, location will speed the process. In a month or two the vinegar should be ready. The clear portion of the vinegar can be poured or siphoned off for use. If another batch is wanted, more of the wine-water mixture can be added to the old culture.

Home Drying of Fruits and Vegetables
by Dale E. Kirk and Carolyn A. Raab

Tasty ready-to-eat snacks and confections are some of the versatile products you can create by drying fruits and vegetables at home. After soaking in water, the rehydrated food can be used in favorite recipes for casseroles, soups, stews and salads. Rehydrated fruits and berries can also make excellent compotes or sauces.

Drying is appealing because the procedure is relatively simple and requires little equipment. Only minimal storage space is needed. Food can be dried in the sun, in the oven, or in a dehydrator.

Drying requires a method of heating the food to evaporate the moisture present, and some means of removing the water vapor formed.

Sun drying utilizes both radiant heat energy and heat transferred to the product from warm air. Natural air currents are usually adequate to carry away the water vapor.

Trays of wood slats, plastic mesh, or aluminum screen may be placed in the sun on support blocks or strips to allow air movement around and through the trays. Galvanized wire is not recommended as a tray material because high-acid foods will react with the zinc coating on the steel wire.

If insects or birds are a problem, a wooden frame can be constructed over the trays to support a plastic mesh or cheesecloth cover. Further protection can be provided by using a totally enclosed frame and a transparent panel to form a solar drying oven.

To dry in the kitchen oven, the thermostat should be set to its lowest temperature (generally about 150° F). Since oven vents provided for removing moisture from roasting and baking are adequate for drying only small quantities of food at one time, the oven door should be left partially opened. For larger loads, the air circulation rate can be increased by placing a household fan outside the oven, directed at one edge of the partially opened oven door.

Dehydrator cabinets may be purchased in many sizes and types. Or they may be built using plans available from State universities or U.S. Department of Agriculture plan services (ask your county Extension office about plans). All cabinets are provided with a heat source and vents for carrying off moist air.

Simpler units may rely on natural convection to carry moist air away, and the heating unit may be limited in output so that the cabinet never exceeds safe drying temperatures near the end of the drying period. This type will be slow in achieving drying temperature if sizable amounts of food are processed at one time.

Trays must be rotated during the processing period to insure even drying. Trays nearest the bottom, exposed to the hottest, driest air, will dry most rapidly.

If the natural convection type cabinet is equipped with a thermostat, it may be fitted with a larger heater. This will provide higher drying temperatures during the early stages but will not give even drying across all trays.

By using a fan to force air across the trays more rapidly, even drying can be obtained across each tray as well as between trays. The forced air system may be used with or without a thermostat.

Dale E. Kirk is a Professor and Agricultural Engineer at Oregon State University, Corvallis. Carolyn A. Raab is Extension Foods and Nutrition Specialist at Oregon State.

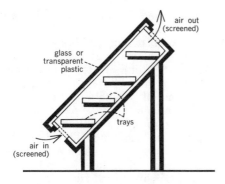

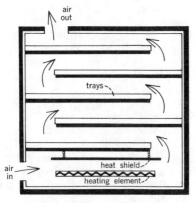

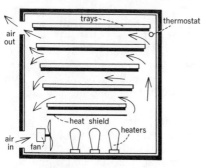

Most food products release moisture rapidly during early stages of drying. This means they can absorb large amounts of heat and give off large quantities of water vapor while remaining at a temperature well below that of the drying air. Maximum drying rates can be achieved by providing a larger, thermostatically-controlled heat source and a fan for circulating air.

To conserve energy and still obtain rapid, even drying across all trays, much of the drying air may be reheated and recirculated. This is particularly effective during the last 70% to 90% of the drying period, when relatively small amounts of water are absorbed by the air as it passes over the partially dried food.

The recirculating system requires either a thermostat or separate switch controls on part of the heating unit to adjust heat output to match the drying load. The amount of air recirculated is determined by the size of the permanent inlet and outlet openings in the box. It can be further controlled by adjusting the door to a partially opened position.

Detailed plans for constructing the recirculation-type drier can be obtained by sending 25¢ to the Western Regional Agricultural Engineering Service (WRAES), Oregon State University, Corvallis, Oreg. 97331 and requesting WRAES Fact Sheet No. 18.

Procedure

Drying is a relatively simple process, but there are a number of recommended techniques. You may need to use a "trial and error" approach to find the drying procedure which works best in a particular situation.

Fruits and vegetables can be dried in pieces or pureed and dried in a thin sheet as a "leather."

The following information summarizes major steps in drying. Detailed instructions are available at county Extension offices. Also, various books on the market give instructions for drying and recipes for using dried food.

Fruits and vegetables selected for drying should be the highest quality

Top, enclosed frame solar drying oven with provision for air movement. Middle, dehydrator with built-in heater relies on air movement to carry off moist air. Bottom, this forced draft dryer can recirculate much of the drying air to conserve energy.

Steps in Drying Fruits and Vegetables

PREPARE
wash, sort
peel, pit/core, slice

PRETREAT
fruits: vegetables:
dip or blanch   blanch

DRY
oven, sun, dehydrator

CONDITION AND STORE
equalize, pasteurize
package and store

---

obtainable—fresh and fully ripened. Wilted or inferior produce will not make a satisfactory dried product. Immature produce lacks flavor and color. Overmature produce may be tough and fibrous or soft and mushy.

Prepare produce immediately after gathering, and begin drying at once. Wash or clean all fresh food thoroughly to remove any dirt or spray. Sort and discard defective food; decay, bruises, or mold on any piece may affect an entire batch.

For greater convenience when you finally use the food, and to speed drying, it is advisable to peel, pit, or core some fruits and vegetables. Smaller pieces dry more quickly and uniformly.

Pretreating

Enzymes in fruits and vegetables are responsible for color and flavor changes during ripening. These changes will continue during drying and storage unless the produce is pretreated to slow down enzyme activity.

Blanching is the recommended pretreatment for vegetables. It helps save some of the vitamin content, sets color, and hastens drying by relaxing tissues. Blanching may also prevent undesirable changes in flavor during storage, and improve reconstitution during cooking.

Steam blanching is preferred because it retains more water-soluble nutrients than water blanching. Blanching times differ, depending on the type of vegetable being dried. Overblanching leads to excessive leaching of vitamins and minerals. Inadequate blanching will not destroy enzymes that cause vitamin loss during drying and storage.

Many light-colored fruits (especially apples, apricots, peaches, nectarines, and pears) tend to darken during drying and storage. To prevent this darkening, the fruit may be pretreated by blanching or by a suitable dip, but effectiveness of pretreatment methods varies.

Fruit may be dipped in one of the following:
—A solution of table salt
—A solution of ascorbic acid. Commercial antioxidant mixtures containing ascorbic acid may also be used, but often are not as effective as pure ascorbic acid.

Fruits may be steam-blanched. However, blanched fruits may turn soft and become difficult to handle.

Sirup blanching may help retain the color of apples, apricots, figs, nectarines, peaches, pears and plums. A sweetened candied product will result.

Fruits with tough skins (grapes, prunes and small dark plums, cherries, figs, and some berries) may be water-blanched to crack the skins. This will allow moisture inside to surface more readily during drying.

Before drying pretreated food, remove any excess moisture by placing the food on paper towels or clean cloths. Drying trays should be loaded with a thin layer of food as directed. If needed, clean cheesecloth can be spread on the trays to prevent food pieces from sticking or falling through.

The amount of food being dried at

one time should not exceed that recommended by instructions.

Drying

A temperature of 135° to 140° F is desirable for dehydrator and oven drying. Moisture must be removed from the food as fast as possible at a temperature that does not seriously affect the food's flavor, texture, color, and nutritive value.

If the initial temperature is too low or air circulation insufficient, the food may undergo undesirable microbiological changes before it dries adequately.

If the temperature is too high and the humidity too low, as when drying small loads in the oven, the food surface may harden. This makes it difficult for moisture to escape during drying.

Oven or dehydrator drying should continue without interruption to prevent microbial growth.

To promote even drying, rotate trays occasionally and stir food if necessary.

Drying time varies according to fruit or vegetable type, size of pieces, and tray load. Dehydrator drying generally takes less time than oven drying. Sun drying takes considerably more time.

Before testing foods for desired dryness, remove a handful and cool for a few moments. Foods that are warm or hot seem softer, more moist, and more pliable than they will when cooled.

Foods should be dry enough to prevent microbial growth and subsequent spoilage. Dried vegetables should be hard and brittle. Dried fruits should be leathery and pliable. For long term storage, home dried fruits will need to be drier than commercially dried fruits sold in grocery stores.

Conditioning and Storing

Fruits cut into a wide range of sizes should be allowed to "sweat" or condition for a week after drying to equalize the moisture among the pieces before placing in long term storage. To condition, place fruit in a non-aluminum, non-plastic container and put in a dry, well-ventilated and protected area. Stir the food gently each day.

Dehydrated foods are free of insect infestation when removed from the dehydrator or oven. However, sun-dried foods can be contaminated and should be treated before storage. Insects or their eggs can be killed by heating dried food at 150° F for 30 minutes in the oven. An alternative is to package the food and place it in the home freezer for 48 hours.

Dried foods should be thoroughly cooled before packaging. Package in small amounts so that food can be used soon after containers have been opened.

Pack food as tightly as possible without crushing into clean, dry, insect-proof containers. Glass jars or moisture-vapor proof freezer cartons or bags (heavy gage plastic type) make good containers. Metal cans with fitted lids can be used if the dried food is first placed in a plastic bag.

Label packaged foods with the packaging date and the type of food.

Store containers of dried foods in a cool, dry, dark place. Check food occasionally to insure that it has not reabsorbed moisture. If there is any sign of spoilage (off-color or mold growth), discard the food. Food affected by moisture, but not moldly, should be used immediately or reheated and repackaged.

All dried foods deteriorate to some extent during storage, losing vitamins, flavor, color, and aroma. However, low storage temperatures prolong storage life, and dried foods may be frozen for long term storage.

Dried foods can be reconstituted by soaking, cooking, or a combination of both, and will resemble their fresh counterparts after reconstitution. However, dried foods are unique and should not be expected to resemble a fresh product in every respect.

Drying does not render the food free of bacteria, yeasts, and molds. Thus, spoilage could occur if soaking is prolonged at room temperature. Refrigerate if soaking for longer than 1 to 2 hours.

To conserve nutritive value, use the liquid remaining after soaking and cooking as part of the water needed in recipes.

One cup of dried vegetables reconstitutes to about 2 cups. To replace the moisture removed from most vegetables, barely cover them with cold water and soak 20 minutes to 2 hours. Cover greens with boiling water. To cook, bring vegetables to a boil and simmer until done.

One cup of dried fruit reconstitutes to about 1½ cups. Add water just to cover the fruit; more can be added later if needed. One to eight hours are required to reconstitute most fruits, depending on fruit type, size of pieces, and water temperature. (Hot water takes less time). Oversoaking will produce a loss of flavor. To cook reconstituted fruit, cover and simmer in the soak water.

Dried or reconstituted fruits and vegetables can be used in a variety of ways.

Use dried fruit for snacks at home, on the trail, or on the ski slopes. Use pieces in cookies or confections.

Serve reconstituted fruit as compotes or as sauces. It can also be incorporated into favorite recipes for breads, gelatin salads, omelets, pies, stuffing, milkshakes, homemade ice cream and cooked cereals.

Add dried vegetables to soups and stews or vegetable dishes. Use as dry snacks or dip chips.

Include reconstituted vegetables in favorite recipes for meat pies and other main dishes, as well as gelatin and vegetable salads.

Powdered vegetables in the dried form make a tasty addition to broths, raw soups, and dressings.

Some vitamin breakdown occurs during drying and storage of dried fruits and vegetables. Ascorbic acid (Vitamin C) is the vitamin most likely to be lost.

Nutritive losses can be kept to a minimum by:

—Blanching the correct length of time

—Packaging dried foods properly and storing containers in a cool, dry, dark place

—Checking dried foods periodically during storage to insure that moisture has not been reabsorbed

—Eating dried foods as soon as possible

—Using liquid remaining after reconstitution in recipes

Storage of Home-Preserved Foods
by Ralph W. Johnston

Proper storage of home-preserved foods, especially of home-canned products, and close scrutiny before serving are essential. If proper storage requirements are not met, home-preserved foods may lose their quality or spoil.

Homemakers should observe some simple techniques for checking home-canned foods before serving them. This will help prevent consumption of food that could cause the rare but extremely dangerous food poisoning called botulism.

Most canned foods are highly perishable yet do not require refrigeration until opened. Unlike frozen foods, they are unaffected by power interruptions or mechanical failures.

However, the hazard of botulism must always be kept in mind. Although botulism is rare, it results in a high death rate of about 65 percent among its victims. Yet it is an easy problem to avoid. Botulism results when home-canned foods are improperly processed. Under these conditions, the spore (a seed-like structure which is highly heat-resistant) of a soil bacterium called *Clostridium botulinum* may survive.

If the food product is low in acidity, as with peas, corn, or beans, the spore can germinate (sprout) and grow during storage at room temperatures. As *Clostridium botulinum* grows, it produces a powerful poison that when ingested can cause severe illness or death. Most cases of botulism in the United States stem from home-canned foods.

The home canner can avoid botulism primarily by following pre-scribed, reliable processing instructions such as those given in USDA Home and Garden Bulletin No. 8, *Home Canning of Fruits and Vegetables*. If you don't have reliable processing instructions, don't attempt home canning. If you have these instructions, read them before and during home canning and do not take short cuts or modify the instructions.

Do not use processing instructions of neighbors or relatives; although frequently given with the best of intentions, they may contain modifications that are inadequate and dangerous. Remember that past safe history of a relative's processing procedure is no guarantee of future safety. Botulism doesn't always occur even in inadequately processed home-canned foods.

After home-canned foods have cooled they are ready to be stored until needed. At this point, the home canner should make his first quality control and safety check, just as commercial canners do.

Jar lids should be examined. If the center of the lid is not depressed or is loose, refrigerate the product immediately and serve at the next meal. Before serving, boil low acid products for 10 minutes. Check all jars for cracks; if they are found, treat jars the same way as those with loose lids.

Observe cans for any evidence of leakage around seams; again if leakage is observed, refrigerate the cans immediately, serve at the next meal, and boil for 10 minutes before serving.

During this first integrity check on home-canned foods, it is unlikely that swelling of the cans or foaming in the jars will be noticeable, because of the short lapse of time since processing. But the first check can easily detect

---

Ralph W. Johnston is Chief, Microbiology Staff, Meat and Poultry Inspection Program, Food Safety and Quality Service.

loose lids, cracked jars and leaking seams on cans.

The next step is to store home-canned products. Proper storage will protect the products from loss of quality and in some cases from spoilage. Store canned foods in a clean, cool, dry area away from bright light —particularly sunlight—and in an area where the foods will not freeze or be exposed to high temperatures. Under these conditions, the products will remain at high quality for at least a year.

Excessive dampness will rust cans or metal lids. If this condition becomes severe, leakage will occur and the product will spoil. Freezing causes expansion of the product and the jar lid may loosen, the jar may crack, or can seams may be stressed. This can lead to leakage and food spoilage.

Fred Farout

When foods are preserved by heating, as in home canning or commercial canning, the heating process is designed to destroy all normal spoilage bacteria that can grow under usual storage conditions, and all bacteria capable of causing human harm. The products are called "commercially sterile" but are not always truly sterile.

A group of bacteria produce extremely heat-resistant spores that can only germinate and grow at high storage temperatures such as those above 103° F. These bacteria often survive both the home and commercial canning process. Even though present, they normally are of little concern from the viewpoint of spoilage and no concern at all from the standpoint of human health. However, if canned foods are stored in attics or near hot water pipes or in any other area where the temperature will exceed 102° F at any time, these heat-loving bacteria (called thermophilic) can grow and spoil the product.

As a rule of thumb, home-canned foods will remain high in quality for one year if properly stored. After a year, loss of quality may occur.

Containers for home-preserved foods are designed to resist any chemical reactions between the product and the containers. However, some products—particularly high acid ones like tomatoes—will slowly react with the metal in the can or the jar lid. Corrosion and container failure may follow during subsequent storage. This action occurs from the inside out and can take place even under good storage conditions.

Jars should be dated when stored, and used within a year from the processing date. Always rotate stock on the shelves so as to use the oldest container first, and can no more units of any single product than you can use in a year.

The last and perhaps most important quality control steps are the final inspection and serving procedures.

After removing the product from storage, carefully inspect the container, and in the case of jars the

Boil home-canned low-acid foods 10 minutes before tasting or serving.

visible contents. This should be done before opening.

If a can or jar lid shows any sign of swelling (bulging) or leakage of product, do not open the container. If a jar lid is loose or the contents of a jar are foamy or otherwise visibly abnormal, do not open. When any of these defects are noted, place the whole container in a heavy plastic bag and tie the top securely. Place this in doubled paper bags with heavy packing of newspapers. Tape or tie the top securely, place in a lidded garbage can, then wash your hands thoroughly.

Not all spoiled or leaking home-canned foods contain the deadly botulism toxin but some do, so extreme caution in disposal is necessary.

If a defective product is found, all of that product prepared at the same time should be removed from storage and similarly inspected.

Never taste the contents of a suspect product. Under certain circumstances, a spoonful of "off" unheated, suspect product has been known to kill.

Finally, bring all home-canned vegetables to a rolling boil after opening and before tasting. Heating makes any odor of spoilage more noticeable. Again, if an odor of spoilage is noted, destroy the product with caution. If the product is normal, cover the pan and continue to boil at least 10 minutes before serving. Only after these precautions are taken are home-canned vegetables safe to taste and serve.

Home Frozen Foods

A plus for home freezing is that slight variations in following directions do not result in a botulism hazard. The bacterium that causes botulism cannot grow in the freezer. Proper freezing prevents the growth of microorganisms that cause spoilage and those that can cause illness.

Besides the initial cost of the freezer itself, energy costs are significant. Utilize the freezer fully to keep the energy costs per unit as low as possible. Fill the freezer when foods are least expensive, use the products as needed, and be careful to use the oldest products first.

Take care not to overload the freezer. If you pack it too tightly with containers of warm food, the freezer will be unable to remove the heat fast enough and spoilage from bacterial growth can result.

To avoid this, freeze foods soon after they have been packed; put no more unfrozen food into a home freezer than will freeze within 24 hours. Usually, this will be about 2 or 3 pounds of food to each cubic foot of capacity.

For quickest freezing, place packages against freezing plates or coils and leave a little space between packages so air can circulate.

Small excesses of product destined for freezing can be held in the refrigerator until the first load is frozen. If a large excess of product exists, chill and carry it in an insulated box or bag as soon as possible to a locker plant.

After freezing, packages may be stored close together. Store them at 0° F or below in order to retain the highest quality for the longest time.

Prolonged storage of frozen foods results in slow loss of quality. The rate of this loss differs with various foods. To maintain high quality, obtain information on recommended storage periods for the foods you freeze. This may be obtained from your county Extension office or from USDA Home and Garden Bulletin No. 10, *Home Freezing of Fruits and Vegetables*.

Storage periods are recommended to guarantee food quality only. If these periods are exceeded, taste may be affected but as long as the product has been kept at 0° F or below there is no question of safety.

The homemaker's greatest concern with a home freezer is mechanical or power failure, which can result in food losses. Some but not all of these can be avoided. Freezers are very dependable mechanical devices yet they do fail. Most failures develop after 5 or more years of use.

The homeowner should clean dust from coils of the freezer once or twice each year in strict accordance with the instruction manual for the unit. At this time watch for any changes that have occurred. Have a dealer or repairman check unusual noises or excessive running.

Air circulation around the coils should not be covered or blocked in any way. Check the plug itself for a firm fit. If the plug is loose in the receptacle, it may fall or be easily bumped out without notice. Replace loose plugs. Better yet, some hardware stores sell clips that clamp the plug in by means of the screw that holds the receptacle plate onto the outlet.

Freezer owners should know where the closest commercial freezer is, in case of an extensive failure. Check your home freezer after thunderstorms or power failures, since freezers have been known to be damaged occasionally when power falls or surges.

Don't Open

A well packed freezer will hold the product for many hours even if the unit is not operating. Normally, power failures are short in duration and no food thawing results. If the power is off, do not open the freezer as this will hasten thawing. Telephone or otherwise determine when the power will be turned on again.

Sometimes freezer failure is discovered only when a homemaker goes to the freezer to get something. If this occurs, condition of the food should be determined immediately. Discard all foods that are thawed and warm, since extensive bacterial growth may have taken place.

Foods may be saved if they remain frozen; or if they are thawed but very cold, about 40° F, and have been held no longer than 1 or 2 days at refrigerator temperatures after thawing. Bacteria grow only slowly in thawed but cold foods. Prompt refreezing of thawed cold foods will lower the quality but not result in spoilage or danger. If you have doubt as to whether the foods are cold or warm, throw them out as the safest course.

Once condition of the foods is determined, plan fast for the next step. If the freezer cannot be repaired quickly, make arrangements to move the food to a commercial locker plant or another freezer. To do so, package the products closely together in paper bags. Place these in cardboard cartons lined and covered with newspapers for insulation, and transfer them immediately.

Another way to save the freezer load is to use dry ice in the freezer itself. Dry ice must be handled with gloves to prevent burns. Also keep in mind that carbon dioxide gas evolves as dry ice evaporates, and can cause unconsciousness if allowed to concentrate.

When transporting dry ice, leave a car window open at least several inches. If you use dry ice in the freezing compartment, make sure a nearby window is cracked open. When packing dry ice into a freezer, figure on 25 to 50 pounds to do the job. Don't break up the ice any more than necessary.

To summarize, frozen foods are seldom involved in food spoilage or food poisoning. Even so, mechanical devices occasionally fail, and freezer owners should have prearranged plans for such an emergency. Preventive maintenance will help reduce the likelihood of failure. If a failure results in food becoming thawed and warm, discard it for safety.

Storing Fresh Fruit and Vegetables
by Anton S. Horn and Esther H. Wilson

Many fruits and vegetables can be stored fresh. But the home gardener must gather them at proper maturity and observe correct temperature, humidity, ventilation, and cleanliness rules.

Basements or outdoor cellars can serve as temporary storage for some produce. A cellar mostly below ground is best for root vegetables. It can be run into a bank and covered with 2½ feet or more of soil. Sometimes outdoor root cellars are made with a door at each end. Combining the outdoor storage cellar with a storm shelter in the event of tornadoes or other needs may be a satisfactory solution.

Modern basements are generally too dry and warm for cool moist storage. However, a suitable storage room may be built by insulating walls and ceiling and ventilating through a basement window. You may ventilate by extending a ventilating flue from half of the window down almost to the floor. Cover the other half of the window with wood and the outside openings of the ventilator with a wire screen for protection against animals and insects.

Keep the room cool by opening the ventilators on cool nights and closing them on warm days. If properly cooled, the room temperature can be controlled between 32° and 40° F during winter. To maintain the humidity, sprinkle water on the floor when produce begins to wilt. A slatted floor and slatted shelves will provide floor drainage and ventilation. A reliable thermometer is needed for operation of any home storage room.

Anton S. Horn is Extension Horticulturist, University of Idaho, Boise. Esther H. Wilson is Extension Nutrition Specialist, University of Idaho, Moscow.

A cool corner in the basement, a back room of a small house with no basement, or a trailer may be suitable. One lady we know uses part of a closet built into the outside corner of a bedroom. It is also possible to adapt storage sheds in carports by insulating and proceeding as outlined earlier.

Pits and trenches or mounds may be used for storage if a root cellar is not available or basement storage is impractical. Also, you may bury a barrel, drainage tile, or galvanized garbage can upright, with four inches of the top protruding above ground level. This will keep potatoes, beets, carrots, turnips, and apples through winter. For convenience, place the produce in sacks or perforated polyethylene bags of a size to hold enough for a few days. Then you can easily take out fruits and vegetables as needed.

Place the barrel on a well drained site, and make a ditch so surface water will be diverted and not run into the container. A garbage can has a good lid, but for a drainage tile or barrel a wooden lid may have to be built. The lid should be covered with straw, and a waterproof cover of canvas or plastic placed over the straw.

Requirements of fruits and vegetables differ. Controlled cold storage or refrigerated storage are best.

Good references are *Storing Vegetables and Fruits in Basements, Cellars, Outbuildings, and Pits*, USDA Home and Garden Bulletin No. 119, and bulletins on this subject prepared by your State Extension service. Your county Extension office may have the bulletins. This office may also be able to tell you how to obtain plans for a fruit and vegetable storage room, or a storm and storage cellar.

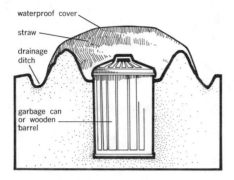

Brief notes on specific storage problems follow:

With proper care, hard-rind varieties of winter pumpkins and squash will keep for several months. Harvest before frost, and leave on a piece of stem when you cut them from the plants.

Store only well-matured fruits that are free of insect damage and mechanical injuries.

Pumpkins and squash for long-term storage keep better when cured for 10 days at 80° to 85° F. If these temperatures are impractical, put the pumpkins and squash near your furnace to cure them. Curing hardens the rinds and heals surface cuts. Bruised areas and pickleworm injuries, however, cannot be healed.

After curing pumpkins and squash, store them in a dry place at 55° to 60° F. If stored at 50° or below, pumpkins and squash are subject to damage by chilling. At temperatures above 60°, they gradually lose moisture and become stringy.

Acorn squash keep well in a dry place at 45° to 50° F for 35 to 40 days. Do not cure acorn squash before storing them. They turn orange, lose moisture, and become stringy if cured for 10 days at 80° to 85° or if stored at 55° or above for more than 6 to 8 weeks.

A dark green rind at harvest indicates succulence and good quality.

Do not store pumpkins and squash in outdoor cellars or pits.

Parsnips, Salsify, Horseradish can be left undug (stored) in the ground.

These vegetables withstand freezing, but alternate freezing and thawing damages them. If you store them in the ground, mulch lightly at the end of the growing season. Keep them covered until outdoor temperatures are consistently low. Then remove the mulch to permit thorough freezing. After they have frozen, mulch deep enough to keep them frozen.

Root Crops

Root crops such as beets, carrots, celeriac, kohlrabi, rutabagas, turnips, and winter radishes should not be put in storage until late fall. Root crops keep best between 32° and 40° F. They require high humidity to prevent shriveling. Continued storage at 45° causes them to sprout new tops and become woody.

Large and overmature root crops may become tough and stringy in storage. Small and immature root crops probably will shrivel.

Dig root crops when the soil is dry and the temperature consistently low. Prepare them immediately for storage. Cut the plant tops about a half inch above the crown. Beets will bleed unless 2 to 3 inches of the top is left. You may wash the roots if you let them dry again before storing. Do not expose them to drying winds, and be sure they are cool when put in storage.

Prevent bitterness in carrots by storing them away from fruits such as apples, which give off volatile gases while ripening.

Turnips and rutabagas give off odors, so don't store them in your basement. Find a separate spot, or store them with other root crops and vegetables in an outdoor cellar or pit.

Some fruits and vegetables can be stored outdoors in a partially buried galvanized garbage can or wooden barrel.

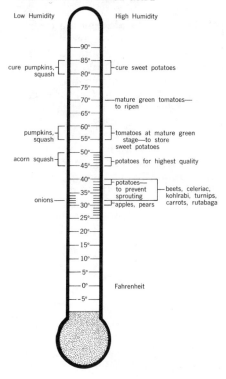

VEGETABLE—FRUIT TEMPERATURE STORAGE GUIDE

Turnips may be left in the garden longer than most other crops. They withstand hard frosts, but are damaged by alternate freezing and thawing. All other root crops can be stored together in your basement storage room.

Root crops keep their crispness longer when bedded in layers of moist sand, peat, or sphagnum moss. However, perforated polyethylene bags and box liners are easier to use than bedding. Root crops can be stored in crates or boxes in moist air, but they gradually lose moisture and quality unless polyethylene liners are used. Carrots and beets may be stored in 10-gallon crocks or any container that will prevent excessive shriveling.

Quick dipping of dried and trimmed turnips, rutabagas, or parsnips in wax will prevent shriveling. Float a layer of jelly-type paraffin on top of a kettleful of heated water which is deep enough to cover the vegetable. Dip room temperature vegetables quickly through the layer of wax.

For a thinner, harder wax film add a little salt and 10 to 20 percent clean beeswax.

Potatoes are the principal root crops you will probably store. Potatoes are eaten from the time they are of sufficient size for early use until storage time, and during storage when the vines have fully ripened.

If potatoes are harvested before maturity the skin may flake off easily. They are all right for immediate use, but not for storage. Immature potatoes shrink badly, bruise easily, and will not keep well very long.

For storage, potatoes should be allowed to mature and develop a thick skin. When the tops lie down the tubers should be mature enough for storage.

Dig potatoes carefully to avoid bruises, for better storage life.

Handle newly dug potatoes with care until the surface has dried or cured a few hours or more. You can keep them in baskets or slatted crates in single layers at first.

Store sound mature tubers in darkness at a minimum relative humidity of 95 percent and 45° to 48° F for highest quality. For very long storage keep at a temperature of 38° to 40° to prevent sprouting. The starch changes to sugar if potatoes are held below 45°. Potatoes may not show any external effect from exposure to these lower temperatures, but sometimes darkened tissue will be seen if the potato is cut and exposed to air.

Light causes considerable "greening" in potatoes. The green portion contains an undesirable substance that gives a bitter flavor.

Sweet potatoes that are well matured, carefully handled, properly cured, and stored at 55° to 60° F can be kept until April or May.

Sweet potatoes are easily bruised and cut. Handle them carefully and as little as possible. Put them directly in storage containers at harvest.

Cure freshly dug sweet potatoes by holding them about 10 days under moist conditions at 80° to 85° F. In the absence of better facilities, sweet potatoes can be cured near your furnace. To maintain high humidity during curing, stack storage crates and cover them with paper or heavy cloth. If the temperature near your furnace is between 65° and 75°, the curing period should last 2 to 3 weeks. After curing, move the crates to a cooler part of your basement or house where a temperature of about 55° to 60° can be maintained.

In houses without central heating, sweet potatoes can be kept behind a cookstove or around a warm chimney. If you keep sweet potatoes this way, wrap them in fireproof paper (to slow down temperature changes) and store them in boxes or barrels.

Sweet potatoes are subject to damage by chilling. Do not store them at 50° F or below.

Outdoor pits are not recommended for storing sweet potatoes, because dampness of the pits encourages decay.

Tomatoes

Even though the home canner has canned plenty of tomatoes, it may be desirable to keep some of the fresh fruit. Keep tomatoes in the garden as long as possible. You can protect against early fall frosts by covering the plants with burlap or old carpets in the evening when frost is predicted. Polyethylene may also be used but injury will occur wherever it touches the plant.

In the summer, tomatoes should be harvested when fully vine-ripened for best quality. Pick when the color is a dark red in red varieties. During fall when frost is likely, mature green fruit can be picked and it will develop a red color when kept in a fairly warm place. The fruit is in the "mature-green" state if the tissues are gelatinous or sticky when the tomato is cut and the tomato interior is yellowish. Immature green tomatoes don't ripen satisfactorily.

To check your judgment, cut a tomato in half that you feel is mature green. If the pulp that fills the compartments is jelly-like, it is mature green. The seeds are dragged aside easily by the knife and not cut through. In immature green tomatoes, seeds are easily cut through and the jelly-like pulp has not yet developed. Usually you can recognize the mature green ones by their glossiness, less hairiness, and more whitish green color.

You can pick mature-green fruits and bury them in deep straw or place in a room where the temperature is 60° to 70° F. The tomatoes will ripen over a period of 3 or 4 weeks. Sunlight is not needed to ripen green-ripe tomatoes, so don't bother to put them on window sills. They ripen satisfactorily in the dark. Generally, tomatoes store best at 55° to 60° and ripen at 70° or room temperature.

You can wash the mature green fruits in a weak solution of household bleach and then wrap in paper to store and ripen.

Some people pull up the vines just before frost and hang them in the basement or garage to ripen their fruit.

Onions

Harvest onions for storage when the neck of the plant dries down, the tops have fallen over, and the roots are dry and have stopped growing from the stem plate. At that time the outer scales of the bulb are drying out and do not cling tightly (outer scales of yellow-skinned varieties change to a darker color).

Pull the onions by hand and lay in a windrow to cure with the tops placed over the bulbs to prevent sunscald. Onions may also be cured in an open shed. Remove onions with thick neck (seeders) before storage and discard all diseased bulbs.

After curing, place onions in open-slatted crates or burlap bags for further field curing or drying. Then place in storage. You may use either common storage or refrigerated storage.

Low temperatures in storage reduce shrinkage due to moisture loss and stop disease development. Keep the humidity as low as possible. Good management of ventilation is important. Ventilate storage early in the morning.

Onions held in cold storage should be placed there immediately after curing. A temperature of 32° F is ideal and will keep onions dormant and relatively free of rot. If sprouts grow it indicates too high a temperature, poor curing, or immature bulbs. If you have root growth the humidity is too high. The humidity should be 65 to 70 percent.

Do not store onions with produce that is likely to absorb the odor. Onions stand slight freezing, but do not handle or move them until they thaw. You can store onions in a dry, well ventilated attic or unheated room. Maintain as near 32° F as you can and keep as dry as possible. You can hang open-mesh bags, about half full, from overhead hooks or nails. Slatted half full crates of onions may be stacked on cross bars.

Apples, Pears

Chemical changes take place in the ripening process of apples and pears. This activity is called respiration. Starch changes to sugar, acids and insoluble pectins decrease, and volatile gases are given off. This continues until the fruit becomes overripe and mealy. During this ripening process oxygen is consumed from the air, and water and carbon dioxide are produced and heat is generated.

You may slow respiration by cooling fruit as rapidly as possible after picking. The sooner this is done the longer the fruit will keep.

Research indicates that when apples are stored at 30° F, about 25 percent more time is required for them to ripen than at 32°. Stored at 40°, the rate of ripening is about double that at 32°. At 60° the rate is close to three times that at 40°, and at 85° the softening and respiration rates have been found to be about double those at 60°. This emphasizes the importance of cooling quickly and keeping cold. The average freezing point of apples is about 28° or 29°.

Most apple varieties keep best at a temperature of 30° to 32° F and a relative humidity of 85 to 88 percent. However, McIntosh, Yellow Newton, and Rhode Island Greening apples do best at 35° to 38°. This prevents internal browning and brown core.

Pears can be stored ideally at 30° to 31° F. The highest freezing point for pears is about 29°. Since pears are likely to shrivel, keep the humidity at 90 percent. Most pears won't ripen satisfactorily for eating at the above temperatures. They should be taken out of storage and ripened between 65° and 70°. This is ideal for Bartletts.

Bartlett pears ripen faster than apples. If you store pears too long they will not ripen properly. Don't store Bartletts after 3 months or Anjou longer than 6 months.

Maintaining desired temperatures for home storage of apples and pears may be difficult. If you must settle for 40° F or even higher, you won't get the good results you would if you refrigerate at the optimum temperatures. Sometimes cold storage facilities are available where you may store your fruit for a set price per container.

Don't mix windfalls (fruits that have dropped to the ground) with

Home Storage Chart

	Where to store	Storage conditions Temperature (F)	Humidity	Storage period
Vegetables:				
Beans and peas, dried	Any cool, dry place	32–40	Dry	Many Years
Beets	Storage cellar or pit	32–40	Moist	Fall-winter
Cabbage	Storage cellar or pit	32–35	Moist	Fall-winter
Carrots	Storage cellar or pit	32–40	Moist	Fall-winter
Celery	Roots in soil in storage cellar	32–40	Moist	Fall-winter
Onions	Any cool, dry place	As near 32 as possible	Dry	Fall-winter
Parsnips	Leave in ground or put in storage cellar	32–40	Moist	Fall-winter
Potatoes	Storage cellar or pit	45–48	Moist	Fall-winter
Pumpkin, winter squash	Unheated room or basement	55–60	Dry	Fall-winter
Rutabagas	Storage cellar or pit	32–40	Moist	Fall-winter
Sweet Potatoes	Unheated room or basement	55–60	Dry	Fall-winter
Tomatoes (green or white)	Unheated room or basement	55–60	Dry	1-6 weeks
Turnips	Storage cellar or pit	32–40	Moist	Fall-winter
Fruits:				
Most apples	Fruit storage cellar	30–32	Moist	Fall-winter
McIntosh, Yellow Newton & Rhode Island Greening	Fruit storage cellar	35–38	Moist	Fall-winter
Grapes	Fruit storage cellar	31–32	Moist	4-6 weeks
Pears	Fruit storage cellar	30–31	Moist	Fall-winter
Peaches	Fruit storage cellar	32	Moist	2 weeks
Apricots	Fruit storage cellar	32	Moist	2 weeks

fruit you pick from the tree. Windfalls are overripe and give off ethylene gas which speeds ripening of picked fruit.

Desirable temperatures may be possible in refrigerator hydrator drawers for small quantities.

An extra refrigerator can be used to store fruit, but do not take the shelves out. When it is empty, for safety reasons take off the doors.

Other Fruit

Storing fresh cherries, peaches, and apricots very long is difficult. Refrigerate as close to 32° F as possible. Peaches ripen well at 65° to 85° and refrigerate well in hydrators for as long as 4 weeks. Peaches may be stored in walk-in refrigerators in larger quantities.

Grapes are generally not adapted to long storage. Concord grapes may be stored 4 to 6 weeks at 31° to 32° F. Catawba and Delaware varieties can be held 8 weeks. Vinifera table grapes such as Emperor and Ribier will keep 3 to 6 months at 30° to 31°.

Since apples, pears, grapes, and other fruit absorb odors from potatoes, onions, and other vegetables, store them separately.

Some kitchen garden herbs, such as chives and parsley, may be potted and cared for as house plants. These plants will supply flavoring and garnishing to enhance wintertime meals.

Trying to predict exactly how long your fruits and vegetables can be stored is next to impossible. Much depends on condition of the product and how successful you are in maintaining correct temperatures and humidity. Generally, you can keep parsnips and carrots all winter, late potatoes 6 to 8 months, cabbage 3 months, onions 6 to 10 months, and pumpkins, squash, root crops, and tomatoes 3 to 6 months.

Cleanliness. One last precaution: Keep the storage areas clean and free of decaying fruit and vegetables; otherwise, molds and bacteria will spread to your sound produce.

If you store nuts (especially peanuts), soybeans, other dry beans or peas, make every effort to prevent growth of molds. Moisture, temperature, and time are necessary to promote their growth. A harmful toxin may be produced if mold growth is allowed to progress. It is important that storage areas be regularly checked so as to avoid this type of contamination.

Discard all produce that shows any sign of decay.

Insects, rats, and other pests can spread disease and are unwanted guests in any food storage area. To escape these undesirables:

—Build them out. Close all cracks and use adequate screening over all openings
—Prevent trash piles from accumulating
—Keep the storage area clean
—Control rats inside and outside. (Seek the advice of your county Extension office or a sanitarian)
—Destroy any infested food
—Remove all containers at least once a year. Wash them and air dry in the sun
—Remember that good housekeeping practices apply to all places where food is stored

For Further Reading:

U.S. Department of Agriculture, *Storing Vegetables and Fruits in Basements, Cellars, Outbuildings, and Pits*, H&G Bul. No. 119, on sale by Superintendent of Documents, U.S. Government Printing Office, Washington, D.C. 20402. 40¢.

Resurgence of Community Canneries

by F. Aline Coffey and Roger Sternberg

A community cannery is a self-help facility equipped for preparing and heat processing food. People bring in produce from their gardens and through their own efforts preserve it for future use.

Community canneries began during the late 1800's in response to the desire of families to work together to preserve their food for the off-season. At the end of World War II there were over 3,800 community canneries in the United States. Most of these wartime canneries were subsidized, but after the war the monies ceased. Growth of the food industry, development of freezing techniques, and the lack of subsidy led to a decline of the canneries.

Today there is a resurgence of interest in establishing community canning centers. This has been influenced by the cost of food, a marked increase in the concern for nutrition, and gardening activities.

A community cannery promotes the preservation of seasonal garden surpluses for consumption during the nonproductive season. It encourages small farmers and nonfarm individuals to produce more food, thereby promoting self-sufficiency for families. It enables families who do not own recommended food preservation equipment to use safe and reliable equipment and techniques.

Availability of nonseasonal foods on a year-round basis can result in a better diet for families, especially if the center incorporates nutrition education classes as part of its program. People who grow their own food may make substantial savings in their food budget. The community cannery creates a social atmosphere of friendly, cooperative work leading to tangible results, and promotes a feeling of self-reliance.

Most of the community canneries in the country have been organized by Community Action Agencies or similar community organizing groups. Individuals, food co-ops, and other groups have successfully set up canneries, but it is recommended that people wanting to establish a canning center contact a community organizing agency. Normally, these agencies have professional people who will work on such a project. They have experience in writing proposals and are aware of potential funding sources.

Support for the canning center can be enhanced by making a special effort to include a diverse membership on a board of directors for the center.

Farmers, low-income people, business people, contractors, Extension personnel, community organizers, local officials, and members of the clergy are all potential supporters and advisors for the cannery.

Although organizing a community cannery requires a lot of work, this need not be a roadblock to initiating the project. It takes many hours to plan the canning operation, draft proposals, develop community support, locate a site, and to select, purchase and install equipment. Because this can easily be a full-time job for one person, efforts should be made to hire a coordinator. In many instances, paid community organizers, Vista volunteers, and home economists have provided valuable assistance in completing the work.

Preparation for and organization of

F. Aline Coffey is Foods and Nutrition Specialist, Vermont Extension Service, University of Vermont, Burlington. Roger Sternberg is Project Coordinator, Bread and Law Taskforce, Montpelier, Vt.

the cannery are the foundation of the project. At least six months should be set aside for organizing.

Points to Consider

Here are some questions to consider before starting a community center:

How many people will commit themselves to organizing a center?

How much time will they give?

How much support can be expected from the community, town officials, local growers?

How many community and family gardens are in the area?

How near are the community gardens to the cannery site?

Is the site near a well-travelled route?

Is parking available?

Can the canning center exist merely to provide a service to the community, or will the cannery have to become involved in a commercial venture?

If some food processed at the cannery is to be sold, are local farmers willing to contract with the cannery to supply it with produce? How close are these farmers to the cannery?

Is a building available for canning purposes (for example, some old creamery)?

If so, what is the size of the building? What is its condition?

Are there cement floors and walls constructed so they can be washed down daily?

Top left, tomatoes in the canning process, at a community cannery. Bottom, sweet corn ready to have kernels removed for canning. Right, at later stage, liquid is poured over hot-packed corn. Note one advantage of a community cannery is that quantities of food can be processed in a few hours.

Is there room for storage, a walk-in cooler?
Is the sewage system adequate?
Does the building have existing equipment that could be put to use?
Is a dependable supply of potable water available?
What is the minimum water pressure and is it constant?
Is the water "hard?" If so, what is the analysis?
What type of electricity is available?
What is the cost of electricity per KWH and demand rate for 240 volt, 3 phase, 60 cycle?
What is the availability and cost of gas (natural or LP) or of fuel oil?
What is the number of families expected to participate? How many are low-income families?
What are the principal foods to be canned?
If it is anticipated that some products will be processed for sale, what will those products be?
Is there a market for the "for-sale" items?
Will canning supplies such as jars, lids, screw bands, tin cans be available? Can they be purchased at wholesale prices?
Is at least one person who is knowledgeable in food preservation methods available to supervise the cannery?
What will be the charge for processing a pint or a quart of food?
Will low-income people be able to pay this amount?
Are funds available to subsidize the canning of food for low-income people?
It is important to obtain a site easily accessible to the public. Selectmen, property owners, realtors should be approached for potential sites. Usually the center has limited funds, and it is difficult and takes time to locate an appropriate building with low-cost rent.
In times of a strained economy and high cost of property maintenance, the business community may be hesitant to provide low-cost housing for the site.

Establishng the faclity in a publicly owned building, such as a school, is a solution in many communities. These canneries are a part of the public school's physical plant and have traditionally been operated under supervision of the vocational agriculture and home economics teachers, using school funds.

In recent years, some schools have wanted to close canneries for several reasons: Lack of operating capital, limited use, lack of interest or knowhow on the part of participants and teachers. With the resurgence of interest in canning, many new cannery ventures are located in schools but are now funded separately from school budgets.

If the cannery is the result of a community endeavor, adjoining small towns could appropriate funds sufficient to set up and man a center. Such a proposal would have to be presented to the town governing bodies. This points up the need for ample planning time. Devising means to allow the cannery to remain open year-round would favor obtaining a site other than in a public building.

Major Costs

Cost of organizing a community cannery is influenced by its size and scope of operation. Expenses can be broken down into these major areas:
- Purchase and installation of equipment
- Building renovation
- Rent
- Labor
- Utilities
- Jars or cans
- Produce
- Miscellaneous costs (office supplies, freight, postage, insurance, cleaning supplies, maintenance)

At least two companies manufacture community canning equipment (Ball Corp. and Dixie Canner Equipment Co.). Prices start at $4,300 for a single-unit operation, and go up to $20,000 for a large center. This does not include the price of a steam boiler, which costs between $3,000 and $5,000. By fabricating some of its own equipment, and by buying used equipment from canning and restaurant equipment suppliers, the cannery can reduce some of its purchase costs substantially.

Installation of the canning equipment and the steam boiler needs to be done by a licensed plumber or steam fitter, or be closely supervised by such a person.

Renovation of a building and installation of the canning equipment can cost between $4,000 and $8,000, including labor costs. Cost can be reduced by soliciting volunteer labor from local craftsmen. The organizers can handle much of the renovation, such as painting, carpentry and cement work. Teams of vocational students may be willing to take on the site renovations as part of their school training.

Salaries for employees can be paid from the cannery's operating budget. Labor costs can be reduced if the workers are already salaried employees provided by other food-related agencies. The cannery can also be an ideal training site for participants in the Comprehensive Employment and Training Act (CETA) and can be staffed successfully in this way.

Regulations

Food and Drug Administration regulations regarding food processing do not apply to community canning centers if they are not involved in interstate commerce. In June, 1976, FDA issued "Suggested Minimum Guidelines for Community Canning Operations" to protect the safety of the consumer.

Environmental regulations that apply to the centers must be carefully followed. Although these regulations are usually not hard to follow, they often mean a possibly unplanned-for expense to the cannery. It may be necessary to apply for a variance to zoning regulations. Cannery supervisory boards should have a working knowledge of all requirements of State and Federal agencies that regulate health, environment, fire, safety, plumbing, electricity, and public building codes.

Sites for the centers should have sewage and draining systems that meet demands of the centers. This would mean a septic system and leach fields, or a municipal sewage system, the latter being the easiest and least expensive method of disposal. Solid waste produced by the center is termed "clean," and effluent from the processing could be put through a strainer, piped out of the center, and then deposited into a leach field.

To maintain high standards of cleanliness and safety, at least one supervisor should be on duty whenever the cannery is in operation. The person in charge must have a thorough knowledge of every aspect of food processing.

The Food and Drug Administration requires that a "certified registered canner" be in attendance only when low-acid foods are processed to be sold. An FDA-approved course is offered by the National Canners Association for commercial cannery personnel in various sections of the country. The cost would involve a registration fee of approximately $125 plus expenses. At present, the course content is geared chiefly toward industry. A shift to a more practical approach would be of greater help to community cannery personnel.

Cannery supervisors and attendants can participate in food preservation classes and demonstrations provided by the Extension Service. When canneries are equipped with commercial food preservation centers, representatives of the manufacturing companies are available for technical information to the cannery staff. Manufacturers may also provide the cannery with a complete operations manual, processing charts, and recipes.

Skills Needed

Cannery supervisors will benefit by employing people to work at the cannery who can provide or learn such skills as:

Bookkeeping/accounting—to keep records of input and outflow of goods and money; to pay bills.

Management—to oversee the flow of food through the center in an efficient manner for smooth operation of the plant.

Maintenance and repair—to maintain equipment and housing in operational condition.

Purchasing/supply—to ensure a supply of materials such as jars and lids.

Sales—to manage sales of surplus retail products if these are processed at the plant.

Public relations—to advertise and promote knowledge of canning centers; to handle complaints and problems of patrons.

Technical—to provide detailed information on processing techniques, food, nutrition, and gardening.

A form of recordkeeping on all foods processed at the plant is essential. This kind of information would include such data as name of person doing the processing, the date, specific food, number of jars, method of processing, time in and time out, and an identification number for foods processed for sale by the cannery.

Canning centers may be incorporated as independent nonprofit cooperatives with a board of directors as the policymaking body. By being organized in conformity with the traditional farmers' cooperative structure, the centers receive special tax considerations. Incorporation on a nonprofit basis is a requirement of many funding sources. The cooperative structure also lends itself to a tighter knit organization, with members feeling they are part of the organization, responsible for its affairs, and willing to pitch in and help if there is some work that needs a few extra hands.

The community cannery should have general liability insurance to cover injuries sustained by the workers or persons using the canning center. Products liability insurance is unnecessary for the cannery operated solely to provide a service to the community. For the cannery that sells commercially, products liability insurance should be obtained.

Hours, Fees

A community cannery should be available to all people interested in preserving food. Ideally, canneries are open during daytime and evening hours. Weekend hours are a possibility. When canneries are limited to processing vegetables and fruits, at

Sealer in a community cannery.

William E. Carnahan

least 6 months of potential operation are lost in certain sections of the country. If at all possible, canneries should be operated to process a wider range of foods such as jams, jellies, pickles, preserves, meat, fish, poultry.

A processing fee is usually set for use of the canning equipment, ranging from 5¢ to 10¢ for pints to 10¢ to 15¢ for quarts. These prices do not include the cost of jar, lid, screw band, or any canning supplies such as salt, vinegar, sugar, spices that may be sold at wholesale prices at the cannery. An additional charge of 50¢ per hour is common for the use of a pulper-juicer and steam-jacketed kettle.

It may be a financial hardship for some low-income families to meet these costs, but they can be given the opportunity to exchange work time at the cannery for payment. Families of limited resources might leave off a percentage of their processed high-acid foods to be sold by the cannery. Sponsoring agencies may apply for grants, such as might be available from Title XX of the Social Security Act, in an effort to subsidize canning costs for low-income families.

To date, no community canneries are completely economically self-sufficient, so far as we know. There are centers in the South which do enough community canning to pay for all their expenses except salaries. To become self-supportive, some canneries are now developing specialty products to be sold commercially. Organic-health food distributorships and food co-ops are often a good market for community cannery processed foods.

The future of community canneries depends on continued interest in home gardening and food preservation, and concern for proper nutrition. Undoubtedly, the cost of food in the marketplace will also be a contributing factor.

Tomato juice being processed at a community cannery. Since cannery is operated by State of Virginia, only charge is for cans used.

377

Questions and Answers on Food Preservation
by Carole Davis and Annetta Cook

Consumers frequently have questions regarding home food preservation practices. Below are some questions commonly asked, and the answers. They concern canning fruits, vegetables, pickles, and jellies, and freezing fruits and vegetables.

Why is open-kettle canning not recommended?

In open-kettle canning, food is cooked in an ordinary kettle, then packed into hot jars and sealed. The food is not processed after packing in the jars.

Open-kettle canning is unsafe because temperatures reached are not high enough to destroy all the spoilage organisms that may be in low-acid foods, such as meat and vegetables, other than tomatoes.

Spoilage bacteria may also enter the food when it is transferred from kettle to jar, making it undesirable as well to can other foods such as fruits, pickles, preserves, and jams by this method.

Why is oven canning unsafe?

Jars may explode, causing personal injury or damage to the oven. Also, temperatures obtained in the oven are not high enough to insure adequate destruction of spoilage organisms in low-acid foods.

Times specified for boiling-water-bath processing of foods do not apply to oven processing since the rate of heat penetration would be different in the oven and the products could easily be underprocessed.

Should jars and lids be sterilized before canning?

Carole Davis and Annetta Cook are Food Technologists with the Consumer and Food Economics Institute, Agricultural Research Service.

No, not when the boiling-water-bath or pressure-canner method is used, because the containers and lids are sterilized during processing. But be sure jars and lids are clean.

Why is no liquid added when tomatoes are canned?

Because tomatoes provide their own juice if pressed gently when packed raw, or when heated before packing hot into jars.

Is it safe to add celery, green pepper, and onion to tomatoes when canning them?

No. Adding other vegetables lowers the acidity of tomatoes. Acidity helps protect against the growth of botulinum bacteria, which can produce a fatal toxin in canned foods. Specific recipes, times, and temperatures determined scientifically for vegetable mixtures need to be used for their safe canning.

Why is headspace important in canning?

Headspace—the distance between the surface of food and the underside of the lid—allows for expansion of solids or bubbling up of liquid during processing. If headspace is not adequate, some food in the container will be forced out, leaving food particles or sirup on the sealing surface and preventing a seal.

When too much headspace is allowed, some air may remain in the jar after processing, causing food at the top of the jar to darken.

What causes jars to break in a canner?

Breakage can occur for several reasons: (1) Using commercial food jars rather than jars manufactured for home canning, (2) Using jars that have hairline cracks, (3) Putting jars directly on bottom of canner instead of on a rack, (4) Putting hot food in cold jars, or (5) Putting jars of raw or unheated food directly into boiling water in the canner, rather than into hot water (sudden change in temperature—too wide a margin between temperature of filled jars and water in canner before processing).

What causes liquid to be lost from jars during processing?

Loss of liquid may be due to packing jars too full. Headspace must be allowed between the top of the food and lid as specified in the instructions for each food. Food expands when processed, so headspace must be adequate or liquid will be forced out of the jar.

Liquid may be lost if the canner's pressure fluctuates during processing. Lowering the pressure too suddenly after processing may also cause liquid to be lost. Pressure canner should be removed from the heat and allowed to cool normally at room temperature.

If liquid is lost from jars during processing, can more be added to fill them again?

No, because if the jar is opened and liquid added, this would allow bacteria to enter the jar and you would need to process again.

Loss of liquid does not cause food to spoil, though food not covered by the liquid may darken.

Why does fruit sometimes float in the jar after canning?

Fruit may float because it is packed too loosely, sirup is too heavy, or because some air remains in tissues of the fruit after heating and processing.

How do you test the seal on home-canned foods?

After jars have cooled, check two-piece lids by pressing the center of the flat metal lid; if lid is down and will not move, jar is sealed.

For porcelain-lined caps, check seal by turning each jar partly over in your hands. If no leakage occurs, the jar is sealed.

What causes lids not to seal?

If food has not been sufficiently heated, a vacuum may not be drawn on the jar of food and the lid will fail to seal. Presence of food particles or sirup on the jar rim could also prevent obtaining an airtight seal. Each jar rim should be wiped clean of all food and sirup before putting the sealing lid in place.

If food has been packed too tightly in the jars or if sufficient headspace has not been allowed, expansion of the food during heating could force sirup or food out of the jar, thus causing poor contact between the lid and jar.

Why should flat metal disks and rubber rings be used only once?

Depressions in the rubber compound made when the lid or ring was first used can prevent obtaining an airtight seal if used a second time.

Why should metal bands be removed after jars have cooled?

If bands are not removed soon after cooling, moisture between the ring and jar may cause the ring to rust, thus making later removal of the bands difficult. The band is no longer necessary after the jar has cooled because the seal has been provided by the flat metal lid with sealing compound and the vacuum created during cooling.

What causes canned foods to change color?

Darkening of food at the top of the jar may be caused by oxidation due to air in the jar, or by too little heating or processing so that enzymes are not destroyed. Overprocessing may cause discoloration of foods throughout the jar.

Pink and blue colors sometimes seen in canned pears, apples, and peaches are caused by chemical changes in the fruit coloring matter.

Iron and copper from cooking utensils or from water in some areas may cause brown, black, and gray colors in some foods.

Why do undersides of metal lids sometimes discolor?

Natural compounds in some foods, particularly acids, corrode metal and make a dark deposit on the underside of jar lids. This deposit on lids of sealed, properly processed canned foods is harmless.

Is it safe to use canned foods which have been frozen as the result of storing them in an unheated storage area?

Freezing does not cause the food to spoil unless the seal is damaged or the jar broken. If the jar is no longer sealed, the food may still be safe to eat if the jar is not broken and the food is still frozen and has not been subjected to thawing and refreezing.

Remove the frozen canned food from jars as carefully as possible. The food may need to thaw slightly to ease its removal from jars, but it should be left in as large blocks as can be removed through the jar opening.

Examine jars for breaks and hairline cracks. If any are found, discard food from those jars. If no cracks are found, food may be transferred from jars into freezer bags or containers and stored in the freezer, or it may be kept in the refrigerator for use within a day or two.

Home-canned foods which have been frozen may be less palatable due to texture changes than properly stored canned foods. Do not recan home-canned foods which have been frozen.

How do you protect canned foods against freezing?

Wrap the jars in paper or cover them with blankets. However, if the storage area temperature is expected to be below freezing (32° F) for more than a day or two, move the food to a warmer storage area.

What does mold on canned food indicate?

It means the jar has not sealed and the food is spoiled. Even if mold appears to be only on the surface, discard all food in the container because parts of the mold may not be visible in the food.

Is it safe to can foods without salt?

Yes. Salt is used for flavor only in canned vegetables and is not necessary for safe processing. Since the characteristic flavor and texture of pickled vegetables depend on salt, do not omit this ingredient from recipes for pickles and relishes.

What kind of salt should be used in pickling? Why?

Use pure granulated salt. Uniodized table salt can be used, but materials added to the salt to prevent caking may make the brine cloudy. Do not use iodized table salt because it may darken pickles.

What type of vinegar should be used for making pickles? Can it be diluted?

Use cider or white distilled vinegar of 4% to 6% acidity (40 to 60 grain). Do not use vinegar of unknown acidity. Do not dilute vinegar unless the recipe so specifies. If a less sour product is preferred, add sugar rather than decrease vinegar.

Why should pickles be processed in a boiling-water bath?

Pickles require heat treatment to destroy organisms that cause spoilage, and to inactivate enzymes that may affect flavor, color, and texture.

Heat processing in a boiling-water bath is recommended for all pickle products. There is always danger of spoilage organisms entering the food when it is transferred from kettle to jar. This is true even when the utmost caution is observed and is the reason open-kettle canning is not recommended.

Why should plastic containers not be used when brining pickles?

Vegetables being pickled undergo physical as well as chemical changes during brining or fermentation. As a result of these changes, the plastic may be affected, causing undesirable compounds to be formed or leached from the plastic.

For fermenting or brining pickles, use a crock or stone jar, unchipped enamel-lined pan, or large glass jar, bowl, or casserole.

What causes pickles to be hollow?

Hollowness in pickles generally results from poorly developed cucumbers, holding cucumbers too long before pickling, too rapid fermentation, too strong or too weak a brine during fermentation.

What causes jelly to be too soft?

Too much juice in the mixture, too little sugar, mixture not acid enough (overripe fruit), or making too big a batch at one time.

What makes jelly tough?

Mixture was cooked too long to reach jellying stage because too little sugar was used in proportion to the pectin and acid in the juice.

What makes crystals form in jelly?

Crystals throughout the jelly may be caused by too much sugar in the jelly mixture, or cooking the mixture too little, too slowly, or too long. Crystals on top of jelly that has been opened and allowed to stand are due to evaporation of liquid.

Tartrate crystals in grape jelly may occur if juice has not been allowed to stand overnight and then strained through a double thickness of cheesecloth before preparing jelly.

Is a one-door refrigerator-freezer combination suitable for freezing and storage of frozen fruits and vegetables?

It may be difficult to obtain the recommended temperature of 0° F or below for freezing and storing foods in this style freezer without freezing food in the refrigerator part as well.

Recommended storage times are severely reduced if a freezer does not maintain 0° F or below. If freezer temperatures are above 10°, do not store frozen food for more than several weeks.

Can containers for commercial foods, such as cottage cheese, margarine, milk, yogurt, ice cream, or sour cream, be used for freezing fruits and vegetables?

Waxed cardboard cartons which previously contained dairy products are not sufficiently moisture-vapor-resistant to use for packaging foods to be frozen.

Plastic commercial-food containers are suitable if they can be tightly sealed and do not become brittle and crack at low temperatures, thus exposing the food to the air.

Can citric acid or lemon juice be used to help prevent fruit from turning dark during freezer storage?

Although these products can be used as anti-darkening agents, neither is as effective as ascorbic acid. Often the quantity of citric acid or lemon juice needed to prevent darkening is so large that natural flavors are masked or the fruit becomes too sour.

Why is it necessary to wash and blanch vegetables before they are frozen?

Washing removes dirt and some of the bacteria from vegetables. Freezing inhibits the growth of bacteria, but does not kill them. Thus it's important that the food, as well as all surfaces it touches, be kept clean so that the number of bacteria on the food is held to a minimum. Bacteria can grow on food if the temperature rises during freezer storage, and when food is thawed.

Except for green peppers and mature onions, vegetables must be blanched to destroy enzymes which could cause undesirable changes in flavor, texture, and color during freezer storage.

Why can green peppers and mature onions be frozen without blanching?

Unlike other vegetables, green peppers and onions do not lose quality during freezer storage if their enzymes are not destroyed by blanching before freezing.

Green peppers frozen without heating are better suited for use in uncooked foods than are blanched peppers. Some of the characteristic flavor of onions is lost if this vegetable is blanched before freezing.

Why is corn which is frozen on-the-cob blanched for longer times than cut corn?

Longer blanching of corn frozen on-the-cob is necessary so that enzymes present in the cob will be destroyed. Otherwise, enzymes in the unheated cob can cause undesirable flavor changes in the corn kernels.

Corn frozen off-the-cob needs only to be blanched just enough to destroy enzymes in the kernels.

Can vegetables be blanched by steaming instead of by heating in boiling water?

The following vegetables may be heated in steam: broccoli, mushrooms, pumpkin, winter squash, and sweetpotatoes.

To steam these vegetables, put 1 to 2 inches of water in a large kettle; bring water to a boil. Add a basket containing a single layer of prepared vegetable; keep the basket at least 1 inch above the water. Cover kettle and start timing.

Steam broccoli 3 minutes; sliced mushrooms, 3 minutes; whole mushrooms (less than 1-inch diameter), 5 minutes. Steam pumpkin, winter squash, or sweetpotatoes until tender.

Is it necessary to make an adjustment in blanching times for vegetables at altitudes above sea level?

At altitudes 5,000 feet or more above sea level, heat vegetables 1 minute longer than the time given in directions for the vegetable being blanched.

What can be done to prevent food from thawing if the freezer should stop running?

Never open the freezer unnecessarily. A fully-loaded freezer will usually remain cold enough to keep food frozen for 2 days if the door is not opened; a half-loaded freezer may not stay cold enough more than a day.

If power cannot be restored or the freezer cannot be fixed before the food would start to thaw, use dry ice. If dry ice is obtained shortly after the failure has occurred and the freezer is fully loaded, 25 pounds of dry ice should keep a 10-cubic-foot freezer at temperatures below freezing for 3 to 4 days; if freezer is less than half-full, for 2 to 3 days.

Place dry ice on boards or heavy cardboard on top of the packages of food. Handle dry ice carefully—never with your bare hands. Wear gloves to prevent burns.

Another alternative is to move food to a neighbor's freezer or to a freezer locker plant where space can be rented.

Can vegetables and fruits which have thawed be refrozen?

Frozen foods that have thawed may be safely refrozen if they still contain ice crystals or if they are still cold—about 40° F—and have been held no longer than 1 or 2 days at refrigerator temperatures (32° to 40°) after thawing.

Since thawing and refreezing reduces the quality of fruits and vegetables, use refrozen foods as soon as possible to save as much of their eating quality as you can.

Food Preservation Glossary
compiled by Annetta Cook and Carole Davis

Acid food—Food with a pH of 4.6 or below. An acid food can be safely processed in a boiling-water bath for specified times. Includes most fruits, tomatoes, and pickled vegetables.

Anaerobes—Bacteria capable of growing without air, as in a sealed container of canned food.

Blanching—Heating vegetables by immersion in boiling water, steaming, sauteing, or stewing to inactivate enzymes capable of causing quality changes in foods during freezer storage.

Boiling-water-bath Canner—A large kettle with lid, rack, and cover; must be deep enough to allow jars to be covered with 1 to 2 inches of water and still have additional height for water to boil actively. Suitable for processing acid foods.

Botulism—Foodborne illness caused from eating canned foods containing the toxin produced by *Clostridium botulinum*, an anaerobic bacterium. This organism can grow and produce toxin in sealed jars of canned foods that are improperly processed.

Canning—Preserving food in airtight rigid containers. Micro-organisms are destroyed by heat-processing containers of food at the temperature and time specified for each food. It is essential to follow reliable canning instructions exactly to insure a safe canned product that is free from botulism-causing bacteria and spoilage organisms.

Cold Pack—Raw, unheated food packed into canning containers and covered with boiling sirup, juice, or water.

Dehydrator—A device which removes moisture, a dryer.

Enzymes—Proteins involved in plant growth processes including maturation and ripening. Enzymes can cause loss of quality in food if they remain active during storage. They are destroyed by canning, or by blanching vegetables before freezing.

Freezer Burn—Small, white, dehydrated areas which occur on improperly wrapped frozen foods. This condition is harmless, but if extensive can cause food to become tough or lose flavor.

Freezing—Preserving food by storing at low temperatures. The recommended temperature for freezer storage is 0° F or below.

Headspace—The space between the top of food in a container and the container lid or closure.

Hot Pack—Food heated in sirup, water or steam, or juice, and packed hot into canning jars.

Hydrator (Vegetable Crisper)—A drawer-like section in refrigerators which protects fresh fruits and vegetables from drying out during refrigerator storage.

Low-acid Food—Food with pH above 4.6. A low-acid food requires processing at high temperature under pressure to destroy micro-organisms and insure a safe canned product. Includes all vegetables except tomatoes.

Micro-organism—Includes bacteria, molds, and yeasts, which when present in food can cause spoilage and even food poisoning. Therefore, they must be destroyed in canning foods or their growth prevented in freezing and drying foods.

Moisture-vapor-proof—Packing materials that prevent loss of moisture from foods during freezer storage. Examples include glass, rigid plastic, and metal freezer containers.

Moisture-vapor-resistant—Packing materials that protect foods from moisture loss during freezer storage. Examples include freezer wraps—paper, plastic, or foil—plastic bags, waxed freezer cartons.

Molds—Microscopic fungi which form air-borne spores (seeds) that may alight on food and grow into cottony mats or fuzz. Some molds or their end products may be harmful, and moldy vegetables or fruit should not be canned. Molds are destroyed by proper canning, but they may develop in leaky containers. Food from leaky containers or any canned food showing mold growth should be discarded without tasting.

Open-kettle Canning—Procedure whereby food is cooked in an ordinary kettle, then packed into hot jars and sealed. Jars of food receive no additional heat processing. This is a dangerous practice as spoilage organisms may enter the jar during the transfer of food from kettle to jar. In low-acid foods,

---

Annetta Cook and Carole Davis are Food Technologists with the Consumer and Food Economics Institute, Agricultural Research Service.

temperatures obtained are not hot enough to insure destruction of all spoilage organisms that may be present in the food.

Pack—Designates how food is packed into containers. Specifies the temperature of food when packed into jars for canning, or the method of sweetening fruits for freezer packs.

Pectin—A substance occurring naturally in many fruits which causes the juice to thicken or gel after heating if the proper proportions of sugar and acid are present. Natural pectins are more prevalent in underripe fruit than mature or overripe fruit. Some fruits have enough natural pectins to make high quality jams and jellies. Others require addition of commercial pectins which are made from either citrus peel or apples.

pH—Measure of acidity of a product. The lower the pH the higher the acidity.

Preserve—To maintain the quality of food for consumption at a later time. Accomplished by canning, freezing, pickling, drying, or making jelly, jams, or preserves.

Processing—Heating food in closed canning jars to insure destruction of micro-organisms so the canned food will remain unspoiled and safe to eat.

Acid foods—such as fruits, tomatoes, and pickles—and jams and preserves are safely processed in a boiling-water-bath. Low-acid vegetables (all vegetables except tomatoes) require processing at higher temperature by use of a pressure canner or pressure saucepan. Times required to insure an adequate process are specified in canning directions for each food.

Sirup Pack—Fruit is packed for freezing in a sugar sirup made by dissolving sugar in water.

Steam-pressure Canner—A large, heavy metal pan having a tight-fitting cover which is fitted with safety valve, steam vent or petcock, and a gage—either weighted or dial. Used for processing low-acid foods under pressure at high temperatures in order to insure their safety.

Steam-pressure Saucepan—Smaller than a canner. If equipped with a gage to maintain pressure at 10 pounds, it is suitable for processing food in pint jars.

Sugar Pack—Sugar is added directly to fruit and mixed gently to draw juice from fruit before packing into freezer containers.

Unsweetened Pack—Fruit packed for freezing without any sweetening added. It may be packed dry or covered with water.

Photography

William E. Carnahan of the Extension Service acted as visual coordinator on the 1977 Yearbook Committee, obtaining photos from a wide variety of sources, and also contributing more photos to the book than anyone else.

Others who helped round up photos for the book include Allan Stoner of the Agricultural Research Service, assistant chairman of the Committee; and three Committee members—Raymond Brush, American Association of Nurserymen, Robert Falasca, American Seed Trade Association, and Evelyn Johnson, Extension Service.

George A. Robinson, Office of Communication, worked on photographic aspects of the book until he left the U.S. Department of Agriculture (USDA) in 1976.

The Agriculture Department is most grateful to the individuals and organizations contributing photos to this book.

Prints of many of the black and white USDA photos may be obtained from the Photography Division, Office of Communication, Room 536-A, U.S. Department of Agriculture, Washington, D.C. 20250. Generally a charge is made to cover expenses. Duplicate slides of some of the color photos also may be ordered. Color slide sets of plant pests, with photography by Clemson University, can be purchased from USDA.

Names of photographers, where known, appear with black and white photos in the book. In such cases no page numbers are given in the photo credits below. Black and white USDA photos are not listed in the credits.

Color photos in the front of the book carry numbers, which are used in the credits to identify them. These numbers are preceded with a "C" in the credits, as C3. Other numbers in the credits refer to page numbers in the black and white part of the book, as 22.

Photographers with no affiliation or address listed were working for USDA.

E. Blair Adams, Washington State University
All America Selections, C3, 137 (top)
William Aplin, SUNSET—Joy of Gardening
Ball Corp., C70, 297
Bernardin, Inc., Home Canning Guide, 314, 315, 329, 342
Milo Burnham, Mississippi State University
Burpee Seeds, 148 (right), 170

William E. Carnahan, C23, C24, C39, C68, C71, C72, last color photo
Chevron Chemical Co., C15, C16, C22
Glenn Christiansen, SUNSET—Joy of Gardening
Clemson Agricultural College, 77 (left)
Clemson University Extension Service, C9, C13, C40–C50, C52–C54, C61, C63—C66
University of Connecticut, 26 (both photos), 31 (top)

E. I. duPont de Nemours and Company, 29 (bottom)
Eastern Grape Grower Magazine, 261 (both)
Ferry Morse Seed Co., 167 (bottom), 168 (bottom)
Wayne Fogle, Foothill Community College
Frank Foster, S. D. Warren Co., a division of Scott Paper Co., first color photo
D. H. Fritts, Montana State University
University of Georgia Extension, 77 (right)
Stan Griffin, D.C. Extension Service, C28
Joseph Harris Seed Co., Inc., 149, 164 (left), 168 (top)
A. S. Harvey, Tennessee Agricultural Experiment Station
Kevin Hayes, Pennsylvania State University
Albert W. Hellenthal, T. M. Duche' Nut Co., C14
Lonnie Hodges, University of Kentucky, C67
International Harvester Co., 211
R. A. Jaynes, Connecticut Agricultural Experiment Station, New Haven
Kansas State University, C31, C34
Robert Knight, C56, C57
John Kucharski, C69
N. S. Mansour, Oregon State University, C33
Charles Marr, Kansas State University, C29
University of Maryland—Vegetable Research Farm, 135

Merchants Publishing Co., C21, C59
The National Future Farmer, 67 (both photos)
National Garden Bureau, 87, 137 (bottom)
Don Normark, SUNSET—*Joy of Gardening*, second color photo
Terence O'Driscoll, C5–C8, C10–C12, C17–C20, C25, C26, C27, C30, C32, C35, C37, C38, C51
John O'Neill, National Garden Bureau
Oregon Agricultural Experiment Station, 5, 11 (both photos), 138, 155 (top), 160 (all photos), 161
Organic Gardening and Farming, 19, 81 (bottom)
Pennsylvania State University, 59 (all photos)
Sallie Peyton, Woodbury, Tenn.
Norman A. Plate, SUNSET—*Joy of Gardening*
George A. Robinson, C73
Vincent E. Rubatzky, University of California
Paris Trail, New York State Agricultural Experiment Station, Geneva
Art Walker, National Canners Association
S. D. Warren Co., a division of Scott Paper Co., C4
Darrow M. Watt, SUNSET—*Joy of Gardening*
Roger Way, New York State Agricultural Experiment Station, Geneva
University of Wisconsin Extension Service, 267

Index

Abraham, George, 102-110
Abraham, Katy, 102-110
Acid: acetic, 355; ascorbic, 305, 337; content in food, 340; tannic, 354
Acidification, 302
Acidity: 54, 119, 323; food chart, 313; food processing, 328; fruit and vegetable chart, 324; wine, 351-352
Aeration: 56, 68, 79, 113; blueberries, 281
Agricultural Extension Service, 90
Alcohol: denatured, 341; in wine, 351-354
Alkalinity, 54
Almond, 53, 284-290
Altitude: 45; corrections, 329; pressure canners, 327-329
Aluminum, 59, 64
American Horticultural Society, 91
Amino acids, 58
Ammonium nitrate, 120, 134
Ammonium sulfate: 120; cucurbits, 190
Anderson, Kelton L., 54-60
Annonaceous fruits, 295
Annuals, 48
Anthracnose: 73; walnuts, 289; watermelon, 191
Ants, 53
Aphids: 74-75, 82, 118; apples, 251; attacking peppers, 145; chard, 164; cole crops, 135; collards, 166; cucurbits, 192-193; root crops, 159; tomatoes, 144
Appalachians, 38
Apple maggot fly, 73, 251
Apple scab, 73
Apples: 41, 53, 246-252; freezing, 336; pest control, 73, 251; storage, 252, 369; varieties, 246-247
Apricots: 53, 253-259; pest control, 73
Arthurs, Kathryn L., 126-132
Artichoke: Jerusalem, 49, 238-240
Ascorbic acid: 305; fruit preparation, 337
Asparagus: 8-10, 12-13, 104, 109, 123; environment needs, 197; harvesting, 200-201; perennial vegetable, 196-201
Assassin bugs, 77
Aster leafhopper, 74
Atemoya, 295
Atlantic Ocean, 45
Autoxidation, 304-305
Avocado: 53, 294; freezing, 336

Bacillus thuringiensis, 77, 83
Bacteria: 81, 84, 128; dried food, 360; food spoilage, 301, 320; thermophilic, 362
Bacteriocide, 75
Baggett, J. R., 157-162
Balcony, 126
Ball Corp., 375
Bamboo, 18
Banadyga, Albert A., 163-170
Bananas, 53, 294
Bark: 5, 20, 32, 56-67, 80; potting soil, 128
Barrels: 18, 102; container growing, 126-127; crushing grapes, 352
Bartelli, Lindo J., 54-60
Bartok, John W., Jr., 24-32
Basement: 70, 86-87; storing produce, 365
Basil, 219
Bates, Earl M., 33-37
Bats, 52
Beans: 45, 47-49, 53, 68, 83, 86, 105, 171-177; bush, 172-173; container grown, 126, 131; how to can green, 333; lima, 7-10, 12-13, 104, 123, 174-177; pest control, 74, 174; pole, 4, 104, 109, 172-173; snap, 8-10, 12-13, 104, 123, 171-172
Bearce, Bradford C., 49-51
Bee, Honey, 53
Beetles: 53, 74, 77; asparagus, 200; Mexican bean, 83
Beets: 7-13, 20, 39, 46-48, 70, 82, 109, 123; container grown, 126, 131; easy to grow, 160; pest control, 74
Berries: 15, 18, 272-278; freezing, 336; pruning, 274; varieties, 272
Berry mold, 74
Berry rot, 74
Biennials, 48
Birds: 49, 52, 76-78, 107; berries, 277; netting, 263
Blackberries: 2, 53, 272-278; pest control, 73, 277
Black knot, 73
Black leg, 166
Black rot: 73; sweet potatoes, 216
Black-sapote, 296
Black walnut, 284-290
Blackwell, Cecil, 2-14
Blanching: pretreatment for drying, 358
Blight, 83
Blossom-end rot, 144
Blueberries: 2, 53, 60; home garden, 279-283; pruning, 283; varieties, 280-281
Boiling water bath: 301; altitude corrections, 329; pickles and relishes, 347-348
Bone meal, 78
Borers, 73
Boron, 49, 60
Botulism: 301, 313, 319; home-preserved foods, 361
Bouwkamp, John C., 212-216
Boysenberries, 272-278
Brambles, 73, 272-278
Broadbeam, 53
Broadcast spreader, 25
Broccoli: 7-10, 12-13, 104-105, 109, 123; cole crops, 133-138; pest control, 74, 135; varieties, 136
Brown rot: 73-74; almonds, 289; nectarines, 253
Brussels sprouts: 8-10, 12-13, 104-105, 109, 123; cole crops, 133-138; pest control, 74, 135; varieties, 138
Bryant, M. Douglas, 41-43
Budding, 51
Bulbs, 51
Burlap: 70, 87; storing onions, 369
Burnham, Milo, 228-244
Bush fruits, 272-278
Butter: herb, 223
Butternut, 284-290

Cabbage: 4, 7-10, 39, 62, 86-87, 109, 123; Chinese, 8-10, 12-13, 109, 123; cole crops, 133-138; pest control, 74, 135; varieties, 136-137
Cabbage loopers, 83
Cabbage worm: 74; cole crops, 135; collards, 166
Cactus fruits, 295-296
Calamodin, 296
Calcium: 4, 49, 107, 120; asparagus, 201; deficiency, 144; sweet potato, 212
Calories: sweet potatoes, 213; wine, 350
Calyx, 49
Cane blight, 73
Cane fruits, 272-278
Canneries: commercial, 302, 323; community, 372-377
Canners: boiling water bath, 314-315, 317, 329; gage types, 325-326; pressure, 315-318, 323-327, 329-330; types of pressure controls, 326
Canning: 305-307; Beginner's Guide, 313-319; boiling water bath, 323; community canneries, 372-377; fruits and vegetables, 328-333; jars, 314-

387

315; packing methods, 316–317; pickles, 348; pressure canners, 323–327; storage, 318, 361–362
Cantaloupes, 7, 45, 104, 109, 123
Capsaicin, 140
Carambola, 294
Carbohydrates, 47
Carbon dioxide, 47–49, 62, 65, 68
Carob, 294
Carrots: 7–10, 12–13, 104–105, 109, 123; a garden favorite, 161; container grown, 126, 131; pest control, 74
Cascade Mountains, 34, 36
Caterpillars, 77
Cattley guava, 77
Cauliflower: 7–10, 12–13, 104–105, 109, 123; cole crops, 133–138; pest control, 74, 135; varieties, 137–138
Cayenne, 140
Celeriac: 8–10, 12–13, 104, 109; culture, 228–229; uses, 160
Celery: 8–10, 12–13, 104, 109, 123, 149–150; growing conditions, 149; harvesting, 150; pest control, 74
Cellars: 69, 87; storing root vegetables, 365
Cercospora leaf spot, 145
Charcoal, activated, 18
Chard: 7–10, 12–13, 102, 104–105, 109, 123; harvesting, 164–165; pest control, 74; Swiss, 82, 104–105, 109, 123; varieties, 163
Chayote, 53, 230–231
Chemical controls, 75–76
Cherimoya, 295
Cherries: 41, 53, 253–259; pest control, 73; varieties, 255
Chervil, 8–10, 12–13
Chicory: 151; witloof, 8–10, 12–13
Chili, 140
Chinese cabbage: 8–10, 12–13, 104, 109, 123; cole crops, 133–138; varieties, 138
Chinese chestnut, 284–290
Chinese gooseberry, 53
Chinkapin chestnuts, 284–290
Chives: 8–10, 12–13, 123, 155–156; uses, 219
Chlorophyll, 47–48
Chromosomes, 50–51
Citrus, 18, 41, 43, 296
Clarification: wine, 353–354
Clay, 55
Cleome, 20
Climate: 17, 33–46; information sources, 36
Climatic tables: Northeast, 39; Northwest, 35; Squtheast, 44; Southwest, 42
Clostridium botulinum: 301–302, 313, 323, 328; improperly processed foods, 361
Clostridium perfringens, 298
Clubroot, 133
Codling moth, 73
Coffey, F. Aline, 372–377
Coldframes, 61–62, 86, 111
Cold pack, 330
Cole crops: 34, 133–138; pest control, 135

Collards: 7–10, 12–13, 104, 109, 123; harvesting, 166; pest control, 74; varieties, 165
Colorado potato beetle, 83
Combs, O. B., 47–49
Community canneries: 372–377; fees, 376–377; points to consider, 373–374; skills needed, 376
Community gardens, 20–23
Compost, 5, 80–81, 119
Conduction, 324–325
Consumer Information Center, 89
Container gardening: 126–132; harvesting, 132; mulching, 130; recommended vegetables, 131–132; watering, 129–130
Containers: dried food, 359; freezing, 321, 335; gardens, 18–20, 112–113, 126–132; home canning, 314–315; jelly, 342; subtropical fruit, 293–294; wine, 353
Convection, 324–325
Cook, Annetta, 334–339, 378–384
Coolers, 61, 66
Cooperative Extension Service, 31, 71, 89–90
Copper, 49, 60
Copper napthenate, 18, 61
Corn: 7–10, 12–13, 102, 104–105, 109, 123; a garden favorite, 181–186; climatic needs, 182; container grown, 126, 131; diseases, 184; harvesting, 185–186; hybrids, 183; insects, 185; sirup, 340–341; types, 181
Corn earworm, 75
Cottonseed meal, 78
Cotyledons, 50
County Extension Service, 90
Cover crops, 79–80, 84–85
Cowpea curculio, 75
Cress: 82, 151; growing indoors, 151; upland, 8–10, 12–13, 123
Crops: cole, 133–138; cover, 79–80, 84–85; cucurbit, 187–195; root, 157–162; rotation, 107
Cucumbers: 7–10, 12–13, 104–105, 109, 123, 187–195; container grown, 126, 131; nutrient needs, 189; pest control, 74; pickling types, 345; planting chart, 190; resistant varieties, 192
Cucumber beetles, 83, 192
Cucurbit crops: 187–195; nutrient needs, 189; planting chart, 190; varieties, 192
Cultivate, 80
Cultivation: 37; berries, 274; hand, 102, 119; potatoes, 210
Cultivator, 24, 110
Curculio, 73–74
Currants: 53, 272–278; freezing, 336
Cuttings, 51, 67
Cutworms: 74, 82, 124; attacking peppers, 145

Damping off: 117; peppers, 145
Dasheen, 231–232
Dates, planting, 8–10, 12–13
Davis, Carole, 328–333, 378–384
Deer, 77

Dehydrator cabinets, 356
Desiccation, 305
Dew, 77
Diamondback, 74
Dill, 219
Diseases: cane fruits, 277–278; plant, 3, 14, 45, 84, 117
Disk harrow, 28
Dixie Canner Equipment Co., 375
Dormancy, 50
Downey, Isabelle, 345–349
Downy mildew: collards, 166; cucurbits, 191
Drainage: 3–4, 38, 54, 56, 103, 113; container gardening, 126
Draper, A. D., 279–283
Drive shaft, 28
Drought, 3, 15, 37
Drying: food quality, 303, 309; fruits and vegetables, 356–360; oven, 356; procedure, 357–358
Dunn, Charlotte M., 320–322
Duster, 30, 86
Dwarf trees, 247–248

Earworms, 75
Earthworms, 76, 106
Eggplant: 7–10, 12–13, 102, 104–105, 109, 132; fertilizing, 141; harvesting, 145–146; transplanting, 142–143; weed control, 144
Egyptian onion, 220
Ellison, J. Howard, 196–201
Embryo, 47, 50
Emerson, Barbara H., 89–92
Endive: 8–10, 12–13, 82, 123; types, 150
Endosperm, 50
Energy, 47
English walnut, 284–290
Environmental Protection Agency, 72
Enzymes: food, 320–321; fruits and vegetables, 304, 358
Equator, 45
Equipment: garden, 24–32; making pickles, 346–347; power, 110; storage, 69, 86
Erosion, 17–18, 45, 58, 79, 119
Escabeche, 140
Escarole, 150
European corn borer, 75

Fans: 65–66, 70; food drying, 356–357
Fats, 47
Faucet, 15, 23
Feijoa, 294
Fences: 4, 20, 61, 68; electric, 77; growing cucumbers, 102; keeping clean, 76; temporary, 85
Fermentation: 302; wine, 350–353
Fertilizer: 31–32, 58, 60, 80–81, 119; asparagus, 198; blueberries, 282–283; cane fruits, 273; chard, 163; cole crops, 134; commercial, 120; container gardens, 130; cucurbits, 189–190; nut crops, 288–289; okra, 224–225; potatoes, 208; recommended for corn, 183; rhubarb 202; stone fruit trees, 256–257

388

strawberries, 267-268; subtropical fruits, 292
Fiber: fruits and vegetables, 305
Fiberglass, 64
Figs: 53, 294; freezing, 336; Indian, 295-296
Filberts, 284-290
Fining: wine, 354
Firewood, 16
Flea beetles: 75, 83; attacking peppers, 145; chard, 164; cole crops, 135; potatoes, 210; root crops, 159; tomatoes, 144
Flooding, 17
Floricanes, 272, 274
Flowers: 15, 47-48, 51-52, 71, 102, 106, 110; nut crops, 286; shedding, 142
Fluorescent lamps, 67-68, 111
Fog, 17
Fogle, Harold W., 253-259
Foliage, 76, 78
Fontenot, J. F., 224-227
Food: acidity chart, 313; canning, 305-307, 328-333; causes of spoilage, 298-299; costs of home grown, 310; determining safe processing time, 324; drying, 303, 308-309; economics of preservation, 310-312; freezing, 308, 320-322, 334-339, 363; pickling, 345-349; preservation, 298-303; preservation glossary, 383-384; quality losses, 304-309; quality of frozen, 320; raw, 304-305; rehydrated, 356; salad vegetables, 147-151; storing fresh, 365-371; storing preserved, 361-364; thawing frozen, 338; using pressure canner, 323-327
Food and Drug Administration, 375
Four o'clock, 20
Freezer: home, 32, 69-70, 321; power failure, 364; temperature, 334
Freezing: containers, 321; garden's harvest, 334-339; home, 320-322; packaging material, 334-335; quality factors, 308; retaining vitamins, 322
French endive, 151
French Sorrel, 222
Frost: 4, 17, 34, 36, 49, 84, 86-87, 108, 116; danger, 124-125; fruit trees, 251, 257
Frost, Doris Thain, 217-223
Frost-free days: representative areas, 35, 39, 42, 44
Fruit: annonaceous, 295; canning, 328-333; drying, 356-360; pest control program, 73-74; preparing for freezing, 336-337; processing, 331; storing fresh, 365-371; thawing, 338; trees, 246-259
Fruitworm, 74
Fungicides: 45, 73-76; cole crops, 135
Fungus, 81, 84
Fusarium wilt: 83, 107; sweet potatoes, 215; tomatoes, 144

Gage: canner types, 325-326; 330

Galletta, G. J., 279-283
Gardening: containers, 126-132; glossary of terms, 93-99; sources of information, 89-92, 110
Gardens: community, 3, 20-23; container, 18-20, 102-103, 105-106; front yard, 20; planning, 102-110; sites, 15-23; tools and equipment, 24-32; Victory, 2
Garlic: 8-10, 12-13, 104-105, 109, 123; planting, 154-155; uses, 220
Garrison, Stephen A., 196-201
Gel, 340
Geotropic, 48
Germination, 47, 68, 114-115
Globe artichoke, 232-233
Glossary: Food Preservation, 383-384; Gardener's, 93-99
Gomez, Ricardo E., 41-43
Gooseberries: 53, 272-278; freezing, 336; pest control, 73
Gourd: 187-195; planting chart, 190
Grafting, 51
Grapes: 260-264; freezing, 336; leading fruit crop, 350; pest control, 73, 263; raisin, 53; three main groups, 350; varieties, 260-262; wine, 264
Grapefruit, 296
Gravity, 48, 52-53, 57
Great Lakes, 38
Great Plains, 4, 17, 37-38
Greenhouses: 37, 61, 63-67; dimensions, 63-64; hobby-size, 111; types, 64; ventilation systems, 65-66
Greens, 163-170
Grubs, 75
Guanabana, 295
Guava, 294
Gulf of Mexico, 45

Hail, 42
Hamilton, Louise W., 304-309
Hardin, N. Carl, 49-51
Hardpan, 4
Harlequin bug: 74; collards, 166
Harrows, 28
Hayes, Kirby, 340-344
Hazelnuts, 284-290
Headspace: canning, 306; freezing, 335
Heat: 300-301; methods of transfer, 324-325; pickle products, 348; sterilization, 306-307
Heaters, 61, 66
Hedgerow: 68, red raspberry, 273
Herbicides: 18, 75; cole crops, 135; strawberry plants, 268
Herbs: 106, 217-223; container grown, 126, 131; freezing, 218; harvesting, 218; pickling, 346; recipes, 223
Hickory nuts, 284-290
Hill, Robert G., Jr., 265-271
Hoe, 24, 61, 80, 86, 102, 110, 121
Honey, 340-341
Honeydew melon, 53
Horn, Anton S., 365-371
Hornworms, 75
Horsepower, 27-28

Horseradish: 8-10, 12-13, 49; growing, 233-235
Horticulture, Directory of American, 91
Hose, garden, 15, 28-29
Hotbeds: 61-63, 86, 111; heating cable, 62
Hot caps, 36, 124
Hot pack, 330
Humidity: 41, 43, 45, 51, 63, 65, 67, 70; food drying, 359; storage, 365
Humus, 5, 14, 57, 81, 84, 106
Husk tomato, 235-236
Hydrogen, 49
Hypocotyl, 50

Indian fig, 295-296
Insecticides: 53, 73-76; cole crops, 135
Insects: 3, 14, 45, 49, 52, 84; beneficial, 77; blueberries, 282; cane fruits, 277-278
Irish famine, 205
Iron, 49, 60
Irrigation: 34, 57, 80; blueberries, 281; cole crops, 135; cucurbits, 190; fruit trees, 251; stone fruit trees, 257; strawberries, 270; trickle, 29-30, 42, 271-275

Jaboticaba, 294
Jalapeno, 140
Jam: 340-344; berry, 272
Japanese persimmon, 296
Japanese walnut, 284-290
Jars: canning, 314-315, 329; crock or stone, 346; jelly, 342
Jaynes, Richard A., 284-290
Jelly: 340-344; berry, 272; herb, 223; problem prevention chart, 343-344; remaking, 344
Jelmeter, 341
Jerusalem artichoke, 49, 238-240
Johnston, Ralph W., 361-364
Johnstone, Bruce, 147-151
Jones, L. G., 224-227
Judkins, Wesley P., 78-83
Juglone, 106
Juice: grape, 350; jelly making, 341

Kale: 7-10, 12-13, 104-105, 109, 123; harvesting, 167; pest control, 74; standard varieties, 166
Kimbrough, W. D., 224-227
Kirk, Dale E., 356-360
Kiwi apple, 53
Kiwi, Yangtao, 294
Klippstein, Ruth N., 310-312
Knight, Robert J., Jr., 291-296
Kochia, 20
Kohlrabi: 7-10, 12-13, 104-105, 109, 123; cole crops, 133-138; pest control, 74, 135; varieties, 138
Kraft, John M., 171-180
Kuhn, Gerald D., 304-309
Kumquat, 296

Labeling: canned foods, 318; dried food, 359; frozen foods, 322, 337; pickles, 348

389

Lacewing, 77
Lady bugs, 77
Leaching, 45
Lead arsenate, 18
Leafhoppers: 73–75; lettuce, 149
Leaf miners: 75, 118; attacking peppers, 145; tomatoes, 144
Leaf roller, 73–74
Leaf scorch, 74
Leaf spot, 74–75
Leeks: 7–10, 12–13, 105, 123; varieties, 156
Lemon: 296; juice, 340, 344
Lettuce: 7–10, 12–13, 104–105, 109, 123, 147–149; pest control, 74; varieties, 126, 132; pest control, 74; varieties, 148
Lids: canning, 314, 329; pickling, 347
Light, 48, 67–68
Lime: 4, 60, 107, 119–120; for acid soil, 133; fruit varieties, 296
Limestone, 81, 107
Loam, 55, 106
Longan, 294–295
Loopers, 74
Loquat, 295
Lovage, 220
Lychee, 295
Lygus bugs, 74

Macadamia, 53, 286
Magnesium, 4, 49, 107, 119–120
Malstrom, Howard L., 284–290
Manganese, 49, 59, 60
Mango, 53, 295
Mansour, N. S., 157–162
Manure: 23, 31, 56–57, 78, 106; green, 119; rhubarb, 202
Marmalades, 340–344
Marr, Charles W., 84–88
Martynia, 236
McGregor, S. E., 51–53
McGrew, J. R., 260–264, 350–355
Meiners, Jack P., 171–180
Melons: 19–20, 34, 39, 49, 62, 102; pest control, 74
Meta, 352–354
Metcalf, Homer N., 228–244
Metric table, 100
Mexican bean beetles, 83
Mice: 4, 7; berries, 277; stone fruit trees, 258
Mildew: celery, 150
Minerals: 5, 47; fruits and vegetables, 305; wine, 350
Mineral salts, 49
Minges, Philip A., 133–138
Mints, 18, 221
Mites: 77; on apples, 251
Moldboard plow, 28
Molds: 313, 320; dried food, 360
Moles, 4
Molybdenum, 49
Mosaic blight, 83
Mosiac virus disease: on peppers, 145
Moths, 53, 73, 83
Mulches: 5, 32, 41–42, 78–79, 87, 110; fruit trees, 251; herbs, 217; organic, 125; plastic, 14,
32, 102, 125, 271; root crops, 159; subtropical fruits, 292–293
Mulching: 11, 14, 37, 43, 80, 108, 125; blueberries, 281; cane fruits, 275; container gardening, 130; control fruit rot, 144; cucurbits, 190; strawberry planting, 268
Mushrooms, 236
Muskmelons: 8–10, 12–13, 104–105, 109, 187–195; harvesting, 194; planting chart, 190
Mustard: 7–10, 12–13, 104–105, 109, 123; origin, 167; pest control, 75; varieties, 168

National Canners Association, 375
National Weather Service, 36
Nectar, 52
Nectarines: 53, 253–259; varieties, 255
Nematicide, 75, 145
Nematodes: 45, 83; damage to tomatoes, 145; okra, 227; sweet potatoes, 216
Netting, wire, 31
New England, 38
New Zealand spinach: 169; harvesting, 170
Niacin: sweet potatoes, 212
Nitrogen: 5, 32, 49, 57, 60, 78–80, 84, 107; deficiency, 5, 58; fertilizer, 5; leafy vegetables, 120; okra, 224; sufficient amounts, 134
Nuclei, 58
Nutrients: 4–5, 40, 48, 58, 76, 78, 84–85, 107; pickles, 345
Nutgrass, 45
Nuts: 284–290; fertilizing, 288–289; harvest, 290; pecan varieties, 285; Persian walnut varieties, 287; storing, 290, 371

Okra: 7–10, 12–13, 104, 109, 123, 224–227; harvesting, 226; pest control, 75; varieties, 225
Oleoresin, 140
Onions: 7–10, 12–13, 104–105, 109, 123; Egyptian, 220; sets, 152; storing, 368–369; transplants, 153
Orchard, 2, 15, 18, 25, 60
Organic: gardening, 78–83; matter, 56, 106–107, 120; wastes, 57
Oriental chestnut trees, 25
Ovary, 49, 51–52
Ovules, 49
Oxygen: 47–49, 54; fruits and vegetables, 304

Paprika, 140
Paraffin: 342; dipping vegetables, 367
Parsley: 7–10, 12–13, 104–105, 109, 123; uses, 221
Parsnips: 7–10, 12–13, 104–105, 109, 123; varieties, 161
Parthenocarpic, 53
Passion fruit, 53, 295
Pathogenic organisms, 78

Patio, 126
Pawpaw, 53
Peaches: 41, 46, 52–53, 253–259; freezing, 336; how to can, 333; pest control, 73; varieties, 254–255
Peach leaf curl, 73
Peanuts: 236–238; storing, 371
Pears: 41, 53, 246–252; pest control, 73, 251; storage, 369; varieties, 246–247
Peas: 7–10, 12–13, 104–105, 109, 123, 177–180; how to freeze, 339; pest control, 75
Peat: moss, 26, 32, 56, 106, 128; pellets, 31, 112; pots, 31, 112
Pea weevil, 75
Pecan: 284–290; trees, 15; varieties, 285
Pectin: commercial, 340–341; liquid and powdered, 341
Pegboard, 24
Pentachlorophenol, 61
Pepo, 187–195
Peppers: 7–10, 12–13, 103–105, 109, 123; container grown, 126; fertilizing, 141; harvesting, 145–146; pest control, 75; transplanting, 142–143; varieties, 140; weed control, 144
Pepper weevil, 75
Perennials, 48
Permeability, 56
Persian melon, 53
Persian walnut: 284–290; varieties, 287
Persimmons: 296; freezing, 336
Pest Control Program, 73–75
Pesticides: 69, 71, 75, 82; precautions, 30–31; sprayer types, 30
Pests: 45, 82; asparagus, 199; control chart, 73–75; management, 71–77; nuts, 289; root crops, 157, 159
Petals, 49
Petcock: pressure canners, 325, 330
pH: 54, 59–60, 79–80, 119, 133; food, 301, 323; herb growing, 217; stone fruits, 254; sweet potatoes, 213
Phloem, 48
Phomopsis rot: eggplant, 145
Phosphoric acid: 78–79; cucurbits, 190
Phosphorus: 32, 49, 58–60, 84, 107; asparagus, 201; commercial fertilizers, 120; sufficient amounts, 134
Photoperiodism, 4
Photosynthesis, 16, 47–48, 50
Phototropic, 48
Pickles: 345–349; classifications, 345; herbs, 219; problems, 349; signs of spoilage, 349
Pickle-worm, 74
Pineapples, 53, 295
Pirate bugs, 77
Pistachio, 284–290
Pistil, 51
Planting: complex art of, 119–125; fall dates, 12–13; Garden Guide, 6; spring dates, 8–10

390

Plants: fruit and vegetable, 47–53; pollination, 51–53; reproduction, 49–51
Plastic: freezer bags, 335; mulches, 14, 32, 102, 125, 271
Plum curculio, 73
Plums: 53, 253–259; freezing, 336; pest control, 73; varieties, 255
Plumule, 50
Pollen, 49, 52
Pollination: apples and pears, 247; plant, 51–53, 102
Polyethylene, 64
Pomegranate, 295
Popcorn, 181
Potash: 78–79, 84–85, 107; commercial fertilizers, 120; cucurbits, 190
Potassium: 32, 49, 59–60; asparagus, 201; sufficient amounts, 134
Potatoes: 7–10, 12–13, 105, 123, 205–211; harvest, 210–211; Irish, 46, 60, 104, 109; pest control, 75, 210; storage, 367; varieties, 207
Potherbs, 163–170
Pots: 31; plastic, 128; red clay, 127
Potting mix, 128
Powdery mildew: 73; apples, 251; cucurbits, 191–192
Praying mantis, 77
Precipitation: representative areas, 35, 39, 42, 44
Preservation: canning, 313–319; drying, 356–360; economics, 310–312; food, 298–303; freezing, 320–322; glossary, 383–384; pickling, 345–349; questions and answers, 378–383
Preserves: 340–344; berry, 272; grape, 350
Pressure canner: 301; low-acid foods, 307, 323–327; types, 325
Primocanes, 272–274
Processing: fruits and vegetables, 331–332; jellied products, 342
Proteins: 47, 58; sweet potatoes, 212
Protoplasm, 58
Prunes, 53
Pruning: fruit trees, 250; shears, 25; stone fruit trees, 256
PTO (power take-off), 27
Pumpkins: 87, 104, 109, 123, 187–195; pest control, 74; planting chart, 190; storage, 195

Quality: home-canned foods, 362; foods for canning, 330; freezing, 335–336; wine, 350
Quince, 53, 246–252

Raab, Carolyn A., 350–360
Rabbits, 31, 77, 258
Raccoons, 31
Rack: canner, 325
Radiation, 324
Radicle, 50
Radishes: 7–10, 12–13, 104–105, 109, 123; container grown, 126, 132; pest control, 75; varieties, 162

Railroad worms, 73
Rainfall, 28, 33, 38, 40–41, 45, 57, 89
Raisins, 350
Rake, 24, 61, 80, 121
Raspberries: 2, 53, 272–278; pest control, 73, 277
Rats, 7
Reasonover, Frances, 313–319
Recommended Daily Allowance (RDA), 212
Records: community canneries, 376; hours spent gardening, 311; keeping, 14, 72, 87
Redwood containers, 127
Refrigerator, 69–70, 86
Relishes, 345–349
Reproduction: plant, 49–51
Respiration: 47; apples and pears, 369; fruits and vegetables, 304
Reynolds, Charles W., 119–125
Rhubarb: 7–10, 12–13, 46, 123; a hardy perennial, 201–204; freezing, 336; varieties, 202
Riboflavin: 305; sweet potatoes, 212
Rocky Mountains, 34
Rodenticide, 75
Rodents, 3–4, 77
Root crops, 157–162; chart of characteristics, 158
Rooting, 51
Roots: 3–4, 47, 49, 78; tuberous, 51
Root maggots: 75; cole crops, 134; collards, 166
Root zone, 58, 62
Rose chafer, 73
Rosemary, 221–222
Ross, David S., 61–70
Rotary tiller, 28
Rotation: crop, 107
Rutabaga: 8–10, 12–13, 39, 70, 123; storage life, 162
Rye, 79, 119, 124

Saccharometer, 351
Sage, 222
Salad Burnet, 222
Salmonellosis, 298
Salsify: 7–10, 12–13, 104, 109; free from cultural problems, 161
Salt: canned foods, 306; pickling, 346
Sand: 20, 42, 55, 62, 70, 83, 106; potting soil, 128
Sauerkraut: 133, 302, 345; problems, 349
Sauls, Julian W., 291–296
Sawdust: 5, 14, 56–57, 79–80, 84, 106; potting soil, 128
Sawfly, 73
Schales, Franklin D., 111–118
Scheel, Dan C., 71–77
Seed: 31, 47, 108; germination, 114–115; herbs, 217; okra, 226; parts, 50; planting, 121; potatoes, 209; sowing, 114; storage, 86; tapes, 121–122
Seedbed, 28
Seeders, 24
Seedlings: 61, 63, 80, 114–116; transplanting, 129

Self-fertile, 52
Self-sterile, 52
Sepals, 49
Serrano, 140
Sewage sludge, 78
Shallots, 8–10, 12–13, 155
Shears: pruning, 25
Sheds: 23–24; storage, 61, 86
Shovels, 24, 86
Sierra Nevada Mountains, 42
Sigman, Catharine C., 340–344
Silt, 55, 106
Sirup: 306; blanching, 358; corn, 340–341; freezing fruit, 336; sweet-sour, 345
Sites: garden, 15–23; selecting, 3
Sludge: sewage, 78
Slugs: 83, 130; cole crops, 135
Slusher, David F., 54–60
Smith, Perry M., 43–46
Snow, 38, 64, 69, 86
Social Security Act, 377
Soil: alkaline, 133; amendments, 76; berries, 273; clay, 4, 106–107; management, 54–60; mixes, 19–20, 64, 103; preparation, 119; sandy, 4, 134; survey, 54–56, 60; test, 4, 80, 87, 89, 119, 133; texture, 43, 55, 106
Soil Conservation Service, 4
Sorrel: 8–10; French, 222
Soursap, 295
Southernpeas, 178–180
Soybeans: 12–13, 104, 109; storing, 371; vegetable, 240–243
Sperm, 50
Sphagnum peat moss, 20
Spices: pickling, 346
Spiders, 77
Spike tooth harrow, 28
Spinach, 4, 7–10, 12–13, 104–105, 109, 123, 168; New Zealand, 7–10, 12–13, 123, 169; varieties, 169
Spores, 49
Sprayers: 69, 86; types, 30
Spring tooth harrow, 28
Sprinkler, 29, 58
Spur blight, 73
Squash: 104–105, 109, 123, 187–195; container grown, 126, 132; pest control, 74; planting chart, 190; resistant varieties, 192
Squash vine borer, 74
Squirrels, 4
Stamens, 49, 52
Stang, Eldon J., 265–271
Staphylococcus, 298
Starter: solutions, 120, 122; vinegar, 355; yeast, 352, 354
Stems, 47, 76
Sternberg, Roger, 372–377
Stigma, 49, 51–52
Stinkbugs, 227
Stirm, Walter L., 38–41
Stomata, 48
Stone fruits: 253–259; storage, 259
Stoner, Alan K., 139–146
Storage: canned foods, 307, 318, 361–362; chart, 370; dried food, 359; fresh fruits and vegetables, 365–371; frozen foods, 337–338, 363–364; home-pre-

391

served foods, 361-364; problems, 366; root crops, 366; vegetables, 86
Straw, 5, 14, 32, 42, 79, 85, 87
Strawberries: 53, 265-271; buying plants, 266; harvesting, 269; how to freeze, 338; irrigation, 270; matted row, 267; pest control, 74, 270; spaced row, 267, two types, 265
Streams, 38
Striped cucumber beetle, 74
Structures, 61-70, 87
Style, 50-51
Subtropical fruit, 291-296
Succession planting: 108; corn, 184; cress, 151; lettuce, 147
Sugar: canned foods, 306; pickling, 346; preserving agent, 340
Sugar-apple, 295
Sulphur, 49, 60, 134
Sulfuric acid, 60
Summer savory, 222
Sun: 3, 16, 34, 54, 62-63, 103, 106; container gardening, 126; drying food, 356; grapes, 260
Sunchoke, 238-240
Swamps, 38
Sweet marjoram, 220-221
Sweet potatoes: 7-10, 12-13, 104, 109, 123; buried treasure, 212-216; growing requirements, 213; storage, 367-368
Swiss chard: 82, 104-105, 109, 123; harvesting, 164-165; varieties, 163
Syrphid flies, 77

Tabasco, 140
Tamarind, 295
Tangelo, 53, 296
Tangerine, 53, 296
Tarragon, 222-223
Temperature: 47, 89; affecting stone fruits, 253; drying food, 359; food preservation, 299-300; representative areas, 35, 39, 42, 44; storage of canned foods, 307
Thermometer: candy, 342; freezer, 322
Thermostats, 61, 67-68, 70
Thiamin, 305
Thyme, 223
Tillers: 61, 110; rotary, 26
Toads, 76, 107
Tomatoes: 7-10, 12-13, 104-105, 109, 123; container grown, 126, 132; cultivated, 139-146; fertilizing, 141; harvesting, 145-146; pest control, 75; storing fresh, 368; transplanting, 142-143; weed control, 144; wire cages, 129
Tomkins, John P., 272-278
Tompkins, Daniel, 201-204
Tools: 61, 64, 69, 110; garden, 24-32, 86
Tope, Nadine Fortna, 323-327
Topsoil, 17, 56, 106, 113
Tractors, 27, 102, 110, 119
Transmissions, 27
Transplants: 31, 111-118, 123-124; asparagus, 198; blueberries, 281; cole crops, 134; hardening, 117; sweet potatoes, 215
Trees: dwarf fruit, 2, 247-248; fruit, 51, 246-259
Trellis: 4, 61, 68, 102; berries, 274; grapes, 262
Tubers, 47, 51
Turnips, 7-10, 12-13, 104, 109, 123; container grown, 126; pest control, 75; varieties, 162
Turnquist, Orrin C., 205-211

Urea, 120
Utzinger, James D., 265-271

Vandemark, J. S., 152-156
V-belts, 27
Vegetables: blanching, 322, 334, 337; canning, 328-333; container gardens, 130-131; drying, 356-360; nutritional value, 6; pest control program, 74-75; planting chart, 104, 123; preparing for freezing, 337; processing, 331-332; salad, 147-151; selecting varieties, 7; small space chart, 105; steaming, 321; storing fresh, 365-371; thawing, 338; transplants, 31; yield chart, 109
Ventilation: 62, 111; greenhouses, 65-66
Vermiculite, 62, 80, 114
Verticillium wilt: 83; cane fruits, 273; eggplant, 145; tomatoes, 144
Villalon, Benigno, 139-146
Vinegar: canned foods, 306; herb, 223; pickling, 346; starter, 355; wine, 355
Vineyard, 2, 350
Vitamins: 47; retaining, 322

Wagner, Philip, 350-355
Walnuts, 284-290
Wann, E. V., 181-186
Wasps: 53, 77; on ripe fruit, 263
Water: 15, 28, 41, 47, 49, 54, 56, 80, 86; amount in canner, 325; brining, 346; container gardening, 129-130; in food, 321; management, 116; table, 56; vapor, 56
Watercress, 243-244
Watering: 58, 108; blueberries, 281; cole crops, 135; strawberry plants, 267; subtropical fruits, 293; systems, 68
Watermelon: 7-10, 12-13, 104-105, 109, 187-195; harvesting, 194; planting chart, 190
Way, Roger D., 246-252
Weather, 3
Webworm, 74
Weeds, 3, 45, 76, 80, 84, 102
Weevils, 7
Weinberger, John H., 253-259
Wheat, 79
Wheelbarrow, 26
Whitaker, Thomas W., 187-195
Whitefly: 118; tomatoes, 144
Wilson, Esther H., 365-371
Wilson, James W., 15-23
Wind, 4, 17, 41, 52
Windbreaks, 4, 17, 69, 85-86, 124
Window boxes, 19, 102, 126, 128
Windscreens: fiberglass, 17, 85
Wine: clarifying, 353-354; coloring matter, 352, 354; fining, 354; making, 350-355; red, 351-352; sugar correction table, 351; sweet, 355; vinegar, 355; white, 354
Wireworms, 75
Witloof chicory, 151
Wolf, Isabel D., 298-303
Woodchucks, 31

Xylem, 48

Yangtao Kiwi, 294
Yeast: 313, 320; cucumber blossoms, 345; dried food, 360; starter, 352, 354; wine fermenting, 350, 352-354

Zinc, 49, 60
Zottola, Edmund A., 298-303
Zucchini, 20
Zygote, 50

LC Card No. 77-600033

For sale by the Superintendent of Documents, U.S. Government Printing Office
Washington, D.C. 20402
Stock Number 001-000-03679-3
Catalog Number A 1.10:977

Metric Counter

WEIGHT

1 Pound (lb.) = .454 Kilograms (kg.)
2 Pounds (lbs.) = 0.9 Kilograms
3 Pounds (lbs.) = 1.4 Kilograms
4 Pounds (lbs.) = 1.8 Kilograms
5 Pounds (lbs.) = 2.3 Kilograms

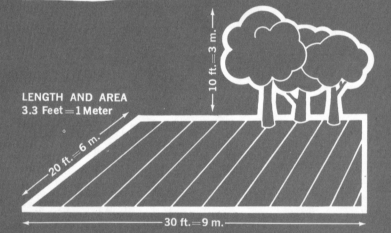

FERTILIZER
5—10—10

40 lbs. = 18 kg.

LENGTH AND AREA
3.3 Feet = 1 Meter

10 ft. = 3 m.
20 ft. = 6 m.
30 ft. = 9 m.

12 Inches = 30.48 Centimeters